Advanced Numerical Methods for Differential Equations

Mathematics and its Applications

Modelling, Engineering, and Social Sciences

Series Editor: Hemen Dutta, Department of Mathematics, Gauhati University

Tensor Calculus and Applications
Simplified Tools and Techniques
Bhaben Kalita

Discrete Mathematical Structures
A Succinct Foundation
Beri Venkatachalapathy Senthil Kumar and Hemen Dutta

Methods of Mathematical Modelling
Fractional Differential Equations
Edited by Harendra Singh, Devendra Kumar, and Dumitru Baleanu

Mathematical Methods in Engineering and Applied Sciences
Edited by Hemen Dutta

Sequence Spaces
Topics in Modern Summability Theory
Mohammad Mursaleen and Feyzi Başar

Fractional Calculus in Medical and Health Science
Devendra Kumar and Jagdev Singh

Topics in Contemporary Mathematical Analysis and Applications
Hemen Dutta

Sloshing in Upright Circular Containers
Theory, Analytical Solutions, and Applications
Alexander Timokha and Ihor Raynovskyy

Advanced Numerical Methods for Differential Equations
Applications in Science and Engineering
Edited by Harendra Singh, Jagdev Singh, S. D. Purohit, and Devendra Kumar

For more information about this series, please visit: https://www.routledge.com/Mathematics-and-its-Applications/book-series/MES

ISSN (online): 2689-0224
ISSN (print): 2689-0232

Advanced Numerical Methods for Differential Equations

Applications in Science and Engineering

Edited by
Harendra Singh
Jagdev Singh
Sunil Dutt Purohit
Devendra Kumar

CRC Press
Taylor & Francis Group
Boca Raton London New York

CRC Press is an imprint of the
Taylor & Francis Group, an **informa** business

First edition published 2021
by CRC Press
6000 Broken Sound Parkway NW, Suite 300, Boca Raton, FL 33487-2742

and by CRC Press
2 Park Square, Milton Park, Abingdon, Oxon, OX14 4RN

Library of Congress Cataloging-in-Publication Data

Names: Singh, Harendra, editor.
Title: Advanced numerical methods for differential equations : applications in science and engineering / edited by Harendra Singh, Jagdev Singh, Sunil Dutt Purohit, and Devendra Kumar.
Description: First edition | Boca Raton : CRC Press, 2021. | Series: Mathematics and its applications : modelling, engineering, and social sciences | Includes bibliographical references and index.
Identifiers: LCCN 2020050789 (print) | LCCN 2020050790 (ebook) | ISBN 9780367473112 (hbk) | ISBN 9781003097938 (ebk)
Subjects: LCSH: Differential equations--Numerical solutions. | Mathematical models.
Classification: LCC TA342 .A374 2021 (print) | LCC TA342 (ebook) | DDC 518/.63--dc23
LC record available at https://lccn.loc.gov/2020050789
LC ebook record available at https://lccn.loc.gov/2020050790

ISBN: 978-0-367-47311-2 (hbk)
ISBN: 978-0-367-56480-3 (pbk)
ISBN: 978-1-003-09793-8 (ebk)

Typeset in Times font
by KnowledgeWorks Global Ltd.

Contents

Preface

This book is prepared for graduate students and researchers working in the area of numerical techniques and mathematical modelling. It describes several useful topics in mathematical modelling having real-life applications in science and engineering. The book consists of 11 chapters and organized as follows:

Chapter 1 presents the numerical approximation of the fractional diffusion equation with a reaction term and discusses the stability and convergence of our numerical discretization, in the context of fractional calculus. The investigations and analysis concern the fractional differential equation described by the Caputo generalized fractional time derivative. The existence and uniqueness of the solution of the fractional diffusion equation are also provided using classical Banach fixed theorem. Graphical representations of the solutions have been presented to illustrate the main results of the chapter.

Chapter 2 investigates the complex and mixed dark-bright travelling wave solutions of the generalized KP-BBM equation. The sine-Gordon expansion method is used to produce dynamical complex wave properties. A wave transform considered is to convert the problems into a simple ordinary differential equation, providing a strong mathematical tool for solving other nonlinear development equations arising in mathematical physics.

Chapter 3 presents recent analytical schemes to construct novel computational solutions of the fractional quantum version of the relativistic energy–momentum relation that is mathematically represented by the nonlinear fractional Klein–Gordon equation. This model describes the spinless relativistic composite particles, like the pion. This process uses a new fractional definition (Atangana–Baleanu derivative operator), which converts the fractional form of the equation to integer-order nonlinear equation; then, by using the obtained analytical solutions to evaluate the initial and boundary conditions, some numerical solutions are discussed. Moreover, the stability property of the obtained solutions is tested to show the ability of our obtained solutions use in the physical experiments. The novelty and advantage of the proposed methods are illustrated by applying this model. Some sketches are plotted to show more about the dynamical behaviour of this model.

Chapter 4 introduces the numerical approximation for the 1-D ultra-relativistic Euler (URE) equations. These equations describe an ideal gas in terms of the particle density n, the spatial part of the four-velocity u, and the pressure p. Due to the inherently nonlinear and complex nature emerging in relativistic flows, the classical Euler equations are not capable of simulating fluid flows near the speed of light. The high-resolution shock capturing cone-grid and wave-front tracking schemes are implemented in order to solve the URE equations. The robustness and efficiency of the proposed schemes are exhibited by the numerical results.

Chapter 5 focuses on numerical treatment of a special type of boundary value problems referred to as singularly perturbed differential equations. The solutions of this type of differential equations exhibit sharp layers and pose challenges for conventional numerical methods. The constructions of layer-adapted meshes, (a) a priori refined meshes and (b) a posteriori refined meshes, has been discussed. The construction of monitor function, which is the main constituent of a posteriori mesh adaptation, has been discussed by considering a priori and a posteriori error estimates. Further, conclude with few numerical experiments for mesh adaptation and their extension in higher dimensions.

Chapter 6 presents non-standard finite difference (NSFD) methods to solve ordinary and partial differential equations. The NSFD methods are easy to use and have more stability than the standard methods. Local truncation of the method is also discussed in brief. Here, numerical computations show that the NSFD scheme is very effective.

Chapter 7 investigates the solutions, from the analytical and numerical point of view, of the set of diffusion equations. The fractional operators presented in the diffusive term of these equations are the Riesz-Welly fractional operators. In particular, the cases related to the reaction processes are analyzed. Therefore, the results obtained from this set of equations show a rich class of behaviours that can be utilized to model several interesting real-life problems.

Chapter 8 presents a local convergence analysis for some families of fifth- and sixth-order iterative methods in order to approximate a locally unique solution of a nonlinear equation in a Banach space setting. Earlier studies have used hypotheses up to the sixth Frechet-derivative of the operator involved, although only the first derivative appears in these methods. We use hypotheses only on the first Frechet-derivative in our local convergence analysis. This way, the applicability of these methods is extended. Moreover, the radius of convergence and computable error bounds on the distances involved are also given in this study. This fact allows us to enlarge the number and type of problems that can be solved by means of using these methods, including differential equations, which are one of the most found problems in optimization and modelling areas. Concrete examples of the theoretical results are also presented in this chapter.

Chapter 9 studies the dynamics modelling of quarantine and isolation. By considering the principles of quarantine and isolation with respect to COVID-19 spread, this chapter derive, extend, analyze, the formulation of the quarantine and isolation scenario with the aim of clarifying the implications that may arise from the process of quarantine and isolation, depending on the day an individual tested positive for COVID-19. This formulation is extended to a fractional differential equation to enable us to present adequate information of our formulation, which may be helpful to the society at large.

Chapter 10 investigates the analytical and numerical solutions of the Atangana conformable fractional (2+1)-dimensional generalized Nizhnik-Novikov-Veselov (GNNV) equations. This system is considered as the isotropic extension of the well-known (1+1)-dimensional KdV equation. Applying four recent analytical schemes for constructing abundant explicit wave solutions of the suggested model allows the employment of the B-spline collection schemes (septic, quantic, cubic) to evaluate the numerical solutions and also to check the accuracy of our obtained solutions via calculating the absolute value of error between exact and numerical solutions. Moreover, the Hamiltonian system properties are used to check the stability of the obtained solutions and to investigate their appropriateness in the model's applications.

Chapter 11 studies the viral dynamics of hepatitis E via semi-analytical hybrid scheme pertaining to homotopy polynomials and the Sumudu transform algorithm. The Caputo fractional derivative is engaged to analyze the dynamics of an HEV model. Employing graphical presentations, the proposed study explores the consequences of the variations of fractional order of a time derivative and time t on susceptible, exposed, infected, and recovered populations. The role of environment is also considered in disease dynamics of the hepatitis E model. This work strongly authenticates the computational strength of the employed scheme. Moreover, the uniqueness and convergence analysis of the method are also established in this study with the aid of fixed point theory of Banach spaces.

About the Author

Harendra Singh is Assistant Professor in the Department of Mathematics, Post-Graduate College, Ghazipur, Uttar Pradesh, India. He did his Master of Science (M.Sc.) in Mathematics from Banaras Hindu University, Varanasi and Ph.D. in Mathematics from Indian Institute of Technology (BHU), Varanasi, India. He has qualified GATE, JRF, and NBHM in Mathematics. He is also awarded post-doctoral fellowship (PDF) in Mathematics from National Institute of Science Education and Research (NISER) Bhubaneswar, Odisha, India. He primarily teaches the subjects like real and complex analysis, functional analysis, abstract algebra, and measure theory in post-graduate level course in Mathematics. His area of interest is Mathematical Modelling, Fractional Differential Equations, Integral Equations, Calculus of Variations, Analytical and Numerical Methods. His works have been published in *Applied Numerical Mathematics, Applied Mathematics and Computations, Applied Mathematical Modelling, Chaos Solitons & Fractals, Numerical Methods for Partial Differential Equations, Physica A, Astrophysics and Space Science, Electronic Journal of Differential Equations, Few-Body System*, and several other peer-reviewed international journals. He has edited a book published by CRC press Taylor and Francis. His 32 research papers have been published in various journals of repute with h-index of 11. His single-authored papers have been published in Applied Mathematics and Computations, Chaos Solitons & Fractals, Astrophysics and Space Science, and other peer-reviewed international journals. He has attained a number of National and International Conferences and presented several research papers. He has also attended Short-Terms Programs and Workshops. He is reviewer of various journals.

Jagdev Singh is Professor in the Department of Mathematics, JECRC University, Jaipur, Rajasthan, India. He did his Master of Science (M.Sc.) in Mathematics and Ph.D. in Mathematics from University of Rajasthan, India. He primarily teaches the subjects like mathematical modelling, real analysis, functional analysis, integral equations, and special functions in post-graduate level course in Mathematics. His area of interest is Integral Transforms, Special Functions, Fractional Calculus, Mathematical Modelling, Mathematical Biology, Fluid Dynamics, Applied Functional Analysis, Nonlinear Dynamics, Analytical and Numerical Methods. He has published three books: Advance Engineering Mathematics (2007), Engineering Mathematics-I (2008), and Engineering Mathematics-II (2013). His works have been published in the *Nonlinear Dynamics, Chaos Solitons & Fractals, Physica A, Journal of Computational and Nonlinear Dynamics, Applied Mathematical Modelling, Entropy, Advances in Nonlinear Analysis, Romanian Reports in Physics, Applied Mathematics and Computation, Chaos*, and several other peer-reviewed international journals. His 120 research papers have been published in various journals of repute with h-index of 30. He has attained a number of National and International Conferences and presented several research papers. He has also attended Summer Courses,

Short-Terms Programs, and Workshops. He is member of Editorial Board of various journals of Mathematics. He is reviewer of various journals.

Sunil Dutt Purohit is Associate Professor of Mathematics at Rajasthan Technical University, Kota, India. He did his Master of Science (M.Sc.) in Mathematics and Ph.D. in Mathematics from Jai Narayan Vyas University, Jodhpur, India. He was awarded University Gold Medal for being topper in M.Sc. Mathematics and awarded Junior Research Fellowship and Senior Research Fellow of Council of Scientific and Industrial Research. He primarily teaches the subjects like integral transforms, complex analysis, numerical analysis, and optimization techniques in graduate and post-graduate level course in engineering mathematics. His research interest includes Special Functions, Basic Hypergeometric Series, Fractional Calculus, Geometric Function Theory, Mathematical Analysis, and Modelling. He has credited more than 150 research articles and four books so far. He has delivered talks at foreign and national institutions. He has also organized a number of academic events. He is a Life Member of Indian Mathematical Society (IMS), Indian Science Congress Association (ISCA), Indian Academy of Mathematics (IAM), and Society for Special Functions and their Applications, Soft Computing Research Society, India (SCRS) and International Association of Engineers (IAENG). Presently, he is general secretary of the Rajasthan Ganita Parishad. He has also contributed in designing and redesigning of syllabus of engineering mathematics B. Tech. course work.

Devendra Kumar is Assistant Professor in the Department of Mathematics, University of Rajasthan, Jaipur, Rajasthan, India. He did his Master of Science (M.Sc.) in Mathematics and Ph.D. in Mathematics from University of Rajasthan, India. He primarily teaches the subjects like real and complex analysis, functional analysis, integral equations, and special functions in post-graduate level course in Mathematics. His area of interest is Integral Transforms, Special Functions, Fractional Calculus, Mathematical Modelling, Applied Functional Analysis, Nonlinear Dynamics, Analytical and Numerical Methods. He has published two books: *Engineering Mathematics-I* (2008) and *Engineering Mathematics-II* (2013). His works have been published in the *Nonlinear Dynamics, Chaos Solitons & Fractals, Physica A, Journal of Computational and Nonlinear Dynamics, Applied Mathematical Modelling, Entropy, Advances in Nonlinear Analysis, Romanian Reports in Physics, Applied Mathematics and Computation, Chaos*, and several other peer-reviewed international journals. His 130 research papers have been published in various journals of repute with h-index of 30. He has attained a number of National and International Conferences and presented several research papers. He has also attended Summer Courses, Short-Terms Programs, and Workshops. He is member of Editorial Board of various journals of Mathematics. He is reviewer of various journals.

1 Stability and Convergence Analysis of Numerical Scheme for the Generalized Fractional Diffusion-Reaction Equation

Ndolane Sene
Laboratoire Lmdan, Département de Mathématiques de la Décision, Université Cheikh Anta Diop de Dakar, Faculté des Sciences Economiques et Gestion, Senegal

CONTENTS

Lead paragraph

- The existence and uniqueness of the fractional differential equation described by the Caputo generalized fractional derivative are investigated.
- The numerical scheme of the fractional diffusion equation with a reaction term is provided.
- The stability and convergence of our numerical discretization are presented.

1.1 INTRODUCTION

The theory of fractional operators is the main contribution to fractional calculus. Fractional calculus is a mathematical field in which the integration and differentiation are done in a non-integer order. Recently, its contribution in sciences and engineering [1–7], in physics modeling [3, 8–14], in epidemiology [15], and in many others fields has been provided. In real-life problems, many physical, mechanical, and biological phenomena are described as differential equations. The differential equations are represented by an integer-order time derivative. Note that modeling the problems with an integer-order time derivative does not take into account the memory effect, whereas modeling using fractional derivatives does take into account this effect. The memory effect is essential in modeling dynamical systems. In other words, the present process depends on past processes. Many phenomena in physics are, like the diffusion processes, fractional. For example, we cite the sub-diffusion process, the super-diffusion process, the ballistic process, the hyper-diffusion process, and others [16–18]. For recent applications of fractional calculus in real-life problems, see in [19–21]. The fractional derivatives play a fundamental role in fractional calculus. We have continuous and discrete types of fractional derivatives. Some of them are with singular kernels: the Riemann-Liouville fractional derivative, the Caputo fractional derivative, the Hilfer fractional derivative, and the generalized form of Riemann-Liouville and Caputo-Liouville fractional derivatives. Their discrete forms have been introduced in many recent investigations, see in [22–24]. The new fractional derivatives introduced in fractional calculus are without singular kernels; here we have the fractional derivative with the Mittag-Leffler kernel and the fractional derivative with the exponential kernel.

In this chapter, we address the physical problems and contribute to the applications of fractional calculus in physics. We mainly study the diffusion-reaction problem in the context of the Caputo generalized fractional derivative. Some investigations related to the numerical schemes of the fractional diffusion equations exist in the literature. In [25], Gao and Sun have proposed the difference scheme for the fractional sub-diffusion equation. In [26], Du et al. have proposed the numerical discretization of the fractional diffusion-wave equation. In [27], Chen et al. have provided a numerical approximation for the Rayleigh-Stokes problem for a heated generalized second-grade fluid. We also note that many works have provided analytical solutions of the fractional diffusion equations. In [21], Sene has given the analytical solution of Stokes' first equation using Fourier and Laplace transformation. In [9], Sene provided the analytical solution of the fractional diffusion equation in two-dimensional spaces and also has provided an analytical solution of the Cateneo-Hristov diffusion equation in terms of the Caputo-Fabrizio fractional derivative. The approximate solutions with the homotopy method can be found in many papers, as in [20].

This chapter is the result of intense research related to the numerical discretization of the fractional diffusion-reaction equation. We mainly investigate the stability and convergence of our proposed numerical discretization. The main objective is to prove that the Caputo generalized fractional derivative can be used in modeling diffusion equations. And the numerical approximation of this derivative, introduced by Shengda et al. in [28], is useful because the numerical schemes of the fractional differential equations are stable and convergent. Our investigations will open a new door in studying the numerical scheme of the fractional differential equations described by the Caputo generalized fractional derivative.

The chapter is structured as follows. In Section 1.2, we recall the most popular fractional derivatives in fractional calculus. In Section 1.3, we address the existence and uniqueness of our proposed model to justify our studies. In Section 1.4, we propose the numerical discretization of the fractional diffusion reaction equation described by the Caputo generalized fractional derivative. In Section 1.5, we study the stability of the proposed numerical discretization. In Section 1.6, we address the convergence of the proposed numerical discretization. Discussion of a graphical representation of the results is given in Section 1.7, and the chapter's conclusion and associated remarks comprise Section 1.8.

1.2 FRACTIONAL DERIVATIVES REVIEW

In this section, we recall the fractional derivatives with and without singular kernels used in fractional calculus. We begin with the Riemann-Liouville fractional derivative and finish with the fractional derivative with Mittag-Leffler kernel. Note that fractional derivatives recalled in this section are those most commonly used in fractional calculus, but there exist many other fractional derivatives similar to those cited.

Definition 1.1: *[29] Let the function defined by $f : [0,+\infty[\longrightarrow \mathbb{R}$. The fractional integral of order α of the function f is defined in the following form*

$$(I^{\alpha}f)(t) = \frac{1}{\Gamma(\alpha)}\int_0^t (t-s)^{\alpha-1} f(s)ds, \tag{1.1}$$

where $\Gamma(...)$ is the gamma function and $t > 0$, and $0 < \alpha < 1$.

The associated fractional derivative was introduced by Riemann and Liouville and is defined as in the following definition.

Definition 1.2: *[29] Let the function defined by $f : [0,+\infty[\longrightarrow \mathbb{R}$. The fractional derivative of order α, of the function f in Riemann-Liouville sense is defined by the following form*

$$(D^{\alpha}f)(t) = \frac{1}{\Gamma(1-\alpha)}\left(\frac{d}{dt}\right)\int_0^t (t-s)^{-\alpha} f(s)ds, \tag{1.2}$$

where $\Gamma(...)$ is the gamma function and $t > 0$, and $0 < \alpha < 1$.

Later, another definition of fractional derivative was proposed by Caputo due to the inconvenience of the first proposed fractional derivative.

Definition 1.3: *[29] Let the function defined by $f : [0, +\infty[\longrightarrow \mathbb{R}$. The Caputo-Liouville fractional derivative of order α, of the function f is defined by the following form*

$$(D_c^\alpha f)(t) = \frac{1}{\Gamma(1-\alpha)} \int_0^t (t-s)^{-\alpha} f'(s)ds, \tag{1.3}$$

where $\Gamma(...)$ is the gamma function and $t > 0$, and $0 < \alpha < 1$.

Recently, the generalized forms of these fractional derivatives were proposed in [29]. They have introduced the following generalizations.

Definition 1.4: *[29–31] Let the function defined by $f : [a, +\infty[\longrightarrow \mathbb{R}$. The generalized integral of order α, $\rho > 0$ of the function f is defined by the following form*

$$(I^{\alpha,\rho} f)(t) = \frac{1}{\Gamma(\alpha)} \int_0^t \left(\frac{t^\rho - s^\rho}{\rho}\right)^{\alpha-1} f(s) \frac{ds}{s^{1-\rho}}, \tag{1.4}$$

where $\Gamma(...)$ is the gamma function and $t > 0$, and $0 < \alpha < 1$.

Definition 1.5: *[29–31] Let the function defined by $f : [0, +\infty[\longrightarrow \mathbb{R}$. The generalized fractional derivative of order α, $\rho > 0$ of the function f in Riemann-Liouville sense is defined by the following form*

$$(D^{\alpha,\rho} f)(t) = \frac{1}{\Gamma(1-\alpha)} \left(t^{1-\rho}\frac{d}{dt}\right) \int_0^t \left(\frac{t^\rho - s^\rho}{\rho}\right)^{-\alpha} f(s) \frac{ds}{s^{1-\rho}}, \tag{1.5}$$

where $\Gamma(...)$ is the gamma function and $t > 0$, and $0 < \alpha < 1$.

Definition 1.6: *[29–31] Let the function defined by $f : [0, +\infty[\longrightarrow \mathbb{R}$. The Caputo-Liouville generalized fractional derivative of order α, $\rho > 0$ of the function f is defined by the following form*

$$(D_c^{\alpha,\rho} f)(t) = \frac{1}{\Gamma(1-\alpha)} \int_0^t \left(\frac{t^\rho - s^\rho}{\rho}\right)^{-\alpha} f'(s)ds, \tag{1.6}$$

where $\Gamma(...)$ is the gamma function and $t > 0$, and $0 < \alpha < 1$.

We can notice all the previous fractional derivatives are with singular kernels. The singular kernel is an inconvenience in many applications. Recently two fractional derivatives without singular kernels have been proposed and find many applications in real-world problems. Here we recall the fractional derivatives with exponential kernel and with Mittag-Leffler kernel.

Definition 1.7: *[32] Consider the function $f : [0, +\infty[\longrightarrow \mathbb{R}$, the fractional derivative with exponential kernel of the function f of order α is defined by the expression*

$$D_\alpha^{CF} f(t) = \frac{M(\alpha)}{1-\alpha} \int_0^t f'(s) \exp\left(-\frac{\alpha}{1-\alpha}(t-s)\right) ds, \tag{1.7}$$

for all $t > 0$, where the order $\alpha \in (0,1)$ and $\Gamma(...)$ is the gamma function.

Definition 1.8: *[32] Consider the function $f : [0,+\infty[\longrightarrow \mathbb{R}$, the fractional derivative with Mittag-Leffler kernel of the function f of order α is expressed in the form*

$$D_{\alpha}^{ABC} f(t) = \frac{AB(\alpha)}{1-\alpha} \int_0^t f'(s) E_{\alpha} \left(-\frac{\alpha}{1-\alpha}(t-s)^{\alpha} \right) ds, \tag{1.8}$$

for all $t > 0$, where the order $\alpha \in (0,1)$ and $\Gamma(...)$ is the gamma function.

Note $M(\alpha)$ and $AB(\alpha)$ represent the normalization term and satisfy $M(0) = M(1) = AB(0) = AB(1) = 1$; see more statements in [32]. In Definition 1.8, the Mittag-Leffler function is represented in the following definition.

Definition 1.9: *[33, 34] We call the Mittag-Leffler function with two parameters a function which is represented by a series and described as follows*

$$E_{\alpha,\beta}(z) = \sum_{k=0}^{\infty} \frac{z^k}{\Gamma(\alpha k + \beta)},$$

where the parameters obey that $\alpha > 0$, $\beta \in \mathbb{R}$ and $z \in \mathbb{C}$.

An important point of note is that when we replace the orders $\alpha = 1$ and $\beta = 1$, we obtain the classical exponential function. That is

$$E_{\alpha,1}(z) = E_{\alpha}(z) = \exp(z).$$

The fractional derivatives shown in this section are the most used in current fractional calculus literature. The discrete forms of the previous fractional derivatives exist, and the reader can find these in [22–24].

1.3 EXISTENCE AND UNIQUENESS VIA BANACH FIXED THEOREM

In this section, we prove the existence and uniqueness of the solution of the fractional diffusion-reaction equation via the Banach fixed theorem. The proof is classic in mathematics and depends on the fractional differential equation under consideration. The following equation describes the fractional differential equation under consideration:

$$D_t^{\alpha,\rho} u(x,t) = \frac{\partial^2 u}{\partial x^2} + \theta(x,t), \tag{1.9}$$

where the initial condition is defined as the following form

$$u(x,0) = f(x), \tag{1.10}$$

and furthermore the function u satisfies the following boundary condition

$$u(0,t) = g(t). \tag{1.11}$$

In Eq. (1.9) the function u represents the density or the concentration of the material, and θ denotes the reaction term in the diffusion process.

In the interests of simplification, let the function be defined by

$$\Pi(u,x,t,\theta) = \frac{\partial^2 u}{\partial x^2} + \theta(x,t). \tag{1.12}$$

The objective of the first step is to prove the function Π is Lipschitz continuous. In this section, the classical norm is used. Applying the Euclidean norm and using triangular inequality to both sides of Eq. (1.12), we obtain

$$\|\Pi(u,x,t,\theta) - \Pi(v,x,t,\theta)\| \le \left\| \frac{\partial^2 u}{\partial x^2} - \frac{\partial^2 v}{\partial x^2} \right\| + \|\theta - \theta\|. \tag{1.13}$$

When we use the condition, the function u is Lipschitz continuous, and we suppose there exists a constant k such that the following relationship is obtained

$$\|\Pi(u,x,t,\theta) - \Pi(v,x,t,\theta)\| \le \left\| \frac{\partial^2 u}{\partial x^2} - \frac{\partial^2 v}{\partial x^2} \right\| \le k\|u-v\|. \tag{1.14}$$

The second step consists of defining the Picard's operator represented in our problem as follows

$$Pu(x,t) = I^{\alpha,\rho}\Pi(u,x,t,\theta), \tag{1.15}$$

where the operator $Pu : H \to H$ and H is closed Banach space. We have to prove the Picard's operator is well defined. We apply the Euclidean norm, and we obtain the following

$$\begin{aligned} \|Pu(x,t) - u(x,0)\| &= \|I^{\alpha,\rho}\Pi(u,x,t,\theta)\|, \\ &\le I^{\alpha,\rho}\|\Pi(u,x,t,\theta)\|, \\ &\le \|\Pi(u,x,t,\theta)\| I^{\alpha,\rho}(1), \end{aligned} \tag{1.16}$$

or $\|\Pi(x,t,\theta)\| \le M$ and $t \le T$, and above Eq. (1.16) can be represented in the form

$$\|Pu(x,t) - u(x,0)\| \le \frac{\rho^{1-\alpha}}{\Gamma(\alpha)}\left(\frac{T^\rho}{\rho}\right)^\alpha M, \tag{1.17}$$

which is equivalent to the well-defined Picard's operator. The last step in the proof consists of providing a condition under which the Picard's operator defines a contraction. Applying the Euclidean norm, we obtain the following relations

$$\begin{aligned} \|Pu(x,t) - Pv(x,t)\| &= \|I^{\alpha,\rho}\Pi(u,x,t,\theta) - \Pi(v,x,t,\theta)\|, \\ &\le I^{\alpha,\rho}\|\Pi(u,x,t,\theta) - \Pi(v,x,t,\theta)\|, \\ &\le \|\Pi(u,x,t,\theta) - \Pi(v,x,t,\theta)\| I^{\alpha,\rho}(1). \end{aligned} \tag{1.18}$$

Using Eq. (1.14), we have the following relation

$$\|Pu(x,t) - Pv(x,t)\| \le \frac{\rho^{1-\alpha}}{\Gamma(\alpha+1)}\left(\frac{T^\rho}{\rho}\right)^\alpha k\|u-v\|. \tag{1.19}$$

We note the Picard's operator is a contraction when the following identity is satisfied

$$\frac{\rho^{1-\alpha}}{\Gamma(\alpha+1)}\left(\frac{T^{\rho}}{\rho}\right)^{\alpha} k<1. \tag{1.20}$$

With the Banach fixed theorem, we conclude the existence and uniqueness of the solution of the fractional diffusion-reaction equation defined by (1.9). Thus, our model is well-defined, and we are now sure the resolution of the problem exists. There exist many methods of getting the solution of Eq. (1.9). We have analytical methods such as the homotopy perturbation method, the Laplace transform, and the Fourier transformation [20,21]. But in this chapter for our problem, we use a numerical scheme.

1.4 NUMERICAL SCHEME OF THE FRACTIONAL DIFFUSION REACTION EQUATION

This section addresses the numerical scheme for the fractional diffusion-reaction equation described by the generalized fractional derivative operator. We adopt the implicit finite difference scheme. The following equation characterizes the fractional differential equation under consideration

$$D_t^{\alpha,\rho}u(x,t)=\frac{\partial^2 u}{\partial x^2}+\theta(x,t), \tag{1.21}$$

where the initial condition is defined as the following form

$$u(x,0)=f(x), \tag{1.22}$$

and furthermore the function u satisfies the boundary condition

$$u(0,t)=g(t). \tag{1.23}$$

The following relation describes the numerical approximation of the Caputo generalized fractional derivative introduced in [28]

$$D_c^{\alpha,\rho}u^n=\sum_{k=1}^{n}\left(u^k-u^{k-1}\right)T_{n,k}, \tag{1.24}$$

where $T_k=\frac{\rho^{1-\alpha}}{\Gamma(2-\alpha)}t_k^{1-\rho}\left[(t_n^{\rho}-t_{k-1}^{\rho})^{1-\alpha}-(t_n^{\rho}-t_k^{\rho})^{1-\alpha}\right]/(t_k-t_{k-1})$. Later, for simplification, the formula defined in Eq. (1.24) is simplified as the following form

$$D_c^{\alpha,\rho}u_n=T_n\left[u^n-u^{n-1}\right]+\sum_{k=1}^{n-1}\left(u^k-u^{k-1}\right)T_{n,k}. \tag{1.25}$$

Equation (1.25) is more useful for a numerical approximation. It will be noted that when the order $\alpha=\rho=1$, we recover the classical approximation of the first integer-order derivative given by

$$D_c^{1,1}u^n=\frac{u^n-u^{n-1}}{t_k-t_{k-1}}. \tag{1.26}$$

The central difference approximation for the second-order derivative respecting the space coordinate is given by

$$\frac{\partial^2 u^n}{\partial x^2} = \frac{u_{j+1}^n - 2u_j^n + u_{j-1}^n}{\Delta x^2} + O\left(\Delta x^2\right). \tag{1.27}$$

The numerical scheme of the fractional diffusion-reaction equation described by the generalized fractional derivative is obtained by combining Eqs. (1.25) and (1.27) and is represented by

$$T_n\left[u^n - u^{n-1}\right] + \sum_{k=1}^{n-1}\left(u^k - u^{k-1}\right)T_k = \frac{\partial^2 u^n}{\partial x^2} + \theta^n + R^n, \tag{1.28}$$

where R^n denotes the residual term. After simplification and calculations, the difference scheme is expressed as the following form

$$u^n = u^{n-1} + \kappa_1\frac{\partial^2 u^n}{\partial x^2} + \kappa_1\sum_{k=1}^{n-1}\left(T_{k+1} - T_k\right)u^{n-k} + T_n u^0 + \kappa_1\theta^n + R^n, \tag{1.29}$$

where $\kappa_1 = T_{n,n}^{-1}$. Our objective in this chapter is to study the stability and convergence of the numerical scheme equation defined in Eq. (1.29).

1.5 STABILITY ANALYSIS OF THE NUMERICAL APPROXIMATION

In this section, we investigate the stability of the numerical scheme proposed for the fractional diffusion equation in the preceding section, and we mainly work with Eq. (1.29). The objective here is to prove the numerical scheme of the fractional differential equation (1.21) under conditions (1.22) and (1.23) is unconditionally stable. First, we show by induction the following inequality

$$\|u^n\|_2 \le \left\|u^0\right\|_2 + \kappa_1\max_{0\le l\le n}\left\|\theta^l\right\|. \tag{1.30}$$

We prove the identity (1.30) by induction. When $n = 1$, into Eq. (1.30), we obtain the following difference scheme

$$u^1 = u^0 + \kappa_1\frac{\partial^2 u^1}{\partial x^2} + \kappa_1\theta^1. \tag{1.31}$$

We multiply both sides of Eq. (1.31) by u^1, giving the following relationships

$$\left(u^1, u^1\right) - \kappa_1\left(\frac{\partial^2 u^1}{\partial x^2}, u^1\right) = \left(u^0, u^1\right) + \kappa_1\left(\theta^1, u^1\right),$$

$$\left(u^1, u^1\right) + \kappa_1\left(\frac{\partial u^1}{\partial x}, \frac{\partial u^1}{\partial x}\right) = \left(u^0, u^1\right) + \kappa_1\left(\theta^1, u^1\right).$$

From Cauchy-Schwarz inequality the following identities

$$\left\|u^1\right\|_2\left\|u^1\right\|_2\leq\left\|u^0\right\|_2\left\|u^1\right\|_2+\kappa_1\left\|\theta^1\right\|_2\left\|u^1\right\|_2,$$
$$\left\|u^1\right\|_2\leq\left\|u^0\right\|_2+\kappa_1\left\|\theta^1\right\|_2\leq\left\|u^0\right\|_2+\kappa_1\max_{0\leq l\leq n}\left\|\theta^l\right\|. \tag{1.32}$$

We suppose Eq. (1.30) is held. Now let us prove the identity

$$\left\|u^{n+1}\right\|_2\leq\left\|u^0\right\|_2+\kappa_1\max_{0\leq l\leq n+1}\left\|\theta^l\right\|, \tag{1.33}$$

is held. We multiply both sides of Eq. (1.33) by u^{n+1}, giving the following relationships

$$\left(u^{n+1},u^{n+1}\right)-\kappa_1\left(\frac{\partial^2u^{n+1}}{\partial x^2},u^{n+1}\right)=\left(u^n,u^{n+1}\right)+\kappa_1\left(\theta^{n+1},u^{n+1}\right)$$
$$+\kappa_1\sum_{k=0}^{n}\left(T_{n,k+1}-T_{n,k}\right)\left(u^k,u^{n+1}\right),$$
$$\left(u^{n+1},u^{n+1}\right)+\kappa_1\left(\frac{\partial u^{n+1}}{\partial x},\frac{\partial u^{n+1}}{\partial x}\right)=\left(u^n,u^{n+1}\right)+\kappa_1\left(\theta^{n+1},u^{n+1}\right)$$
$$+\kappa_1\sum_{k=0}^{n}\left(T_{n,k+1}-T_{n,k}\right)\left(u^k,u^{n+1}\right).$$

We now use the Cauchy-Schwarz equation, giving the following relations

$$\left\|u^{n+1}\right\|_2\left\|u^{n+1}\right\|_2\leq\left\|u^n\right\|_2\left\|u^{n+1}\right\|_2+\kappa_1\left\|\theta^{n+1}\right\|_2\left\|u^{n+1}\right\|_2$$
$$+\kappa_1\sum_{k=0}^{n}\left(T_{n,k+1}-T_{n,k}\right)\left\|u^k\right\|_2\left\|u^{n+1}\right\|_2,$$
$$\left\|u^{n+1}\right\|_2-\left\|u^n\right\|_2\leq\kappa_1\left\|\theta^{n+1}\right\|_2+\kappa_1\sum_{k=0}^{n}\left(T_{n,k+1}-T_{n,k}\right)\left\|u^k\right\|_2. \tag{1.34}$$

Using the assumption $\left\|u^n\right\|_2\leq\left\|u^0\right\|_2+\kappa_1\max\left\|\theta^n\right\|$ and $T_{n,k}>T_{n,k+1}$, we obtain by neglecting the negative term

$$\left\|u^{n+1}\right\|_2\leq\left\|u^0\right\|_2+\kappa_1\max_{0\leq l\leq n+1}\left\|\theta^l\right\|. \tag{1.35}$$

For the second step of the proof, let the terms defined $\xi^n=U^n-u^n$, where U^n represents the exact solution of the fractional diffusion equation (1.21), satisfying the discretization (1.29). We note both U^n and u^n satisfy Eq. (1.30), that is

$$\left\|u^n\right\|_2\leq\left\|u^0\right\|_2+\kappa_1\max_{0\leq l\leq n}\left\|\theta^l\right\|, \tag{1.36}$$

and

$$\left\|U^n\right\|_2\leq\left\|U^0\right\|_2+\kappa_1\max_{0\leq l\leq n}\left\|\theta^l\right\|. \tag{1.37}$$

Making the difference between Eq. (1.37) and Eq. (1.36), and doing manipulations using triangular inequality, the error term satisfies the following identity

$$\|\xi^n\|_2 \leq \|\xi^0\|_2. \tag{1.38}$$

Thus, the numerical approximation (1.29) of the fractional diffusion reaction equation described by the Caputo generalized fractional derivative is unconditionally stable.

1.6 CONVERGENCE ANALYSIS OF THE NUMERICAL APPROXIMATION

The second part of the investigation into the numerical discretization (1.29) of the fractional diffusion reaction equation concerns the convergence of our scheme. In this section, we use the approximate discretization with the error term R^n. We first define the round-off error, denoted by $\varepsilon^n = u^n - U^n$, where u^n denotes the approximate solution and satisfies the discretization (1.29), that is

$$u^n = u^{n-1} + \kappa_1 \frac{\partial^2 u^n}{\partial x^2} + \kappa_1 \sum_{k=0}^{n-1} \left(T_{n,k+1} - T_{n,k}\right) u^k + \kappa_1 \theta^n + R^n, \tag{1.39}$$

and U^n represents the exact solution of the numerical scheme (1.29), that is

$$U^n = U^{n-1} + \kappa_1 \frac{\partial^2 U^n}{\partial x^2} + \kappa_1 \sum_{k=0}^{n-1} \left(T_{n,k+1} - T_{n,k}\right) U^k + \kappa_1 \theta^n. \tag{1.40}$$

The round-off error $\varepsilon^n = u^n - U^n$ follows from the difference between Eq. (1.39) and Eq. (1.40). We obtain the following relationship

$$\varepsilon^n = \varepsilon^{n-1} + \kappa_1 \frac{\partial^2 \varepsilon^n}{\partial x^2} + \kappa_1 \sum_{k=0}^{n-1} \left(T_{n,k+1} - T_{n,k}\right) \varepsilon^k + R^n. \tag{1.41}$$

We repeat the same reasoning as the previous section. We prove by induction the following inequality

$$\|\varepsilon^n\|_2 \leq \left\|\varepsilon^0\right\|_2 + \kappa_1 \max_{0 \leq l \leq n} \left\|R^l\right\|, \tag{1.42}$$

where $\left\|\varepsilon^0\right\|_2 = 0$. In other words, we have to prove the following identity

$$\|\varepsilon^n\|_2 \leq \kappa_1 \max_{0 \leq l \leq n} \left\|R^l\right\|. \tag{1.43}$$

When $n = 1$, into Eq. (1.41), we obtain the following difference scheme

$$\varepsilon^1 = \varepsilon^0 + \kappa_1 \frac{\partial^2 \varepsilon^1}{\partial x^2} + \kappa_1 R^1. \tag{1.44}$$

We multiply both sides of Eq. (1.44) by ε^1, giving the following relationships

$$\left(\varepsilon^1,\varepsilon^1\right)-\kappa_1\left(\frac{\partial^2\varepsilon^1}{\partial x^2},\varepsilon^1\right)=\left(\varepsilon^0,\varepsilon^1\right)+\kappa_1\left(R^1,\varepsilon^1\right),$$
$$\left(\varepsilon^1,\varepsilon^1\right)+\kappa_1\left(\frac{\partial\varepsilon^1}{\partial x},\frac{\partial\varepsilon^1}{\partial x}\right)=\left(\varepsilon^0,\varepsilon^1\right)+\kappa_1\left(R^1,\varepsilon^1\right).$$

We use Cauchy-Schwarz inequality, and the following relationship is held

$$\left\|\varepsilon^1\right\|_2\left\|\varepsilon^1\right\|_2\leq\left\|\varepsilon^0\right\|_2\left\|\varepsilon^1\right\|_2+\kappa_1\left\|R^1\right\|_2\left\|\varepsilon^1\right\|_2,$$
$$\left\|\varepsilon^1\right\|_2\leq\left\|\varepsilon^0\right\|_2+\kappa_1\left\|R^1\right\|_2\leq\left\|\varepsilon^0\right\|_2+\kappa_1\max_{0\leq l\leq n}\left\|R^l\right\|. \tag{1.45}$$

Following this, and the fact that $\left\|\varepsilon^0\right\|_2=0$, the round-off error satisfies at the first step the following identity:

$$\left\|\varepsilon^1\right\|_2\leq\kappa_1\max_{0\leq l\leq n}\left\|R^l\right\|. \tag{1.46}$$

To satisfy Eq. (1.43), let's prove the following identity

$$\left\|\varepsilon^{n+1}\right\|_2\leq\kappa_1\max_{0\leq l\leq n+1}\left\|R^l\right\| \tag{1.47}$$

is satisfied too. We multiply both sides of Eq. (1.41) by ε^{n+1}, giving the following relationships

$$\left(\varepsilon^{n+1},\varepsilon^{n+1}\right)-\kappa_1\left(\frac{\partial^2\varepsilon^{n+1}}{\partial x^2},\varepsilon^{n+1}\right)=\left(\varepsilon^n,\varepsilon^{n+1}\right)+\kappa_1\left(R^{n+1},\varepsilon^{n+1}\right)$$
$$+\kappa_1\sum_{k=0}^{n}\left(T_{n,k+1}-T_{n,k}\right)\left(\varepsilon^k,\varepsilon^{n+1}\right),$$
$$\left(\varepsilon^{n+1},\varepsilon^{n+1}\right)+\kappa_1\left(\frac{\partial\varepsilon^{n+1}}{\partial x},\frac{\partial\varepsilon^{n+1}}{\partial x}\right)=\left(\varepsilon^n,\varepsilon^{n+1}\right)+\kappa_1\left(R^{n+1},\varepsilon^{n+1}\right)$$
$$+\kappa_1\sum_{k=0}^{n}\left(T_{n,k+1}-T_{n,k}\right)\left(\varepsilon^k,\varepsilon^{n+1}\right).$$

Using the Cauchy-Schwarz formula, we have the following relationships

$$\left\|\varepsilon^{n+1}\right\|_2\left\|\varepsilon^{n+1}\right\|_2\leq\left\|\varepsilon^n\right\|_2\left\|\varepsilon^{n+1}\right\|_2+\kappa_1\left\|R^{n+1}\right\|_2\left\|\varepsilon^{n+1}\right\|_2$$
$$+\kappa_1\sum_{k=0}^{n}\left(T_{n,k+1}-T_{n,k}\right)\left\|\varepsilon^k\right\|_2\left\|\varepsilon^{n+1}\right\|_2,$$
$$\left\|\varepsilon^{n+1}\right\|_2-\left\|\varepsilon^n\right\|_2\leq\kappa_1\left\|R^{n+1}\right\|_2+\kappa_1\sum_{k=0}^{n}\left(T_{n,k+1}-T_{n,k}\right)\left\|\varepsilon^k\right\|_2. \tag{1.48}$$

Using the previous assumption $\left\|\varepsilon^n\right\|_2\leq\left\|\varepsilon^0\right\|_2+\kappa_1\max_{0\leq l\leq n}\left\|R^l\right\|$ and $T_{n,k}>T_{n,k+1}$, we obtain by neglecting the negative term the following

$$\left\|\varepsilon^{n+1}\right\|_2\leq\left\|\varepsilon^0\right\|_2+\kappa_1\max_{0\leq l\leq n}\left\|R^l\right\|. \tag{1.49}$$

Following this, and the fact that $\left\|\varepsilon^0\right\|_2 = 0$, the round-off error satisfies at the step $n+1$ the following identity

$$\left\|\varepsilon^{n+1}\right\|_2 \leq \kappa_1 \max_{0 \leq l \leq n+1} \left\|R^l\right\|. \tag{1.50}$$

The convergence of our numerical scheme follows from the fact that the round-off error satisfies the following inequality

$$\|\varepsilon^n\|_2 \leq \kappa_1 \max_{0 \leq l \leq n} \left\|R^l\right\| = \kappa_1 \mathcal{O}(\Delta t). \tag{1.51}$$

1.7 THE GRAPHICS WITH THE NUMERICAL SCHEME

In this section, we illustrate the numerical method used here by solving an example. We consider the fractional diffusion equation described by the differential equation given by the following expression

$$D_t^{\alpha,\rho} u(x,t) = \frac{\partial^2 u}{\partial x^2} + \theta(x,t), \tag{1.52}$$

where the initial condition is defined as the following form

$$u(x,0) = e^x, \tag{1.53}$$

and the function u satisfies the following boundary relation

$$u(0,t) = E_\alpha\left(2\left(\frac{t^\rho}{\rho}\right)^\alpha\right), \tag{1.54}$$

where the function playing the reaction role is the function u

$$\theta(x,t) = u(x,t). \tag{1.55}$$

For the numerical statements, we make the following assumptions: we suppose $t = 1$, $\Delta x = 0.04$, and $N = 100$. The exact solution of the problem is given by the expression

$$u(x,t) = e^x E_\alpha\left(2\left(\frac{t^\rho}{\rho}\right)^\alpha\right). \tag{1.56}$$

In Figure 1.1, we depict the numerical solution generated by our numerical schemes with the order $\alpha = 1$ and $\rho = 0.6; 07; 1$.

In Figure 1.2, we depict with the order $\alpha = 1$ and $\rho = 0.7$ the exact solution (continuous line) and the numerical solution (dotted line). We can observe that solutions are in agreement.

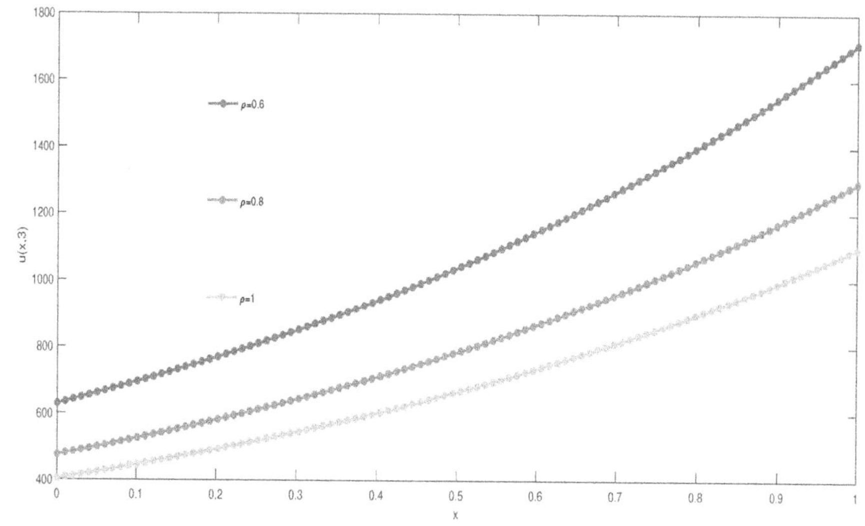

Figure 1.1 Numerical solution for the fractional diffusion equation with reaction.

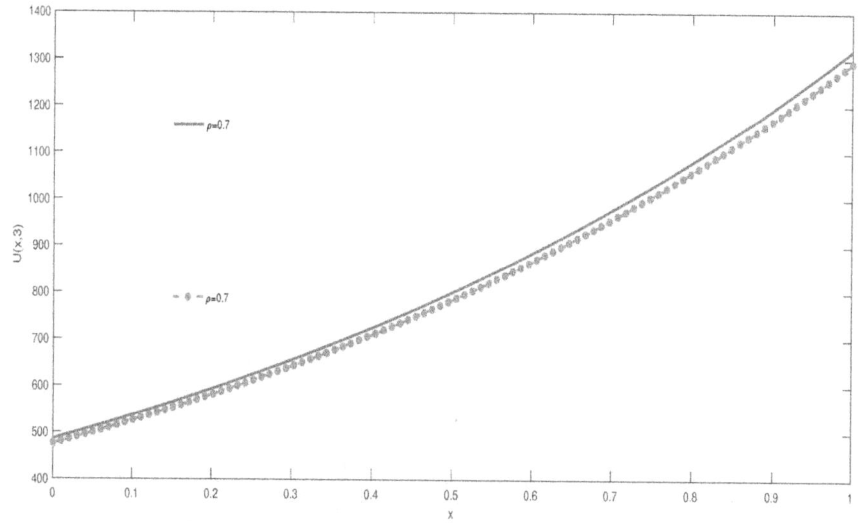

Figure 1.2 Comparison between the exact and the approximate solutions.

1.8 CONCLUSION

In this chapter, we have implemented numerical schemes for the fractional diffusion equation with a reaction term. The particularity of this work is the use of the numerical scheme for the Caputo generalized fractional derivative. The existence and uniqueness of the fractional diffusion-reaction equation have been discussed too. We have also focused on the stability and convergence of the numerical scheme theoretically. Graphical representations of the numerical schemes of the proposed model have been depicted, and we notice the exact solution and the numerical solution are in good agreement.

REFERENCES

1. Baleanu, D., ChengWu, G. and Zeng S. D. (2017), *Chaos analysis and asymptotic stability of generalized Caputo fractional differential equations*, Chaos Soli. Fract., **102**, 99–105.
2. Tasbozan, O., Esen, A., Yagmurlu, N. M. and Ucar, Y. (2013), *A numerical solution to fractional diffusion equation for force-free case*, Abstr. Appl. Anal., **2013**, 6.
3. Sene, N. (2019), *Integral balance methods for Stokes' first, equation described by the left generalized fractional derivative*, Physics, **1**, 154–166.
4. Sene, N. (2020), *Global asymptotic stability of the fractional differential equations*, J. Nonlinear Sci. Appl., **13**, 171–175.
5. Singh, H. (2016), *A new numerical algorithm for fractional model of Bloch equation in nuclear magnetic resonance*, Alex. Eng. J., **55(3)**, 2863–2869.
6. Singh, H. (2018), *Approximate solution of fractional vibration equation using Jacobi polynomials*, Appl. Math. Comput., **317**, 85–100.
7. Singh, H. (2016), *Operational matrix approach for approximate solution of fractional model of Bloch equation*, J. King Saud Univ. Sci., **29(2)**, 235–240.
8. Meerschaert, M. M. and Tadjeran C. (2004), *Finite difference approximations for fractional advection-dispersion flow equations*, J. Comput. Appl. Math., **172**, 65–67.
9. Sene, N. (2019), *Solutions of fractional diffusion equations and Cattaneo-Hristov diffusion models*, Int. J. Appli. Anal., **17(2)**, 191–207.
10. Singh, H. (2018), *An efficient computational method for the approximate solution of nonlinear Lane-Emden type equations arising in astrophysics*, Astrophys Space Sci., **363**, 71.
11. Singh, H., Srivastava, H. M. and Kumar, D. (2017), *A reliable numerical algorithm for the fractional vibration equation*, Chaos Soli. Fract., **103**, 131–138.
12. Singh, H., Kumar, D. and Baleanu, D. (2019), *Methods of Mathematical Modelling: Fractional Differential Equations*, CRC Press (Taylor and Francis Group): Boca Raton, FL ISBN: 978-0-367-22008-2.

13. Owolabi, K. M., Akgul A. and Atangana, A. (2020), *Modelling and analysis of fractal-fractional partial differential equations: Application to reaction-diffusion system*, Alex. Eng. J., 59 (4), 2477–2490.

14. Atangana, A., Akgul A. and Owolabi, K. M. (2020), *Analysis of fractal fractional differential equations*, Alex. Eng. J., https://doi.org/ 0.1016/ j.aej.2020.01.005.

15. Mouaouine, A., Boukhouima, A., Hattaf, K. and Yousfi, N. (2018), *A fractional order SIR epidemic model with nonlinear incidence rate*, Advances in Difference Equations, **2018:160**, 146–156.

16. Santos, M. D. and Gomez, I. S. (2018), *A fractional Fokker–Planck equation for non-singular kernel operators*. J. Stat. Mech. Theory Exp., 123205.

17. Santos, M. D. (2019), *Fractional Prabhakar Derivative in Diffusion Equation with Non-Static Stochastic Resetting*. Physics, **1**, 40–58.

18. Santos, M. D. (2018), *Non-Gaussian Distributions to Random Walk in the Context of Memory Kernels*. Fractal Fract., **2**, 20.

19. Henry, B. I., Langlands, T. A. M. and Straka, P. (2009), *An Introduction to Fractional Diffusion*, World Sci. Rev.

20. Sene, N. (2019), *Homotopy Perturbation ρ-Laplace Transform Method and Its Application to the Fractional Diffusion Equation and the Fractional Diffusion-Reaction Equation*, Fractal Fract., **3**, 14.

21. Sene, N. (2018), *Stokes' first problem for heated flat plate with Atangana–Baleanu fractional derivative*, Chaos Soli. Fract., **117**, 68–75.

22. Abdeljawad, T. (2018), *Different type kernel h-fractional differences and their fractional h-sums*, Chaos Soli. Fract., **116**, 146–156.

23. Abdeljawad, T. and Al-Mdallal, Q. M. (2018), *Discrete Mittag-Leffler kernel type fractional difference initial value problems and Gronwall's inequality*, J. Comput. Appl. Math., **339**, 218–230.

24. Abdeljawad, T. and Baleanu, D. (2016), *Discrete fractional differences with non-singular discrete Mittag-Leffler kernels*, Adv. Diff. Equat., **2016**, 232.

25. Gao, G. H. and Sun, Z. Z. (2011), *A compact finite difference scheme for the fractional sub-diffusion equations*, J. Comp. Phys., **230**, 586–595.

26. Du, R., Yan, Y. and Liang, Z. (2018), *A high-order scheme to approximate the Caputo fractional derivative and its application to solve the fractional diffusion wave equation*. J. Comp. Phys. **376**, 1312–1330.

27. Chen, C. M., Liu, F. and Anh, V. (2008), *Numerical analysis of the Rayleigh–Stokes problem for a heated generalized second grade fluid with fractional derivatives*, Appl. Math. Comput., **204**, 340–351.

28. Shengda, Z., Dumitru, B., Yunru, B. and Cheng, W. G. (2017), *Fractional differential equations of Caputo-Katugampola type and numerical solutions*, Appl. Math. Comput., **315**, 549–554.

29. Fahd J. and Abdeljawad, T. (2018), *A modified Laplace transform for certain generalized fractional operators*, Results in Nonlinear Analysis, **2**, 88–98.

30. Sene, N. (2019), Integral-Balance Methods for the Fractional Diffusion Equation Described by the Caputo-Generalized Fractional Derivative, *Methods of Mathematical Modelling: Fractional Differential Equations*, CRC Press 87.

31. Sene, N. (2020), *Stability analysis of electrical RLC circuit described by the Caputo-Liouville generalized fractional derivative*, Alex. Eng. J., https://doi.org/10.1016/j.aej.2020.01.008.

32. Atangana, A. and Baleanu, D. (2016), *New fractional derivatives with nonlocal and non-singular kernel: theory and application to heat transfer model*, Therm. Sci. **20**, DOI: 10.2298/TSCI160111018A.

33. Erdélyi, A., Magnus, W., Oberhettinger, F. and Tricomi, F. G. (1955), *Higher Transcendental Functions*, McGraw-Hill: New York, NY, **3**, *2018*.

34. Mittag-Leffler, M. G. (1904), Sopra la funzione $E_\alpha(x)$, C. R. Acad. Sci., **13**, 3–5.

2 Studying on the Complex and Mixed Dark-Bright Travelling Wave Solutions of the Generalized KP-BBM Equation

Haci Mehmet Baskonus,[1] *Ajay Kumar*[2], *M.S. Rawat,*[3] *Bilgin Senel,*[4] *Gulnur Yel,*[5] *and Mine Senel*[4]
[1]Department of Mathematics and Science Education, Faculty of Education, Harran University, Sanliurfa, Turkey
[2,3]Department of Mathematics, H.N.B Garhwal University (A Central University) Srinagar, India
[4]Fethiye Faculty of Business Administration, Mugla Sitki Kocman University, Mugla, Turkey
[5]Final International University, Kyrenia Mersin 10, Turkey

CONTENTS

2.1 INTRODUCTION

Nonlinear equations play an important role in a wide variety of areas in engineering and physics, such as heat flow, quantum mechanics, solid-state physics, chemical kinematics, fluid mechanics, optical fibres, bacteria physics, wave proliferation phenomena, and spreading of shoal water waves. Therefore, many different techniques have been suggested to find exact solutions of partial differential equations. For example, the extended tanh method for abundant solitary wave solutions of nonlinear wave equation was proposed by Wazwaz [1]. Zakharov-Kuznetsov-Benjamin-Bona-Mahony (ZK-BBM) equation for compact and noncompact physical structures was proposed by Wazwaz [2]. (2+1)-dimensional (KP-BBM) equation for an explicit solution and stability analysis was proposed by Ganguly and Das [3]. The sine-cosine method for handling wave equation was proposed by Wazwaz [4]. A transformation method was proposed by Yan to establish a relation between linear and nonlinear wave theories [5]. (3+1)-dimensional B-type Kadomtsev-Petviashvili-Boussinesq equation for rational and semi-rational was proposed by Hu et al. [6]. The unstable Schrodinger equation was proposed by Lu et al. [7]. Drinefeld-Sokolov-Wilson equation for travelling wave solution was proposed by Syed and Sadaf [8]. Periodic and solitary wave solutions of Kawahara and modified Kawahara equation by using the sine-cosine method were proposed by Yusufoglu et al. [9]. (2+1)-dimensional nonlinear electrical transmission line equation for Jacobi elliptic function solution was proposed by Zayed and Tala-Tebue [10]. Automated tanh-function method for finding solitary wave solutions to nonlinear evolution equation was proposed by Parkes and Duffy [11]. Complex travelling wave solutions to the Konopelchenko-Dubrovsky model were proposed by Dusunceli [12]. The (4+1)-dimensional Fokas equation for exact travelling wave solutions was proposed by Al-Amr and El-Ganaini [13]. Three-wave resonant interaction, multi-dark, dark solitons, breathers waves, and their interactions and dynamics, was proposed by Zhang et al. [14]. Complex structures of Kundu-Eckhaus equation via improved Bernoulli sub-equation function method were proposed by Bulut and Baskonus [15]. Complex hyperbolic structures to the Lonngren wave equation were proposed by Baskonus et al. [16]. Some asymptotic methods for strongly nonlinear equations were proposed by He [17]. The $(\frac{G'}{G})$ method for travelling wave solution of nonlinear evolution equations was proposed by Zhang and Mingliang [18]. Analytic treatment of two higher-dimensional nonlinear partial differential equation was proposed by Hosseini and Gholamin [19]. (2+1)-dimensional Zakharov-Kuznetsov (ZK) equation for travelling wave solutions was proposed by Khalfallah [20]. A generalized second extended (3+1)-dimensional Jimbo-Miwa equation for conservation laws was proposed by Moleleki et al. [21]. Perturbation theory for the nonlinear wave equation was proposed by Girgis and Biswas [22]. Modified simple equation method and its applications to nonlinear partial differential equations was proposed by Mirzazadeh [23]. On symmetry preserving and symmetry broken bright, dark, and anti-dark soliton solutions of nonlinear Schrödinger were proposed by Priya et al. [24]. Elliptic boundary value problems for finite difference method for a numerical solution were proposed by Jaboob and Pandey [25]. (2+1)-dimensional partially nonlinear distributed coefficient Gross-Pitaevskii equation for analytical matter wave was proposed by Liu [26]. Applications of He's semi-inverse variational method and improved tan (Phi/2)-expansion method (ITEM) to the nonlinear long-short wave interaction system were proposed by Tekiyeh et al. [27]. (2+1)-dimensional soliton equation for new periodic and soliton solutions was proposed by Arbabi and Najafi [28]. Yu and Hong-Cai [29] discussed the explicit solution (2+1)-dimensional nonlinear KP-BBM equation. The (3+1)-dimensional potential Yu-Toda-Sasa-Fukuyama equation for multiple lump solution was proposed by Zhao and He [30]. Other mathematical techniques to nonlinear models can be found in [31–47].

This chapter is organized as follows. In Section 2.2, we present the general properties of the sine-Gordon expansion method (SGEM) in a detailed manner. In section 2.3, the SGEM is applied to the generalized KP-BBM equation and also to obtain well-known complex wave patterns like soliton. The solution process is quite ingenious and all types of nonlinear development equation are investigated. In the last section of the chapter, we present a conclusion.

2.2 THE SGEM

Consider the sine-Gordon equation of the form [6, 19, 48]

$$\frac{\partial^2 u}{\partial x^2} - \frac{\partial^2 u}{\partial t^2} = m^2 \sin(u), \tag{2.1}$$

where $u = u(x,t)$ and m is a constant. The wave variable $\xi = \mu(x - ct)$ transforms the above PDE to the following ODE

$$\frac{d^2U}{d\xi^2} = \frac{m^2}{\mu^2(1-c^2)} \sin(U), \tag{2.2}$$

where $U(\xi) = U$ and ξ, c , stand for the amplitude and velocity of the travelling wave. Equation (2.2) can be reduced into first-order ODE

$$\left(\frac{d(U/2)}{d\xi}\right)^2 = \frac{m^2}{\mu^2(1-c^2)} \sin^2(U/2) + K, \tag{2.3}$$

where K is a constant of integration. K is supposed zero. Let $w(\xi) = \frac{U(\xi)}{2}$ and $b^2 = \frac{m^2}{\mu^2(1-c^2)} \sin^2(U/2)$, then Eq. (2.3) is converted to

$$w' = \frac{dw}{d\xi} = b\sin(w). \tag{2.4}$$

Set $b = 1$ in Eq. (2.4). Then, Eq. (2.4) yields two considerable relationship

$$\sin(w) = \sin[w(\xi)] = \frac{2de^{\xi}}{d^2e^{2\xi}+1} \downarrow d_{=1} = \operatorname{sech}(\xi), \tag{2.5}$$

$$\cos(w) = \cos[w(\xi)] = \frac{d^2e^{2\xi}-1}{d^2e^{2\xi}+1} \downarrow d_{=1} = \tanh(\xi), \tag{2.6}$$

where d is a non-zero constant. The NPDE of the form

$$R(u, u_x, u_t, u_{xx}, u_{tt}, u_{xt,...}) = 0 \tag{2.7}$$

can be converted to an ODE

$$R(U, U', U'', ...) = 0 \tag{2.8}$$

by using a consistent wave transform $U = U(\xi)$, where the transform variable ξ is defined as $\mu(x - ct)$. Then, the trial solution to Eq. (2.8) of the form

$$U(\xi) = \sum_{i=1}^{n} \tanh^{i-1}(\xi)[B_i\text{sech}(\xi) + A_i\tanh(\xi)] + A_0 \tag{2.9}$$

may be written as

$$U(\omega) = \sum_{i=1}^{n} \cos^{i-1}(\omega)[B_i \sin(\omega) + A_i \cos(\omega)] + A_0, \tag{2.10}$$

owing to Eqs. (2.5)–(2.6). The replacement of the forecast solutions Eq. (2.10) into Eq. (2.8) the coefficient of power of $\sin^i(w), \cos^i(w)$ is supposed as zero. Next, solution of the expiration algebraic tract is attempted for the coefficients $A_0, A_1, B_1, \ldots, a, v$. Then, and if they exist, the solutions are generated by using Eqs. (2.5)–(2.6) and ξ.

2.3 APPLICATIONS OF SGEM AND MATHEMATICAL ANALYSIS

In this section, we investigate the generalized KP-BBM equation [3], defined as

$$(u_t + u_x - a(u^n)_x - b(u^n)_{xxt})_x + ku_{yy} = 0. \tag{2.11}$$

In 2008, Wazwaz studied a family of physical properties of Eq. (2.11) in [3, 4], in which the coefficients a, b, and k are constant and non-zero. In addition to the nonlinear conveyance terms $(u^n)_x$, the dispersion term $(u^n)_{xxt}$ is also nonlinear. If $n = 1$, Eq. (2.11) converts to the well-known KP-BBM equation. This equations model dynamical wave properties in a nonlinear spreading tract, and as a result, the BBM equations comprehend fabricate impact. The solutions of the BBM equations exhibition certain soliton-like conduct that is not explicable by any know theory BBM equation involucre matter of the under mentioned type [11], superficies waves of long wavelength in serous, acoustics-gravity waves incompressible flow, hydromagnetic waves in chilled bacteria, and acoustic waves in a harmonious crystal.

The fragile interplay between the conveyance and the actual nonlinear prevalence procreates solitary wave with exact complex wave exemplar solution that is ordinarily articulated by the trigonometric functions sine and cosine.

2.3.1 INVESTIGATION OF GENERALIZED KP-BBM EQUATION

The wave variable

$$u(x, y, t) = U(\xi), \quad \xi = x + y - ct. \tag{2.12}$$

Substituting Eq. (2.12) into Eq. (2.11), then integrate twice with regard to ξ, the following nonlinear differential equation is obtained

$$(1 + k - c)U - a(U^n) + bc(U^n)'' = 0, \tag{2.13}$$

or equivalently

$$(1+k-c)U - a(U^n) + bc(nU^{n-1}U'' + n(n-1)U^{n-2}(U')^2) = 0. \tag{2.14}$$

By integrating the resulting equations and assuming every constants of integration to be zero. Balancing $U^{n-1}U''$ with U in Eq. (2.14) gives

$$(n-1)N + N + 2 = N, \tag{2.15}$$

so that

$$N = -\frac{2}{n-1} \tag{2.16}$$

We use transform to obtain the closed form solution

$$U = V^{-\frac{1}{n-1}} \tag{2.17}$$

to convert Eq. (2.15) into

$$(1+k-c)(n-1)^2V^3 - a(n-1)^2V^2 - bcn(1-2n)(V')^2 - bcn(n-1)VV'' = 0. \tag{2.18}$$

Balancing the term VV'' with V^3 in Eq. (2.18) we find $N+N+2 = 3N$ so that $N = 2$. Using the sine-Gordon expansion method to Eq. (2.10) we get

$$V(w) = B_1\sin(w) + A_1\cos(w) + B_2\cos(w)\sin(w) + A_2\cos^2(w) + A_0. \tag{2.19}$$

Differentiating Eq. (2.19) twice yields

$$V''(w) = B_1\cos^2(w)\sin(w) - B_1\sin^3(w) - 2A_1\sin^2(w)\cos(w) + B_2\cos^3\sin(w) - 5B_2\sin^3(w)\cos(w) - 4A_2\cos^2(w)\sin^2(w) + 2A_2\sin^4(w). \tag{2.20}$$

By substituting Eqs (19–20) into Eq. (2.18) and employing mathematical manipulation, we obtain at a nonlinear algebraic set. We are able to solve this algebraic set with the help of symbolic computational programme, as follows.

Case 1

$A_0 = -iB_2,\ A_1 = 0,\ A_2 = iB_2,\ B_1 = 0,\ a = 4bc,\ k = -1 + c + \frac{6ibc}{B_2}$

$$u_1(x,y,t) = \frac{1}{-iB_2 - \mathrm{sech}[ct-x-y]B_2\tanh[ct-x-y] + iB_2\tanh^2[ct-x-y]}. \tag{2.21}$$

Case 2

$A_0 = \frac{3a}{2-2c+2k};\ A_1 = 0;\ A_2 = \frac{3a}{2(-1+c-k)};\ B_1 = 0;\ B_2 = \frac{3ia}{2-2c+2k};\ b = \frac{a}{4c}$

$$u_2(x,y,t) = \frac{2-2c+2k}{3a - 3ia\,\mathrm{sech}[ct-x-y]\tanh[ct-x-y] + 3a\tanh^2[ct-x-y]}, \tag{2.22}$$

Case 3

$A_0 = -A_2;\ A_1 = 0;\ B_1 = 0;\ B_2 = iA_2;\ b = \frac{aA_2}{6a+4(1+k)A_2};\ c = 1 + k + \frac{3a}{2A_2}$

$$u_3(x,y,t) = \frac{1}{-A_2 + i\text{sech}[x+y-t(1+k+\frac{3a}{2A_2})]A_2\tanh[x+y-t(1+k+\frac{3a}{2A_2})]} \\ +A_2\tanh^2[x+y-t(1+k+\frac{3a}{2A_2})]. \tag{2.23}$$

Case 4

$A_0 = -A_2, A_1 = B_1 = B_2 = 0, c = \frac{a}{16b}, k = -1 + \frac{a}{16b} - \frac{3a}{4A_2},$

$$u_4(x,y,t) = \frac{1}{-A_2 + A_2\tanh^2[\frac{at}{16b} - x - y]}. \tag{2.24}$$

Case 5

$k = -2;\ c = 2;\ A_0 = -\frac{2bn(1+n)}{3(-1+n)^2};\ A_1 = 0;\ A_2 = \frac{2bn(1+n)}{3(-1+n)^2};\ B_1 = 0;$
$B_2 = -\frac{2ibn(1+n)}{3(-1+n)^2};\ a = \frac{2bn^2}{(-1+n)^2}$

$$u_5(x,y,t) = (-\frac{2bn(1+n)}{3(-1+n)^2} + \frac{2ibn(1+n)\text{sech}[2t-x-y]\tanh[2t-x-y]}{3(-1+n)^2} \\ + \frac{2bn(1+n)\tanh^2[2t-x-y]}{3(-1+n)^2})^{\frac{-1}{-1+n}}. \tag{2.25}$$

Case 6

$k = -2;\ c = 2;\ A_0 = -\frac{a(1+n)}{3n};\ A_1 = 0;\ A_2 = \frac{a(1+n)}{3n};\ B_1 = 0;\ B_2 = -\frac{ia(1+n)}{3n};$
$b = \frac{a(-1+n)^2}{2n^2}$

$$u_6(x,y,t) = (-\frac{a(1+n)}{3n} + \frac{ia(1+n)\text{sech}[2t-x-y]\tanh[2t-x-y]}{3n} \\ + \frac{a(1+n)\tanh^2[2t-x-y]}{3n})^{\frac{-1}{-1+n}}. \tag{2.26}$$

Case 7

$k = -2;\ c = 2;\ A_0 = iB_2;\ A_0 = 0; A_1 = -A_2 = iB_2;\ B_1 = -B_2;$
$a = 6iB_2;\ b = 12iB_2;\ n = -\frac{1}{3},$

$$u_7(x,y,t) = (-\text{sech}[2t-x-y]B_2 - iB_2\tanh[2t-x-y] - \text{sech}[2t-x-y] \\ B_2\tanh[2t-x-y] - iB_2\tanh^2[2t-x-y])^{\frac{3}{4}}. \tag{2.27}$$

2.4 CONCLUSIONS

The construction of soliton wave patterns of generalized KP-BBM equations has been investigated via an analytical technique, the SGEM. We obtained the solutions through the hyperbolic and trigonometric-function travelling waves solution of the generalized KP-BBM equation, and the process of selecting the appropriate values for the parameters is shown. The 3D and 2D graph and contour simulations of these solutions are drawn to provide an alternative point of view to the solutions (i.e. Eqs. (2.21)–(2.27) The 3D, 2D, graph, and contour simulations can be seen in Figures 2.1–2.13. We chose $n = 2$ in Cases 1–4 and $n = 2$ in Case 5 for plotting generalized KP-BBM. The proposed method is powerful due to its not requiring linearization, discretization, or disorganization. The acquired results exhibit the confirmation of the algorithm and give it comprehensive applicability to NLPDE. This was enabled by the use of a software program, allowing us to determine solutions to many sets, with less computational work and needing less powerful computers.

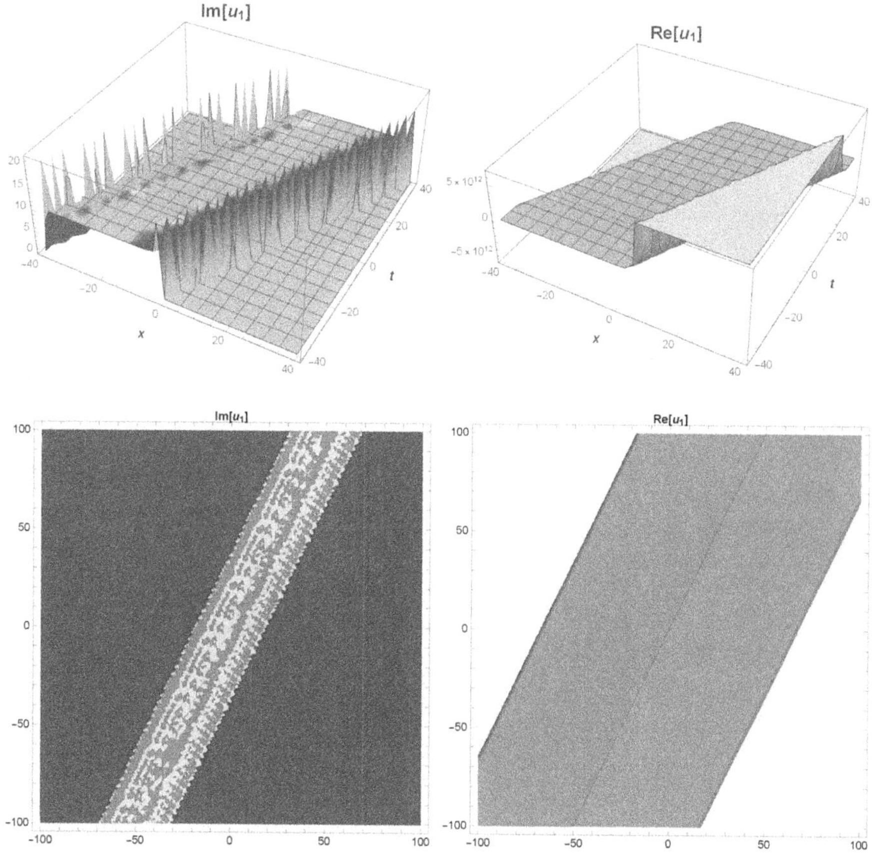

Figure 2.1 The 3D and contour surfaces of the analytical solution Eq. (2.21) by considering the values $B_2 = 0.1, c = 0.5, y = 0.3$.

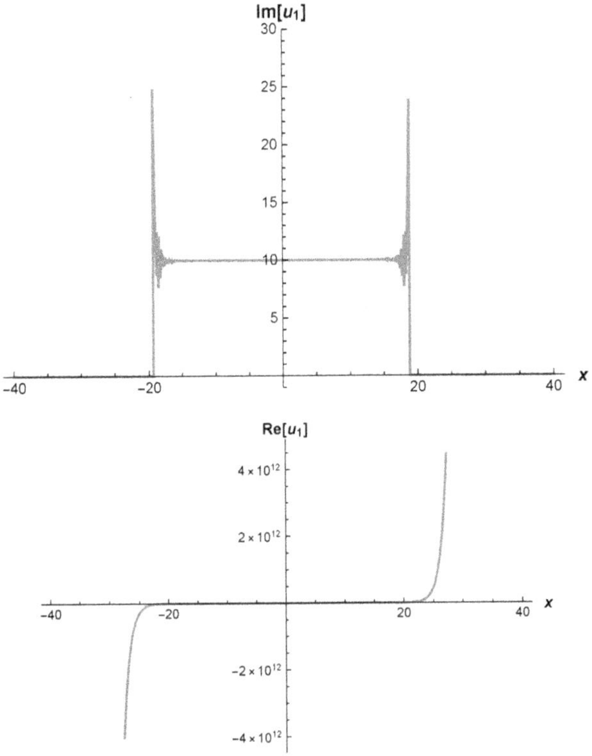

Figure 2.2 The 2D surfaces of the analytical solution Eq. (2.21) by considering the values $B_2 = 0.1, c = 0.5, y = 0.3, t = 0.1$.

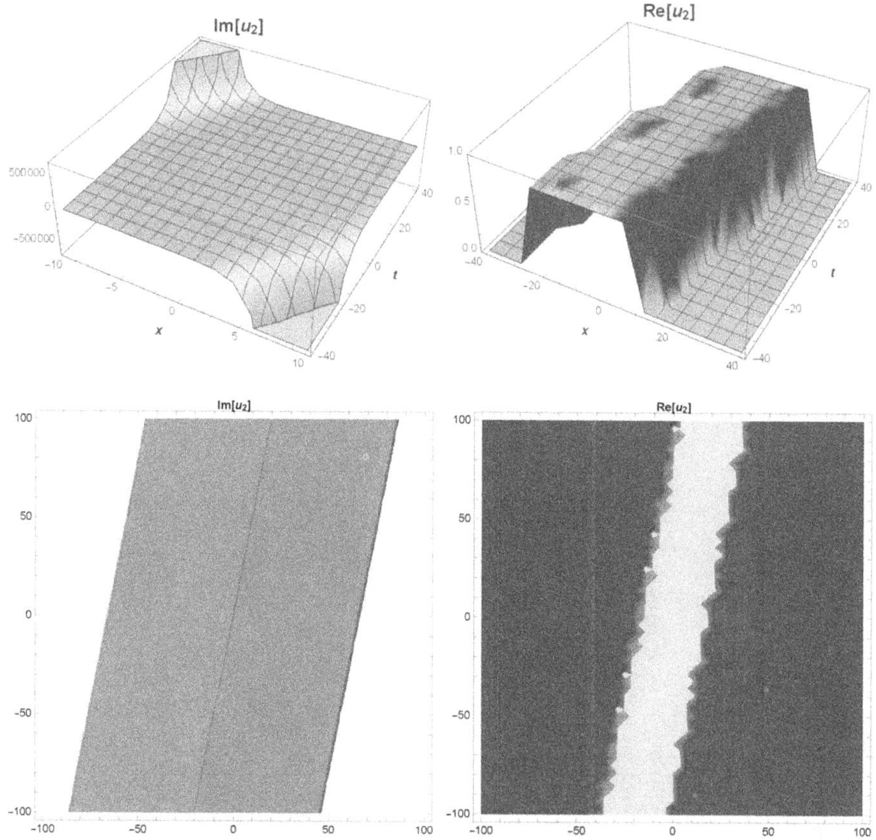

Figure 2.3 The 3D and contour surfaces of the analytical solution Eq. (2.22) by considering the values $k = 0.6, a = 1.1, c = 0.2, y = 0.3$.

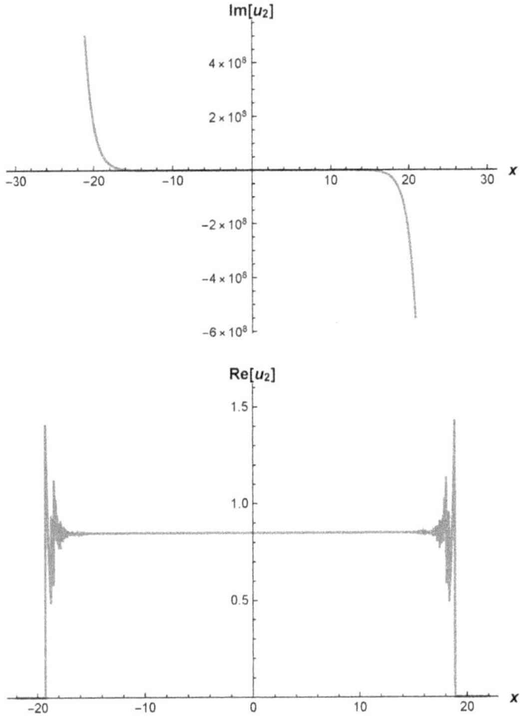

Figure 2.4 The 2D surfaces of the analytical solution Eq. (2.22) by considering the values $k = 0.6, a = 1.1, c = 0.2, y = 0.3, t = 0.6$.

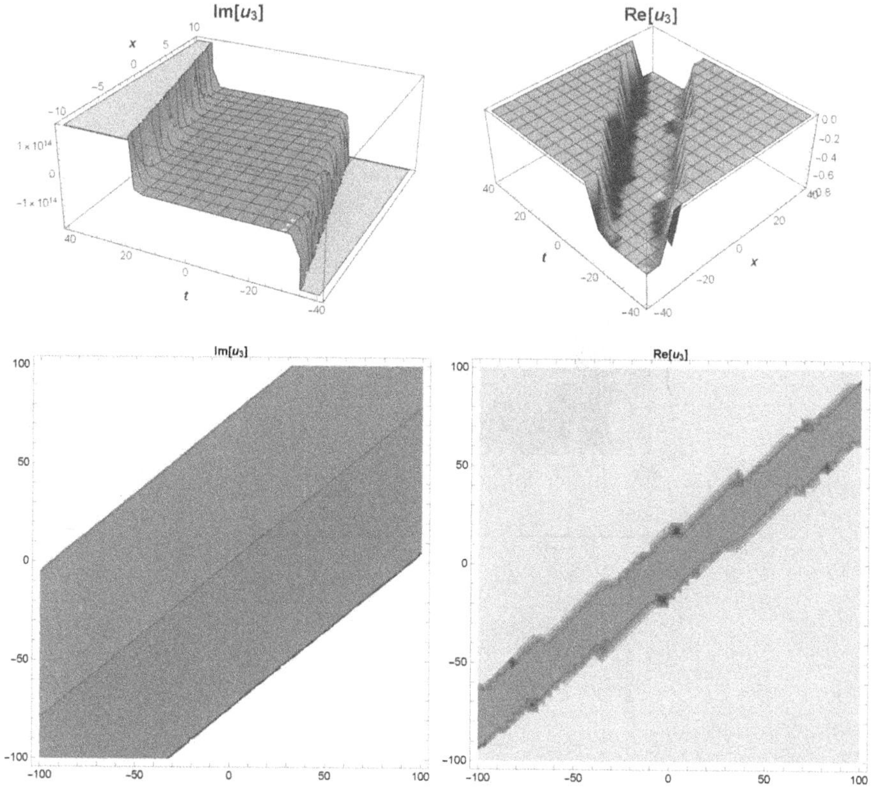

Figure 2.5 The 3D and contour surfaces of the analytical solution Eq. (2.23) by considering the values $A_2 = 2.1, k = 0.1, a = 0.23, y = 0.01$.

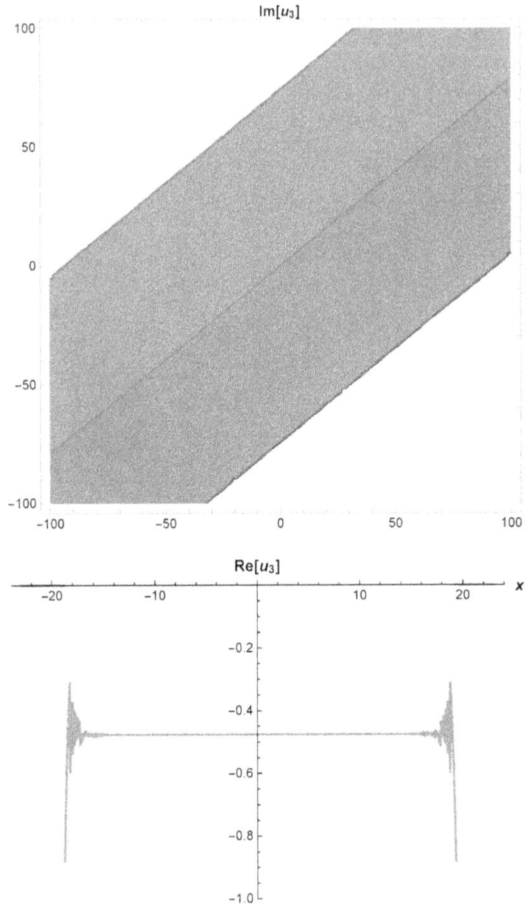

Figure 2.6 The 2D surfaces of the analytical solution Eq. (2.23) by considering the values $A_2 = 2.1, k = 0.1, a = 0.23, y = 0.01, t = 0.21$.

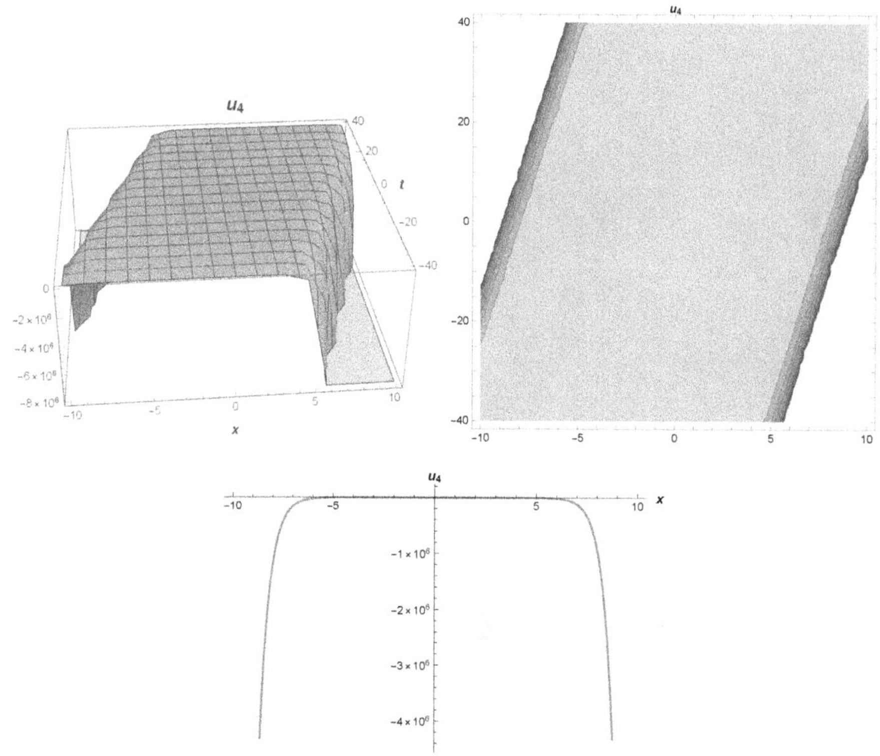

Figure 2.7 The 3D and contour surfaces of the analytical solution Eq. (2.24) by considering the values $A_2 = 2.1, b = 0.23, a = 0.3, y = 0.01, t = 1$.

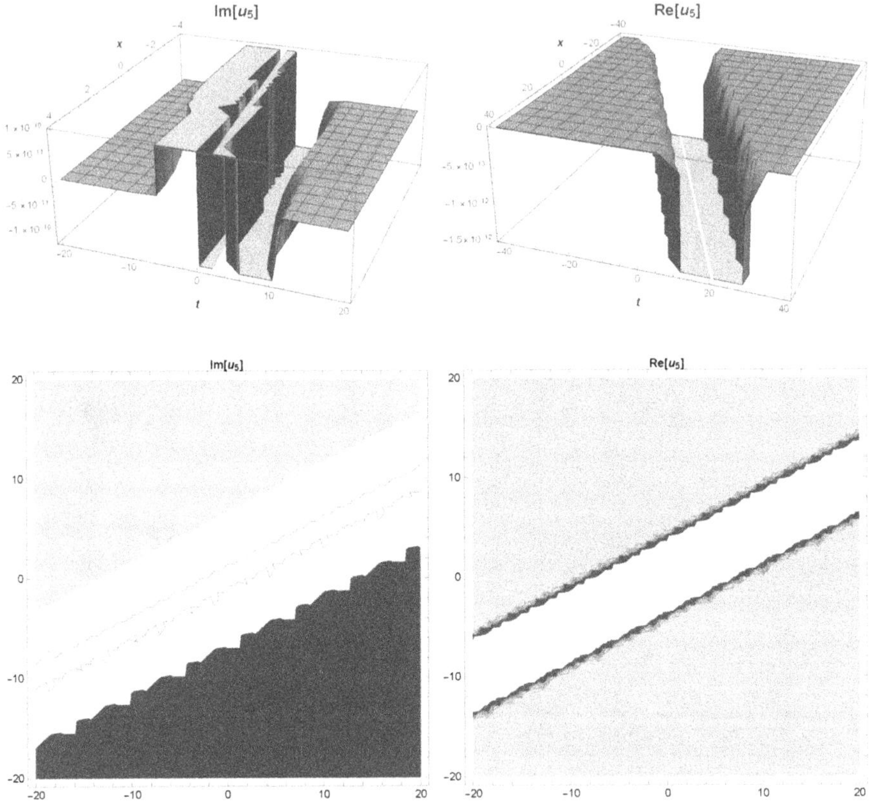

Figure 2.8 The 3D and contour surfaces of the analytical solution Eq. (2.25) by considering the values $k = -2, c = 2, b = 0.5, y = 0.3$.

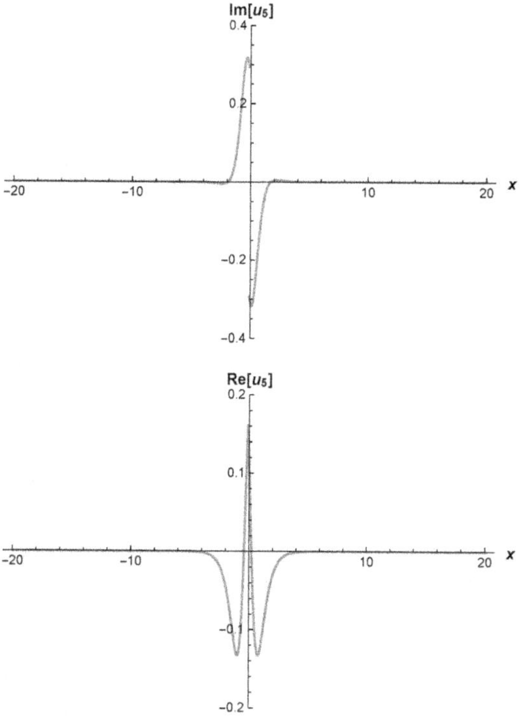

Figure 2.9 The 2D surfaces of the analytical solution Eq. (2.25) by considering the values $k=-2, c=2, b=0.5, y=0.3, t=0.1$.

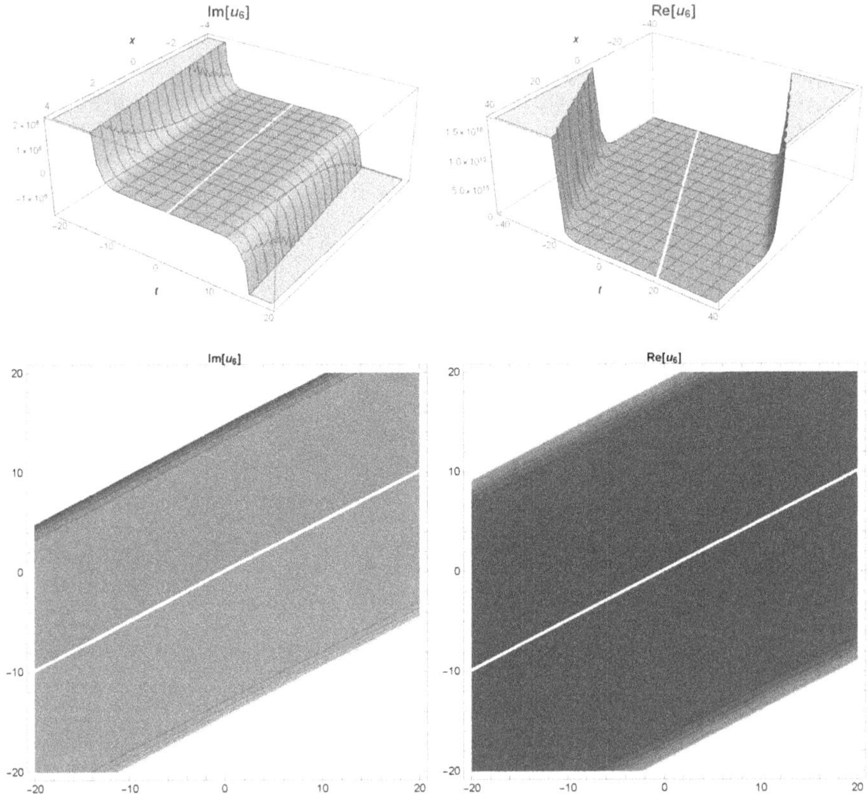

Figure 2.10 The 3D and contour surfaces of the analytical solution Eq. (2.26) by considering the values $k = -2, c = 2, a = 0.5, y = 0.3$.

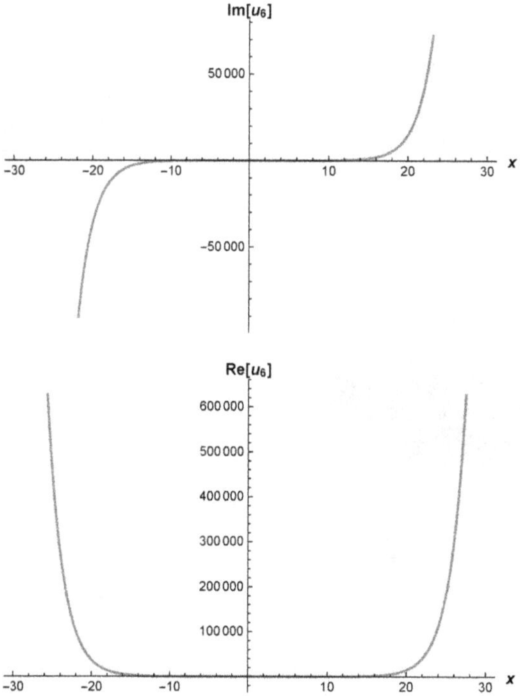

Figure 2.11 The 2D surfaces of the analytical solution Eq. (2.26) by considering the values $k = -2, c = 2, b = 0.5, y = 0.3, t = 0.6$.

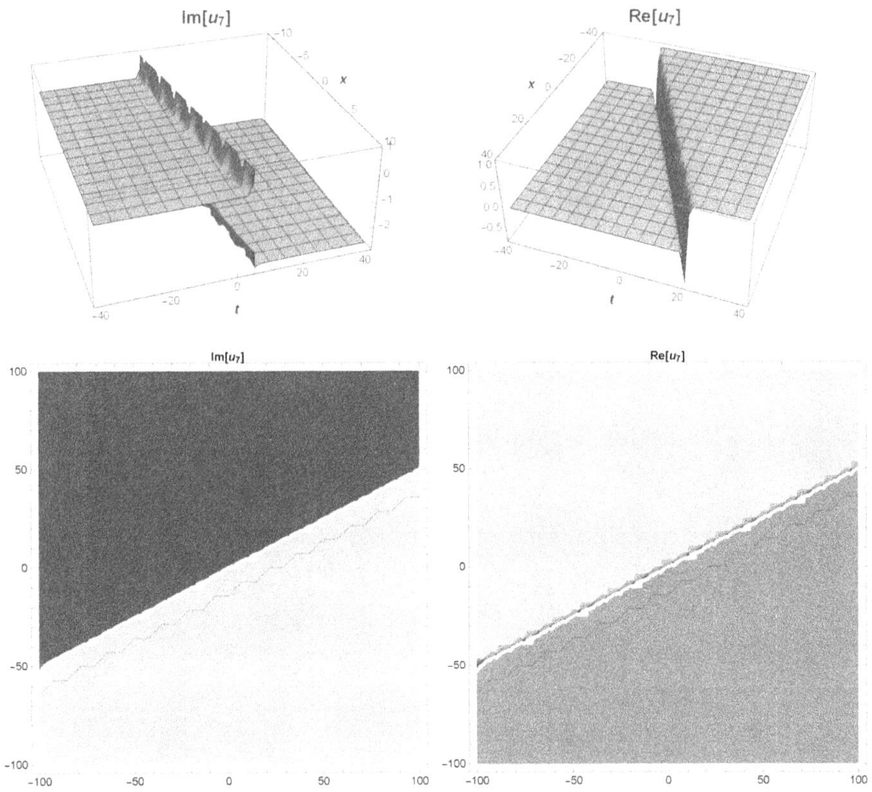

Figure 2.12 The 3D and contour surfaces of the analytical solution Eq. (2.27) by considering the values $k = -2, c = 2, B_2 = 2.1, y = -0.1$.

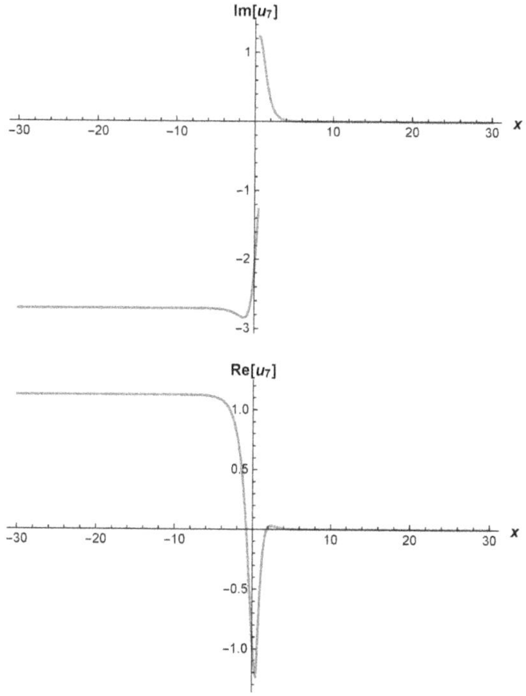

Figure 2.13 The 2D surfaces of the analytical solution Eq. (2.27) by considering the values $k = -2, c = 2, B_2 = 2.1, y = -0.1, t = 0.21$.

REFERENCES

1. Wazwaz, A.M., (2007). The extended tanh method for abundant solitary wave solutions of nonlinear wave equations. Applied Mathematics and Computation, 187, 1131–1142.

2. Wazwaz, A.M., (2005). Compact and noncompact physical structures for the ZK–BBM equation. Applied Mathematics and Computation, 169, 713–725.

3. Ganguly, A., Das, A., (2015). Explicit solutions and stability analysis of the (2 + 1)–dimensional KP-BBM equation with dispersion effect. Communications in Nonlinear Science and Numerical Simulation, PII:S1007–5704(15), 1–27.

4. Wazwaz, A.M., (2004). A sine–cosine method for handling nonlinear wave equations. Mathematical and Computer Modelling, 40, 499–508.

5. Yan, C. (1996). A simple transformation for nonlinear waves. Physics Letters A, 224(1), 77–84.

6. Hu, C.C., Deng, Y.S., Tian, B., Sun, Y., Zhang, C.R., (2019). Rational and semi-rational solutions for the (3+1)-dimensional B–type Kadomtsev Petviashvili Boussinesq equation. Modern Physics Letters B, 950296, 1–20.

7. Lu, D., Seadawy, A.R., Arshad, M., (2009). Applications of extended simple equation method on unstable nonlinear Shrödinger's equations. Physica Scripta, 80, 350–360.

8. Syed, D.M.T., Sadaf, B., (2014). New traveling wave solutions of Drinefeld Sokolov Wilson equation using Tanh and extended Tanh methods. Journal of the Egyptian Mathematical Society, 22(3), 517–523.

9. Yusufoglu, E., Bekir, A., Alp, M., (2008). Periodic and solitary wave solutions of Kawahara and modified Kawahara equations by using sine−cosine method. Chaos, Solitons and Fractals, 37, 1193–1197.

10. Zayed, E.M., Tala-Tebue, E. (2018). New Jacobi elliptic function solutions, solitons and other solutions for the (2+1)-dimensional nonlinear electrical transmission line equation. European Physical Journal-Plus, 133(314), 1–20.

11. Parkes, E.J., Duffy, B.R., (1998). An automated tanh-function method for finding solitary wave solutions to nonlinear evolution equations. Computer Physics Communications, 98, 288–300.

12. Dusunceli, F., (2019). New exponential and complex traveling wave solutions to the Konopelchenko-Dubrovsky model. Advances in Mathematical Physics, 7801247, 1–9.

13. Al-Amr, M.O., El-Ganaini, S., (2017). New exact traveling wave solutions of the (4+1)-dimensional Fokas equation. Computers and Mathematics with Applications, 74(6), 1274–1287.

14. Zhang, G., Yan, Z., Wen, X.Y., (2018). Three-wave resonant interactions: Multi-dark-dark-dark solitons, breathers, rogue waves, and their interactions and dynamics. Physica D: Nonlinear Phenomena, 366, 27–42.

15. Bulut, H., Baskonus, H.M., (2015). On the complex structures of Kundu-Eckhaus equation via improved Bernoulli sub-equation function method. Waves Random Complex Media, 25(4), 720–728.

16. Baskonus, H.M., Bulut, H., Sulaiman, T.A., (2019). New complex hyperbolic structures to the Lonngren-Wave equation by using Sine-Gordon expansion method. Applied Mathematics and Nonlinear Sciences, 4(1), 141–150.

17. He, J.H., (2006). Some asymptotic methods for strongly nonlinear equations. International Journal of Modern Physics B, 20, 1141–1199.

18. Zhang, J.L., Mingliang, X.W., (2008). The $(\frac{G'}{G})$ expansion method and travelling wave solutions of nonlinear evolution equations in mathematical physics. Physics Letters A, 372, 417–423.

19. Hosseini, K., Gholamin, P., (2015). Feng's first integral method for analytic treatment of two higher dimensional nonlinear partial differential equations. Differential Equations and Dynamical Systems, 23, 317–325.

20. Khalfallah, M., (2007). New exact traveling wave solutions of the (2+1) dimensional Zakharov-Kuznetsov (ZK) equation. Analele Stiintifice ale Universitatii Ovidius Constanta, 15(2), 35–44.

21. Moleleki, L.D., Motsepa, T., Khalique, C.M., (2018). Solutions and conservation laws of a generalized second extended (3+1)-dimensional Jimbo-Miwa equation. Applied Mathematics and Nonlinear Sciences, 3(2), 459–474.

22. Girgis, L., Biswas, A., (2010). Soliton perturbation theory for nonlinear wave equations. Applied Mathematics and Computation, 216, 2226–2231.

23. Mirzazadeh, M., (2014). Modified simple equation method and its applications to nonlinear partial differential equations. Information Sciences Letters, 3(1), 1–9.

24. Priya, N.V., Senthilvelan, M., Rangarajan, G., Lakshmanan, M., (2018). On symmetry preserving and symmetry broken bright, dark and antidark soliton solutions of nonlocal nonlinear Schrödinger equation, Physics Letters A, 383, 15–26.

25. Pandey, P.K., Jaboob, S.S.A., (2018). A finite difference method for a numerical solution of elliptic boundary value problems. Applied Mathematics and Nonlinear Sciences, 3(1), 311–320.

26. Liu, Q., (2020). Analytical matter wave solutions of a (2+1)-dimensional partially nonlocal distributed–coefficient Gross–Pitaevskii equation with alinear potential. International Journal for Light and Electron Optics, 163434, 2020.

27. Tekiyeh, R.M., Manafian, J., Baskonus, H.M., Dusunceli, F., (2019). Applications of He's semi-inversevariational method and ITEM to the nonlinear long–short wave interaction system. International Journal of Advanced and Applied Sciences, 6(8), 53–64.

28. Arbabi, S., Najafi, M., (2012). New periodic and soliton solutions of (2+1) dimensional soliton equation. Journal of Advanced Computer Science and Technology, 1(4), 232–239.

29. Yu, Y., Hong-Cai, M., (2010). Explicit solutions of (2 + 1)-dimensional nonlinear KP-BBM equation by using Exp-function method. Applied Mathematics and Computation, 217, 1391–1397.

30. Zhao, Z., He, L., (2019). Multiple lump solutions of the (3+1)–dimensional potential Yu-Toda-Sasa-Fukuyama equation. Applied Mathematics Letters, 95, 114–121.

31. Wazwaz, A.M., (2008). The extended tanh method for new compact and noncompact solutions for the KP–BBM and the ZK–BBM equations. Chaos Solitons and Fractals, 38, 1505–1516.

32. Fan, E., Hon, Y.C., (2003). Applications of extended tanh method to special types of nonlinear equations. Applied Mathematics and Computation, 141, 351–358.

33. Dusunceli, F., (2018). Solutions for the Drinfeld-Sokolov equation using an IBSEFM method. MSU Journal of Science, 6(1), 505–510.

34. Inan, I.E., Kaya, D., (2007). Exact solutions of some nonlinear partial differential equations. Physica A, 381, 104–115.

35. Yoshimura, K., Kobayashi, R., Ohmura, T., Kajimoto, Y., Miura, T., (2016). A new mathematical model for pattern formation by cranial sutures. Journal of Theoretical Biology, 408, 66–74.

36. Islam, M.N., Asaduzzaman, M., Ali, M.S., (2019). Exact wave solutions to the simplified modified Camassa-Holm equation in mathematical physics. AIMS Mathematics, 5(1), 26–41.

37. Taghizadeh, N., Mirzazadeh, M., Paghaleh, A.S., Vahidi, J., (2012). Exact solutions of nonlinear evolution equations by using the modified simple equation method. Ain Shams Engineering, 3, 321–325.

38. Huang, Q.M., (2019). Integrability and dark soliton solutions for a high-order variable coefficients nonlinear Schrödinger equation. Applied Mathematics Letters, 93, 29–33.

39. Raslan, R.K., Evans, D.J., (2005). The tanh function method for solving some important non-linear partial differential equations. International Journal of Computer Mathematics, 82(7), 897–905.

40. Kofane, T.C., Jiotsa, A.K., Fozap, D.C.T., Tala-Tebue, E., (2014). Envelope periodic solutions for a discrete network with the Jacobi elliptic functions and the alternative $(\frac{G'}{G})$expansion method including the generalized Riccati equation. European Physical Journal-Plus, 129(136), 1–20.

41. Serkin, V.N., Ramirez, A., Belyaev, T.L., (2019). Nonlinear-optical analogies to the Moses sea parting effect: Dark soliton in forbidden dispersion or nonlinearity. Optik–International Journal for Light and Electron Optics, 192, 162–928.

42. Wazwaz, A. M., (2005). The tanh function method for solving some important non-linear partial differential equations. International Journal of Computer Mathematics, 82(7), 897–905.

43. Malfliet, W., (2004). The tanh method: A tool for solving certain classes of nonlinear evolution and wave equations. Journal of Computational and Applied Mathematics, (164–165), 529–541.

44. Zhao, Y.M., (2013). F-expansion method and its application for finding new exact solutions to the Kudryashov–Sinelshchikov equation. Journal of Applied Mathematics, 2013(895760), 1–7.

45. Gao, W., Senel, M., Yel, G., Baskonus, H. M., Senel, B., (2020). New complex wave patterns to the electrical transmission line model arising in network system. Aims Mathematics, 5(3), 1881–1892.

46. Gao, W., Ismael, H. F., Husien, A.M., Bulut, H., Baskonus, H.M., (2020). Optical soliton solutions of the nonlinear Schrödinger and resonant nonlinear Schrödinger equation with parabolic law. Applied Science, 10(1), 219, 1–20.

47. Gao, W., Yel, G., Baskonus, H. M., Cattani, C., (2020). Complex solitons in the conformable (2+1)-dimensional Ablowitz-Kaup-Newell-Segur equation. Aims Mathematics, 5(1), 507–521.

48. Yel, G., Baskonus, H.M., Bulut. H., (2017). Novel archetypes of new coupled Konno–Oono equation by using sine–Gordon expansion method. Optical and Quantum Electronics, 49(285), 1–20.

3 Abundant Computational and Numerical Solutions of the Fractional Quantum Version of the Relativistic Energy–Momentum Relation

Mostafa M.A. Khater,[1,2] *Raghda A.M. Attia,*[1,3] *Saud Owyed,*[4] *and Abdel-Haleem Abdel-Aty*[4,5]

[1]Department of Mathematics, Faculty of Science, Jiangsu University, China
[2]Department of Mathematics, Obour Institutes, Cairo, Egypt
[3]Department of Basic Science, Higher Technological Institute 10th of Ramadan City, Egypt
[4]Department of Physics, College of Sciences, University of Bisha, Bisha, Saudi Arabia
[5]Physics Department, Faculty of Science, Al-Azhar University, Assiut, Egypt

CONTENTS

3.1 INTRODUCTION

Fractional calculus is considered as one of the significant branches of science, especially for the phenomena that cannot be represented in the nonlinear partial differential equations (NLPDEs) with an-integer order. The failure of this species to describe some aspects of physics, engineering, and other events is due to the nonlocal property. This property is just found in NLPDEs with the fractional order. The contributions of this type of equations have been manifested in many areas such as physics, mechanical engineering, economics, chemistry, signal processing, food supplement, applied mathematics, quasi-chaotic dynamical systems, hydrodynamics, system identification, statistics, finance, fluid mechanics, solid-state biology, dynamical systems with chaotic dynamical behavior, optical fibers, electric control theory, economics, and diffusion problems. The advantage of this calculus is its ability to provide a great explanation of the nonlocal property of these models since it depends on both the historical and current states of the problem in the contract of the classical calculus, which depends on the current state only. Based on the importance of this kind of calculus, many definitions have been derived, such as conformable fractional derivative, fractional Riemann–Liouville derivatives, Caputo, Caputo–Fabrizio definition, and so on [1–10]. These definitions have been employed to convert the fractional NLPDEs to nonlinear integer-order ordinary differential equation. Then the computational and numerical schemes can be applied to get various types of solutions for these models and the examples of these schemes [11–24].

This chapter investigates the quantum version of the relativistic energy–momentum relation by applying four recent computational and four recent numerical schemes [25–47]. These schemes seek the analytical and numerical solutions of this model to check the accuracy of the obtained solutions by evaluating the absolute value of error between analytical and numerical obtained solutions. Physically, the relativistic–dispersion relation or energy–momentum relation is considered as a relativistic equation relating an object's intrinsic, momentum, and total energy [48–51].

$$\mathscr{E}^2 = (\mathscr{P}\mathscr{C})^2 + \left(\mathscr{M}_0\mathscr{C}^2\right)^2, \tag{3.1}$$

where $\left[\mathscr{E}, \mathscr{P}, \mathscr{C}, \mathscr{M}_0\right]$ represent total energy, a momentum of magnitude, the speed of light, and intrinsic rest mass, respectively. The energy–momentum relation is consistent with the familiar mass–energy relation in both its interpretations, given by

$$\mathscr{E} = \mathscr{M}\mathscr{C}^2, \tag{3.2}$$

where $\mathscr{M}$ is the (total) relativistic mass. Equation (3.1) relates the rest mass $\mathscr{M}_0$ to the total energy. All three equations hold true simultaneously. Moreover, Eq. (3.1) has three special cases, which are given by

1. For $\left[\mathscr{M}_0 = 0\right]$, Eq. (3.1) is usually used for the photons to explain the relation between radiant energy and radiant momentum (causing radiation pressure). This relation was discovered in the 19th century.
2. For the body's speed $\left[\mathscr{V} \lll \mathscr{C}\right]$, Eq. (3.1) takes the following formula

$$\mathscr{E} = \frac{1}{2}\mathscr{M}_0\mathscr{V}^2 + \mathscr{M}_0\mathscr{C}^2, \tag{3.3}$$

   where $\frac{1}{2}\mathscr{M}_0\mathscr{V}^2$ represent the total energy of the body, that is simply its classical kinetic energy.
3. For $\left[\mathscr{V} = 0\right]$, the energy–momentum relation and both forms of the mass–energy relation (mentioned above) all become the same.

We study one of the quantized models of the relativistic energy–momentum relation: the Klein–Gordon equation (Klein–Fock–Gordon equation/Klein–Gordon–Fock equation) [52–56]. This model is a relativistic wave equation, related to the Schrödinger equation [57–61]. Solving this equation investigates the quantum scalar characteristic or pseudo-scalar field's property. Its theoretical relevance is similar to that of the Dirac equation, a relativistic wave equation derived by British physicist Paul Dirac in 1928 and given by [62–64]

$$\left(\mathscr{P}\mathscr{M}\mathscr{C}^2 + \mathscr{C}\sum_{n=1}^{3}\alpha_n\beta_n\right)\Psi = i\hbar\Psi_t, \tag{3.4}$$

where $\left[\mathscr{P}, \mathscr{M}, \mathscr{C}, \alpha, \beta_n\right]$ represent arbitrary constant, the electron of rest mass with space-time coordinates x, t, the speed of light, arbitrary constants, the components of the momentum, $\Psi = \Psi(x,t)$ the wave function for the electron of rest mass with space-time coordinates, and the reduced Planck constant. On the other hand, the KG equation is derived by the physicists Oskar Klein and Walter Gordon, who in 1926 proposed that it describes relativistic electrons. This model is given by

$$\frac{1}{\mathscr{C}^2}\Psi_{tt} - \nabla^2\Psi + \frac{\mathscr{M}^2\mathscr{C}^2}{\hbar}\Psi = 0. \tag{3.5}$$

Equation (3.5) describes the spinless relativistic composite particles, like the pion. In this context, we study the fractional nonlinear formula of Eq. (3.5) that is given by

$$\mathscr{D}_{tt}^{2\alpha}\pounds - \pounds_{xx} + a\pounds + b\pounds^3 = 0, \tag{3.6}$$

where $0 < \alpha < 0$, a, b are arbitrary constants. Additionally, $\pounds = \pounds(x,t)$ represents the spinless relativistic composite particles. Applying the Atangana–Baleanu derivative operator to Eq. (3.6) has the following definition [65, 66].

Definition 3.1: *Atangana–Baleanu fractional operator is given by*

$$^{ABR}\mathscr{D}_{a+}^{\alpha}\mathscr{F}(t) = \frac{\mathscr{B}(\alpha)}{1-\alpha}\frac{d}{dt}\int_a^t \mathscr{F}(x)\,\mathscr{E}_\alpha\left(\frac{-\alpha\,(t-\alpha)^\alpha}{1-\alpha}\right)dx, \tag{3.7}$$

where $\mathscr{E}_\alpha$ is the Mittag–Leffler function, defined by the following formula

$$\mathscr{E}_\alpha\left(\frac{-\alpha\,(t-\alpha)^\alpha}{1-\alpha}\right) = \sum_{n=0}^{\infty}\frac{\left(\frac{-\alpha}{1-\alpha}\right)^n (t-x)^{\alpha n}}{\Gamma(\alpha n+1)}$$

and $\mathscr{B}(\alpha)$ being a normalization function. Thus

$$^{ABR}\mathscr{D}_{a+}^{\alpha}\mathscr{F}(x) = \frac{\mathscr{B}(\alpha)}{1-\alpha}\sum_{n=0}^{\infty}\left(\frac{-\alpha}{1-\alpha}\right)^n {}^{RL}\mathscr{I}_a^{\alpha n}\mathscr{F}(x). \tag{3.8}$$

This definition leads to the following wave transformation

$$\pounds(x,t) = \pounds(\Xi),\ \Xi = x + \frac{(1-\alpha)\,c\,t^{\alpha(-n)}}{B(\alpha)\sum_{n=0}^{\infty}\left(-\frac{\alpha}{1-\alpha}\right)^n\Gamma(1-\alpha n)}. \tag{3.9}$$

This transform converts the fractional nonlinear KG equation into the following ordinary differential equation with an-integer order

$$-a\pounds + b\pounds^3 + \left(c^2-1\right)\pounds'' = 0. \tag{3.10}$$

Balancing the nonlinear term and highest order derivative term yields $(n = 1)$.

We now offer a description of the structure of this chapter. Section 3.2 applies four recent schemes to the fractional nonlinear KG equation to get distinct formulas of computational solutions. Moreover, some sketches are given to show more physical properties of this model. Section 3.3 discusses the stability property of the obtained solutions. Section 3.4, investigates the numerical solutions of this model. Also, some figures are plotted to explain the absolute value of error between exact and numerical solutions. Section 3.6 gives the conclusion of the research.

3.2 ANALYTICAL EXPLICIT WAVE SOLUTIONS

This section applies some recent computational schemes to the fractional quantum version of the relativistic energy–momentum relation (the nonlinear fractional Klein–Gordon equation). The aim of this section is to investigate novel explicit wave solutions to explain more physical properties of this fractional quantum model.

3.2.1 EXTENDED EXP ($-\phi(\Xi)$) EXPANSION METHOD

Applying the extended exp $(-\phi(\Xi))$ expansion method to Eq. (3.10) leads to a general form of solutions according to this technique that takes the following form:

$$\pounds(\Xi) = \frac{\sum_{i=0}^{m} a_i e^{-i\phi(\Xi)}}{\sum_{i=0}^{n} b_i e^{-i\phi(\Xi)}} = \frac{a_1 e^{-\phi(\Xi)} + a_2 e^{-2\phi(\Xi)} + a_0}{b_1 e^{-\phi(\Xi)} + b_0}, \tag{3.11}$$

where $\left[a_0, a_1, a_2, b_0, b_1\right]$ are arbitrary constants. Additionally, $\phi(\Xi)$ is the solution function of the following auxiliary equation:

$$\phi'(\Xi) = \beta + \delta\, e^{\phi(\Xi)} + \frac{\rho}{e^{\phi(\Xi)}}, \tag{3.12}$$

where $\left[\beta, \delta, \rho\right]$ are arbitrary constants. Substituting Eqs. (3.11) and (3.12) into Eq. (3.10), and collecting all terms with the same power of $\left[e^{-i\phi(\Xi)}, i = 0, 1, 2, ...\right]$ leads to a system of algebraic systems. Using Mathematica 12 to solve this system yields the following:

Family I

$$\left[a_0 \to -\frac{i\beta b_0 \sqrt{c^2-1}}{\sqrt{2}\sqrt{b}}, a_1 \to -\frac{i\sqrt{c^2-1}\,(\beta b_1 + 2 b_0 \rho)}{\sqrt{2}\sqrt{b}}, a_2 \to -\frac{i\sqrt{2} b_1 \sqrt{c^2-1}\rho}{\sqrt{b}}, a \to -\frac{1}{2}\left(c^2-1\right)\left(\beta^2 - 4\delta\rho\right)\right],$$

where $\left[b < 0, c \neq \pm 1\right]$. Thus, the explicit wave solutions of the fractional nonlinear KG equation are given under the following conditions with the next formulas:

Case 1 ($\rho = 1$):

For $\left[\beta^2 - 4\delta > 0 \,\&\, \delta \neq 0\right]$

$$\pounds_1(x,t) = -\frac{i\sqrt{c^2-1}\left(\beta\sqrt{\beta^2-4\delta}\tanh\left(\frac{1}{2}\sqrt{\beta^2-4\delta}\times(\eta+\frac{(1-\alpha)\left(ct^{\alpha(-n)}\right)}{B(\alpha)\sum_{n=0}^{\infty}\left(-\frac{\alpha}{1-\alpha}\right)^n\Gamma(1-\alpha n)}+x)\right)+\beta^2-4\delta\right)}{\sqrt{2}\sqrt{b}\left(\sqrt{\beta^2-4\delta}\tanh\left(\frac{1}{2}\sqrt{\beta^2-4\delta}\times(\eta+\frac{(1-\alpha)\left(ct^{\alpha(-n)}\right)}{B(\alpha)\sum_{n=0}^{\infty}\left(-\frac{\alpha}{1-\alpha}\right)^n\Gamma(1-\alpha n)}+x)\right)+\beta\right)}, \tag{3.13}$$

$$\pounds_2(x,t) = -\frac{i\sqrt{c^2-1}\left(\beta\sqrt{\beta^2-4\delta}\coth\left(\frac{1}{2}\sqrt{\beta^2-4\delta}\times(\eta+\frac{(1-\alpha)\left(ct^{\alpha(-n)}\right)}{B(\alpha)\sum_{n=0}^{\infty}\left(-\frac{\alpha}{1-\alpha}\right)^n\Gamma(1-\alpha n)}+x)\right)+\beta^2-4\delta\right)}{\sqrt{2}\sqrt{b}\left(\sqrt{\beta^2-4\delta}\coth\left(\frac{1}{2}\sqrt{\beta^2-4\delta}\times(\eta+\frac{(1-\alpha)\left(ct^{\alpha(-n)}\right)}{B(\alpha)\sum_{n=0}^{\infty}\left(-\frac{\alpha}{1-\alpha}\right)^n\Gamma(1-\alpha n)}+x)\right)+\beta\right)}. \tag{3.14}$$

For $\left[\beta^2 - 4\delta > 0 \,\&\, \delta \neq 0\right]$

$$\pounds_3(x,t) = -\frac{i\beta\sqrt{c^2-1}\coth\left(\frac{1}{2}\beta(\eta+\frac{(1-\alpha)\left(ct^{\alpha(-n)}\right)}{B(\alpha)\sum_{n=0}^{\infty}\left(-\frac{\alpha}{1-\alpha}\right)^n\Gamma(1-\alpha n)}+x)\right)}{\sqrt{2}\sqrt{b}}. \tag{3.15}$$

For $\left[\beta^2 - 4\delta = 0 \,\&\, \delta \neq 0 \,\&\, \beta \neq 0\right]$

$$\pounds_4(x,t) = -\frac{i\sqrt{2}\beta\sqrt{c^2-1}}{\sqrt{b}(\beta(\eta+\frac{(1-\alpha)\left(ct^{\alpha(-n)}\right)}{B(\alpha)\sum_{n=0}^{\infty}\left(-\frac{\alpha}{1-\alpha}\right)^n\Gamma(1-\alpha n)}+x)+2)}. \tag{3.16}$$

For $\left[\beta^2 - 4\delta = 0 \,\&\, \delta = 0 \,\&\, \beta = 0\right]$

$$\pounds_5(x,t) = -\frac{i\sqrt{2}\sqrt{c^2-1}}{\sqrt{b}(\eta+\frac{(1-\alpha)\left(ct^{\alpha(-n)}\right)}{B(\alpha)\sum_{n=0}^{\infty}\left(-\frac{\alpha}{1-\alpha}\right)^n\Gamma(1-\alpha n)}+x)}. \tag{3.17}$$

For $\left[\beta^2-4\delta<0\&\delta\neq 0\&\beta\neq 0\right]$

$$£_6=\frac{i\sqrt{c^2-1}\left(\beta\sqrt{4\delta-\beta^2}\tan\left(\frac{1}{2}\sqrt{4\delta-\beta^2}\times(\eta+\frac{(1-\alpha)\left(ct^{\alpha(-n)}\right)}{B(\alpha)\sum_{n=0}^{\infty}\left(-\frac{\alpha}{1-\alpha}\right)^n\Gamma(1-\alpha n)}+x)\right)-\beta^2+4\delta\right)}{\sqrt{2}\sqrt{b}\left(\beta-\sqrt{4\delta-\beta^2}\tan\left(\frac{1}{2}\sqrt{4\delta-\beta^2}\times(\eta+\frac{(1-\alpha)\left(ct^{\alpha(-n)}\right)}{B(\alpha)\sum_{n=0}^{\infty}\left(-\frac{\alpha}{1-\alpha}\right)^n\Gamma(1-\alpha n)}+x)\right)\right)}, \quad (3.18)$$

$$£_7=\frac{i\sqrt{c^2-1}\left(\beta\sqrt{4\delta-\beta^2}\cot\left(\frac{1}{2}\sqrt{4\delta-\beta^2}\times(\eta+\frac{(1-\alpha)\left(ct^{\alpha(-n)}\right)}{B(\alpha)\sum_{n=0}^{\infty}\left(-\frac{\alpha}{1-\alpha}\right)^n\Gamma(1-\alpha n)}+x)\right)-\beta^2+4\delta\right)}{\sqrt{2}\sqrt{b}\left(\beta-\sqrt{4\delta-\beta^2}\cot\left(\frac{1}{2}\sqrt{4\delta-\beta^2}\times(\eta+\frac{(1-\alpha)\left(ct^{\alpha(-n)}\right)}{B(\alpha)\sum_{n=0}^{\infty}\left(-\frac{\alpha}{1-\alpha}\right)^n\Gamma(1-\alpha n)}+x)\right)\right)}. \quad (3.19)$$

Case 2 ($\beta=0$):

For $\left[\rho>0\&\delta>0\right]$

$$£_8(x,t)=-\frac{i\sqrt{2}\sqrt{c^2-1}\delta\sqrt{\frac{\rho}{\delta}}\cot\left(\sqrt{\delta\rho}(\frac{(1-\alpha)\left(ct^{\alpha(-n)}\right)}{B(\alpha)\sum_{n=0}^{\infty}\left(-\frac{\alpha}{1-\alpha}\right)^n\Gamma(1-\alpha n)}+x+\vartheta)\right)}{\sqrt{b}}, \quad (3.20)$$

$$£_9(x,t)=\frac{i\sqrt{2}\sqrt{c^2-1}\delta\sqrt{\frac{\rho}{\delta}}\tan\left(\sqrt{\delta\rho}(\frac{(1-\alpha)\left(ct^{\alpha(-n)}\right)}{B(\alpha)\sum_{n=0}^{\infty}\left(-\frac{\alpha}{1-\alpha}\right)^n\Gamma(1-\alpha n)}+x+\vartheta)\right)}{\sqrt{b}}. \quad (3.21)$$

For $\left[\rho\delta<0\&\rho<0\right]$

$$£_{10}(x,t)=\frac{i\sqrt{2}\sqrt{c^2-1}\rho\sqrt{-\frac{\delta}{\rho}}\tanh\times\left(\sqrt{-\delta\rho}(\frac{(1-\alpha)\left(ct^{\alpha(-n)}\right)}{B(\alpha)\sum_{n=0}^{\infty}\left(-\frac{\alpha}{1-\alpha}\right)^n\Gamma(1-\alpha n)}+x+\vartheta)\right)}{\sqrt{b}}, \quad (3.22)$$

$$£_{11}(x,t)=\frac{i\sqrt{2}\sqrt{c^2-1}\rho\sqrt{-\frac{\delta}{\rho}}\coth\times\left(\sqrt{-\delta\rho}(\frac{(1-\alpha)\left(ct^{\alpha(-n)}\right)}{B(\alpha)\sum_{n=0}^{\infty}\left(-\frac{\alpha}{1-\alpha}\right)^n\Gamma(1-\alpha n)}+x+\vartheta)\right)}{\sqrt{b}}. \quad (3.23)$$

Case 3 ($\delta = \beta = 0$):

$$\pounds_{12}(x,t) = -\frac{i\sqrt{2}\sqrt{c^2-1}}{\sqrt{b}\left(\frac{(1-\alpha)\left(ct^{\alpha(-n)}\right)}{B(\alpha)\sum_{n=0}^{\infty}\left(-\frac{\alpha}{1-\alpha}\right)^n\Gamma(1-\alpha n)}+x+\vartheta\right)}. \tag{3.24}$$

Family II

$$\left[a_1 \to -\frac{i\sqrt{2}\beta b_1\sqrt{c^2-1}}{\sqrt{b}}, a_2 \to -\frac{i\sqrt{2}b_1\sqrt{c^2-1}\rho}{\sqrt{b}}, b_0 \to \frac{\beta b_1}{2\rho}, a \to \left(c^2-1\right)\right.$$
$$\left.\times\left(\beta^2+2\delta\rho\right) - \frac{3i\sqrt{2}a_0\sqrt{b}\sqrt{c^2-1}\rho}{b_1},\right]$$

where $\left[b < 0, c \neq \pm 1\right]$. Thus, the explicit wave solutions of the fractional nonlinear KG equation are given under the following conditions with the next formulas:

Case 1 ($\rho = 1$):

For $\left[\beta^2 - 4\delta > 0 \,\&\, \delta \neq 0\right]$

$$\pounds_{13}(x,t) = \frac{2\left(a_0 + \frac{2i\sqrt{2}b_1\sqrt{c^2-1}\delta\left(\beta\sqrt{\beta^2-4\delta}\tanh\left(\frac{1}{2}\sqrt{\beta^2-4\delta}(\eta+\xi)\right)+\beta^2-2\delta\right)}{\sqrt{b}\left(\sqrt{\beta^2-4\delta}\tanh\left(\frac{1}{2}\sqrt{\beta^2-4\delta}(\eta+\xi)\right)+\beta\right)^2}\right)}{b_1\left(\beta - \frac{4\delta}{\sqrt{\beta^2-4\delta}\tanh\left(\frac{1}{2}\sqrt{\beta^2-4\delta}(\eta+\xi)\right)+\beta}\right)}, \tag{3.25}$$

$$\pounds_{14}(x,t) = \frac{2\left(a_0 + \frac{2i\sqrt{2}b_1\sqrt{c^2-1}\delta\left(\beta\sqrt{\beta^2-4\delta}\coth\left(\frac{1}{2}\sqrt{\beta^2-4\delta}(\eta+\xi)\right)+\beta^2-2\delta\right)}{\sqrt{b}\left(\sqrt{\beta^2-4\delta}\coth\left(\frac{1}{2}\sqrt{\beta^2-4\delta}(\eta+\xi)\right)+\beta\right)^2}\right)}{b_1\left(\beta - \frac{4\delta}{\sqrt{\beta^2-4\delta}\coth\left(\frac{1}{2}\sqrt{\beta^2-4\delta}(\eta+\xi)\right)+\beta}\right)}. \tag{3.26}$$

For $\left[\beta^2 - 4\delta > 0 \,\&\, \delta \neq 0\right]$

$$\pounds_{15}(x,t) = \frac{2a_0\tanh\left(\frac{1}{2}\beta(\eta+\xi)\right)}{\beta b_1} - \frac{i\sqrt{2}\beta\sqrt{c^2-1}\operatorname{csch}(\beta(\eta+\xi))}{\sqrt{b}}. \tag{3.27}$$

For $\left[\beta^2 - 4\delta = 0 \,\&\, \delta \neq 0 \,\&\, \beta \neq 0\right]$

$$\pounds_{16}(x,t) = \frac{(\beta(\eta+\xi)+2)\left(a_0 + \frac{ib_1\beta^3\sqrt{c^2-1}(\eta+\xi)(\beta(\eta+\xi)+4)}{2\sqrt{2}\sqrt{b}(\beta(\eta+\xi)+2)^2}\right)}{\beta b_1}. \tag{3.28}$$

For $\left[\beta^2-4\delta=0\&\delta=0\&\beta=0\right]$

$$£_{17}(x,t)=\frac{a_0(\eta+\xi)}{b_1}-\frac{i\sqrt{2}\sqrt{c^2-1}}{\sqrt{b}(\eta+\xi)}. \tag{3.29}$$

For $\left[\beta^2-4\delta<0\&\delta\neq 0\&\beta\neq 0\right]$

$$£_{18}=\frac{2\left(a_0+\frac{2i\sqrt{2}b_1\sqrt{c^2-1}\delta\left(-\beta\sqrt{4\delta-\beta^2}\tan\left(\frac{1}{2}\sqrt{4\delta-\beta^2}(\eta+\xi)\right)+\beta^2-2\delta\right)}{\sqrt{b}\left(\beta-\sqrt{4\delta-\beta^2}\tan\left(\frac{1}{2}\sqrt{4\delta-\beta^2}(\eta+\xi)\right)\right)^2}\right)}{b_1\left(\beta-\frac{4\delta}{\beta-\sqrt{4\delta-\beta^2}\tan\left(\frac{1}{2}\sqrt{4\delta-\beta^2}(\eta+\xi)\right)}\right)}, \tag{3.30}$$

$$£_{19}=\frac{2\left(a_0+\frac{2i\sqrt{2}b_1\sqrt{c^2-1}\delta\left(-\beta\sqrt{4\delta-\beta^2}\cot\left(\frac{1}{2}\sqrt{4\delta-\beta^2}(\eta+\xi)\right)+\beta^2-2\delta\right)}{\sqrt{b}\left(\beta-\sqrt{4\delta-\beta^2}\cot\left(\frac{1}{2}\sqrt{4\delta-\beta^2}(\eta+\xi)\right)\right)^2}\right)}{b_1\left(\beta-\frac{4\delta}{\beta-\sqrt{4\delta-\beta^2}\cot\left(\frac{1}{2}\sqrt{4\delta-\beta^2}(\eta+\xi)\right)}\right)}. \tag{3.31}$$

Case 2 ($\beta=0$):

For $\left[\rho>0\&\delta>0\right]$

$$£_{20}(x,t)=\sqrt{\frac{\rho}{\delta}}\tan\left(\sqrt{\delta\rho}(\xi+\vartheta)\right)\left(\frac{a_0}{b_1}-\frac{i\sqrt{2}\sqrt{c^2-1}\delta\cot^2\left(\sqrt{\delta\rho}(\xi+\vartheta)\right)}{\sqrt{b}}\right), \tag{3.32}$$

$$£_{21}(x,t)=\sqrt{\frac{\rho}{\delta}}\tan\left(\sqrt{\delta\rho}(\xi+\vartheta)\right)\left(-\frac{a_0\cot^2\left(\sqrt{\delta\rho}(\xi+\vartheta)\right)}{b_1}+\frac{i\sqrt{2}\sqrt{c^2-1}\delta}{\sqrt{b}}\right). \tag{3.33}$$

For $\left[\rho\delta<0\&\rho<0\right]$

$$£_{22}(x,t)=\frac{a_0\left(-\sqrt{b}\right)\coth\left(\sqrt{-\delta\rho}(\xi+\vartheta)\right)-i\sqrt{2}b_1\sqrt{c^2-1}\delta\tanh\left(\sqrt{-\delta\rho}(\xi+\vartheta)\right)}{\sqrt{b}b_1\sqrt{-\frac{\delta}{\rho}}}, \tag{3.34}$$

$$\pounds_{23}(x,t)=\frac{\begin{array}{c}-a_0\sqrt{b}\tanh\left(\sqrt{-\delta\rho}(\xi+\vartheta)\right)\\ -i\sqrt{2}b_1\sqrt{c^2-1}\delta\coth\left(\sqrt{-\delta\rho}(\xi+\vartheta)\right)\end{array}}{\sqrt{b}b_1\sqrt{-\frac{\delta}{\rho}}}. \tag{3.35}$$

Case 3 $(\delta=\beta=0)$**:**

$$\pounds_{24}(x,t)=\frac{\frac{a_0\rho(\xi+\vartheta)^2}{b_1}-\frac{i\sqrt{2}\sqrt{c^2-1}}{\sqrt{b}}}{\xi+\vartheta} \tag{3.36}$$

where $\left[\xi=\frac{(1-\alpha)\left(ct^{\alpha(-n)}\right)}{B(\alpha)\sum_{n=0}^{\infty}\left(-\frac{\alpha}{1-\alpha}\right)^n\Gamma(1-\alpha n)}+x\right]$.

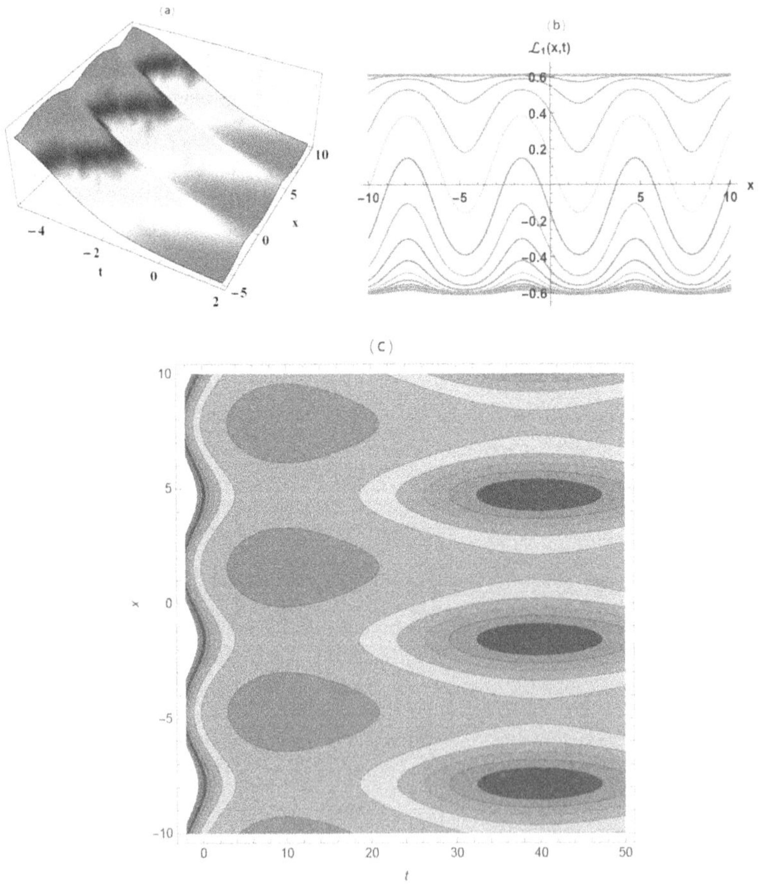

Figure 3.1 Numerical simulation of Eq. (3.13) in three distinct forms.

3.2.2 EXTENDED FAN EXPANSION METHOD

Applying the extended Fan expansion method to Eq. (3.10) leads to a general form of solutions according to this technique that takes the following form:

$$\pounds(\Xi) = \sum_{i=0}^{n} a_i \phi(\Xi)^i = a_1 \phi(\Xi) + a_0, \tag{3.37}$$

where $\left[a_0, a_1\right]$ are arbitrary constants. Additionally, $\phi(\Xi)$ is the solution function of the following auxiliary equation:

$$\phi'(\Xi)^2 = \mu^2 \phi(\Xi)^4 + \phi(\Xi)^2 \left(2\mu\chi + \rho^2\right) + 2\mu\rho\phi(\Xi)^3 + 2\chi\rho\phi(\Xi) + \chi^2, \tag{3.38}$$

where $\left[\mu, \chi, \rho\right]$ are arbitrary constants. Substituting Eq. (3.37) and (3.38) into Eq. (3.10) and collecting all terms with the same power of $\left[\phi(\Xi)^i, i = 0, 1, 2, \ldots\right]$ leads to a system of algebraic systems. Using Mathematica 12 to solve this system yields

$$\left[a_0 \to \frac{\sqrt{\rho^2 - c^2\rho^2}}{\sqrt{2}\sqrt{b}}, a_1 \to \frac{\sqrt{2}\mu\sqrt{-(c^2-1)\rho^2}}{\sqrt{b}\rho}, a \to -\frac{1}{2}\left(c^2-1\right)\left(\rho^2 - 4\mu\chi\right)\right],$$

where $\left[b > 0, c \neq 1\right]$. Therefore, the explicit wave solutions of the fractional nonlinear KG equation are given under the following conditions with the next formulas:

For $\left[\rho^2 - 4\mu\chi > 0, \mu\rho \neq 0 (\mu\chi \neq 0)\right]$

$$\pounds_{25}(x,t) = -\frac{\sqrt{-(c^2-1)\rho^2}\sqrt{\rho^2-4\mu\chi}\tanh \times \left(\frac{1}{2}\sqrt{\rho^2-4\mu\chi}\left(x - \frac{(\alpha-1)ct^{-\alpha}}{B(\alpha)\sum_{n=0}^{\infty}\left(-\frac{\alpha}{1-\alpha}\right)^n\Gamma(1-\alpha n)}\right)\right)}{\sqrt{2}\sqrt{b}\rho}, \tag{3.39}$$

$$\pounds_{26}(x,t) = -\frac{\sqrt{-(c^2-1)\rho^2}\sqrt{\rho^2-4\mu\chi}\coth \times \left(\frac{1}{2}\sqrt{\rho^2-4\mu\chi}\left(x - \frac{(\alpha-1)ct^{-\alpha}}{B(\alpha)\sum_{n=0}^{\infty}\left(-\frac{\alpha}{1-\alpha}\right)^n\Gamma(1-\alpha n)}\right)\right)}{\sqrt{2}\sqrt{b}\rho}, \tag{3.40}$$

$$\begin{aligned}\pounds_{27}(x,t) = &\frac{-1}{\sqrt{2}\sqrt{b}\rho}\left[\sqrt{-(c^2-1)\rho^2}\sqrt{\rho^2-4\mu\chi}\left(\tanh\left(\sqrt{\rho^2-4\mu\chi}\left(x\right.\right.\right.\right.\\ &\left.\left.-\frac{(\alpha-1)ct^{-\alpha}}{B(\alpha)\sum_{n=0}^{\infty}(-\frac{\alpha}{1-\alpha})^n\Gamma(1-\alpha n)}\right)\right) \pm i\,\mathrm{sech}\left(\sqrt{\rho^2-4\mu\chi}\left(x\right.\right.\\ &\left.\left.\left.\left.-\frac{(\alpha-1)ct^{-\alpha}}{B(\alpha)\sum_{n=0}^{\infty}(-\frac{\alpha}{1-\alpha})^n\Gamma(1-\alpha n)}\right)\right)\right)\right],\end{aligned} \tag{3.41}$$

$$
\begin{aligned}
\pounds_{28}(x,t) = & \frac{-1}{\sqrt{2}\sqrt{b}\rho}\left[\sqrt{-(c^2-1)\rho^2}\sqrt{\rho^2-4\mu\chi}\left(\coth\left(\sqrt{\rho^2-4\mu\chi}\left(x\right.\right.\right.\right. \\
& \left.\left. - \frac{(\alpha-1)ct^{-\alpha}}{B(\alpha)\sum_{n=0}^{\infty}\left(-\frac{\alpha}{1-\alpha}\right)^n\Gamma(1-\alpha n)}\right)\right) \pm \operatorname{csch}\left(\sqrt{\rho^2-4\mu\chi}\left(x\right.\right. \\
& \left.\left.\left.\left. - \frac{(\alpha-1)ct^{-\alpha}}{B(\alpha)\sum_{n=0}^{\infty}\left(-\frac{\alpha}{1-\alpha}\right)^n\Gamma(1-\alpha n)}\right)\right)\right)\right],
\end{aligned}
\tag{3.42}
$$

$$
\begin{aligned}
\pounds_{29}(x,t) = & \frac{-1}{2\sqrt{2}\sqrt{b}\rho}\left[\sqrt{-(c^2-1)\rho^2}\sqrt{\rho^2-4\mu\chi}\tanh\left(\frac{1}{4}\sqrt{\rho^2-4\mu\chi}\left(x\right.\right.\right. \\
& \left.\left. - \frac{(\alpha-1)ct^{-\alpha}}{B(\alpha)\sum_{n=0}^{\infty}\left(-\frac{\alpha}{1-\alpha}\right)^n\Gamma(1-\alpha n)}\right)\right)\left(\coth^2\left(\frac{1}{4}\sqrt{\rho^2-4\mu\chi}\left(x\right.\right.\right. \\
& \left.\left.\left.\left. - \frac{(\alpha-1)ct^{-\alpha}}{B(\alpha)\sum_{n=0}^{\infty}\left(-\frac{\alpha}{1-\alpha}\right)^n\Gamma(1-\alpha n)}\right)\right) + 1\right)\right],
\end{aligned}
\tag{3.43}
$$

$$
\begin{aligned}
\pounds_{30}(x,t) = & \left[\sqrt{-(c^2-1)\rho^2}\left(\sqrt{(A^2+B^2)(\rho^2-4\mu\chi)}\right.\right. \\
& - A\sqrt{\rho^2-4\mu\chi}\cosh\left(\sqrt{\rho^2-4\mu\chi}\left(x\right.\right. \\
& \left.\left.\left.\left. - \frac{(\alpha-1)ct^{-\alpha}}{B(\alpha)\sum_{n=0}^{\infty}\left(-\frac{\alpha}{1-\alpha}\right)^n\Gamma(1-\alpha n)}\right)\right)\right)\right] \Big/ \\
& \left[\sqrt{2}\sqrt{b}\rho\left(A\sinh\left(\sqrt{\rho^2-4\mu\chi}\left(x\right.\right.\right.\right. \\
& \left.\left.\left.\left. - \frac{(\alpha-1)ct^{-\alpha}}{B(\alpha)\sum_{n=0}^{\infty}\left(-\frac{\alpha}{1-\alpha}\right)^n\Gamma(1-\alpha n)}\right)\right) + B\right)\right],
\end{aligned}
\tag{3.44}
$$

$$
\begin{aligned}
\pounds_{31}(x,t) = & \left[\sqrt{-(c^2-1)\rho^2}\left(A\sqrt{\rho^2-4\mu\chi}\sinh\left(\sqrt{\rho^2-4\mu\chi}\right.\right.\right. \\
& \times\left.\left(x - \frac{(\alpha-1)ct^{-\alpha}}{B(\alpha)\sum_{n=0}^{\infty}\left(-\frac{\alpha}{1-\alpha}\right)^n\Gamma(1-\alpha n)}\right)\right) \\
& \left.\left. + \sqrt{(B^2-A^2)(\rho^2-4\mu\chi)}\right)\right] \Big/
\end{aligned}
$$

$$\times\left[-\sqrt{2}\sqrt{b}\rho\left(A\cosh\left(\sqrt{\rho^2-4\mu\chi}\left(x-\frac{(\alpha-1)ct^{-\alpha}}{B(\alpha)\sum_{n=0}^{\infty}\left(-\frac{\alpha}{1-\alpha}\right)^n\Gamma(1-\alpha n)}\right)\right)+B\right)\right], \tag{3.45}$$

where A and B are two nonzero constants satisfying $B^2-A^2>0$.

$$£_{32}(x,t)=\frac{\sqrt{-(c^2-1)\rho^2}\left(1-\frac{4\mu\chi}{\rho^2-\rho\sqrt{\rho^2-4\mu\chi}\tanh\left(\frac{1}{2}\sqrt{\rho^2-4\mu\chi}\times\left(x-\frac{(\alpha-1)ct^{-\alpha}}{B(\alpha)\sum_{n=0}^{\infty}\left(-\frac{\alpha}{1-\alpha}\right)^n\Gamma(1-\alpha n)}\right)\right)}\right)}{\sqrt{2}\sqrt{b}}, \tag{3.46}$$

$$£_{33}(x,t)=\frac{\sqrt{-(c^2-1)\rho^2}\left(1-\frac{4\mu\chi}{\rho^2-\rho\sqrt{\rho^2-4\mu\chi}\coth\left(\frac{1}{2}\sqrt{\rho^2-4\mu\chi}\times\left(x-\frac{(\alpha-1)ct^{-\alpha}}{B(\alpha)\sum_{n=0}^{\infty}\left(-\frac{\alpha}{1-\alpha}\right)^n\Gamma(1-\alpha n)}\right)\right)}\right)}{\sqrt{2}\sqrt{b}}, \tag{3.47}$$

$$\begin{aligned}£_{34}(x,t)=&\frac{\sqrt{-(c^2-1)\rho^2}}{\sqrt{2}\sqrt{b}}\left(1-\left[\left(4\mu\chi\cosh\left(\sqrt{\rho^2-4\mu\chi}\right.\right.\right.\right.\\&\left.\left.\left.\times\left(x-\frac{(\alpha-1)ct^{-\alpha}}{B(\alpha)\sum_{n=0}^{\infty}\left(-\frac{\alpha}{1-\alpha}\right)^n\Gamma(1-\alpha n)}\right)\right)\right)\\&\Big/\left(-\rho\sqrt{\rho^2-4\mu\chi}\sinh\left(\sqrt{\rho^2-4\mu\chi}\right.\right.\\&\left.\times\left(x-\frac{(\alpha-1)ct^{-\alpha}}{B(\alpha)\sum_{n=0}^{\infty}\left(-\frac{\alpha}{1-\alpha}\right)^n\Gamma(1-\alpha n)}\right)\right)\\&+\rho\left(\rho\cosh\left(\sqrt{\rho^2-4\mu\chi}\right.\right.\\&\left.\left.\left.\left.\left.\times\left(x-\frac{(\alpha-1)ct^{-\alpha}}{B(\alpha)\sum_{n=0}^{\infty}\left(-\frac{\alpha}{1-\alpha}\right)^n\Gamma(1-\alpha n)}\right)\right)\pm i\sqrt{\rho^2-4\mu\chi}\right)\right)\right]\right),\end{aligned} \tag{3.48}$$

$$\pounds_{35}(x,t) = \frac{\sqrt{-(c^2-1)\rho^2}}{\sqrt{2}\sqrt{b}}\Bigg(1-\Bigg(4\mu\chi\Big/\Bigg(\rho^2-\rho\,\mathrm{csch}\Bigg(\sqrt{\rho^2-4\mu\chi}\Bigg(x - \frac{(\alpha-1)ct^{-\alpha}}{B(\alpha)\sum_{n=0}^{\infty}\left(-\frac{\alpha}{1-\alpha}\right)^n\Gamma(1-\alpha n)}\Bigg)\Bigg) \times\Bigg(\sqrt{\rho^2-4\mu\chi}\cosh\Bigg(\sqrt{\rho^2-4\mu\chi}\Bigg(x - \frac{(\alpha-1)ct^{-\alpha}}{B(\alpha)\sum_{n=0}^{\infty}\left(-\frac{\alpha}{1-\alpha}\right)^n\Gamma(1-\alpha n)}\Bigg)\Bigg)\pm\sqrt{\rho^2-4\mu\chi}\Bigg)\Bigg), \tag{3.49}$$

$$\begin{aligned}\pounds_{36}(x,t) = &\sqrt{-(c^2-1)\rho^2}\Bigg((\rho^2-4\mu\chi)\sinh \\ &\times\Bigg(\frac{1}{2}\sqrt{\rho^2-4\mu\chi}\Bigg(x-\frac{(\alpha-1)ct^{-\alpha}}{B(\alpha)\sum_{n=0}^{\infty}\left(-\frac{\alpha}{1-\alpha}\right)^n\Gamma(1-\alpha n)}\Bigg)\Bigg) \\ &-\rho\sqrt{\rho^2-4\mu\chi}\cosh \\ &\times\Bigg(2\sqrt{\rho^2-4\mu\chi}\Bigg(x-\frac{(\alpha-1)ct^{-\alpha}}{B(\alpha)\sum_{n=0}^{\infty}\left(-\frac{\alpha}{1-\alpha}\right)^n\Gamma(1-\alpha n)}\Bigg)\Bigg)\Bigg) \\ &\Big/\Bigg(\sqrt{2}\sqrt{b}\rho\Bigg(\rho\sinh \\ &\times\Bigg(\frac{1}{2}\sqrt{\rho^2-4\mu\chi}\Bigg(x-\frac{(\alpha-1)ct^{-\alpha}}{B(\alpha)\sum_{n=0}^{\infty}\left(-\frac{\alpha}{1-\alpha}\right)^n\Gamma(1-\alpha n)}\Bigg)\Bigg) \\ &-\sqrt{\rho^2-4\mu\chi}\cosh \\ &\times\Bigg(2\sqrt{\rho^2-4\mu\chi}\Bigg(x-\frac{(\alpha-1)ct^{-\alpha}}{B(\alpha)\sum_{n=0}^{\infty}\left(-\frac{\alpha}{1-\alpha}\right)^n\Gamma(1-\alpha n)}\Bigg)\Bigg)\Bigg)\Bigg).\end{aligned} \tag{3.50}$$

For $\left[\rho^2-4\mu\chi<0,\ \mu\rho\neq 0\,(\mu\chi\neq 0)\right]$

$$\pounds_{37}(x,t) = \frac{\sqrt{-(c^2-1)\rho^2}\sqrt{4\mu\chi-\rho^2}\times\tan\left(\frac{1}{2}\sqrt{4\mu\chi-\rho^2}\left(x-\frac{(\alpha-1)ct^{-\alpha}}{B(\alpha)\sum_{n=0}^{\infty}\left(-\frac{\alpha}{1-\alpha}\right)^n\Gamma(1-\alpha n)}\right)\right)}{\sqrt{2}\sqrt{b}\rho}, \tag{3.51}$$

$$\pounds_{38}(x,t) = -\frac{\sqrt{-(c^2-1)\rho^2}\sqrt{4\mu\chi-\rho^2}\cot\left(\frac{1}{2}\sqrt{4\mu\chi-\rho^2}\left(x-\frac{(\alpha-1)ct^{-\alpha}}{B(\alpha)\sum_{n=0}^{\infty}\left(-\frac{\alpha}{1-\alpha}\right)^n\Gamma(1-\alpha n)}\right)\right)}{\sqrt{2}\sqrt{b}\rho}, \tag{3.52}$$

$$\pounds_{39}(x,t) = \frac{\sqrt{-(c^2-1)\rho^2}\sqrt{\rho^2-4\mu\chi}}{\sqrt{2}\sqrt{b}\rho} \times \left(\tan\left(\sqrt{4\mu\chi-\rho^2}\left(x-\frac{(\alpha-1)ct^{-\alpha}}{B(\alpha)\sum_{n=0}^{\infty}\left(-\frac{\alpha}{1-\alpha}\right)^n\Gamma(1-\alpha n)}\right)\right) \pm \sec\left(\sqrt{4\mu\chi-\rho^2}\left(x-\frac{(\alpha-1)ct^{-\alpha}}{B(\alpha)\sum_{n=0}^{\infty}\left(-\frac{\alpha}{1-\alpha}\right)^n\Gamma(1-\alpha n)}\right)\right)\right), \tag{3.53}$$

$$\pounds_{40}(x,t) = -\frac{\sqrt{-(c^2-1)\rho^2}\sqrt{\rho^2-4\mu\chi}}{\sqrt{2}\sqrt{b}\rho} \times \left(\cot\left(\sqrt{4\mu\chi-\rho^2}\left(x-\frac{(\alpha-1)ct^{-\alpha}}{B(\alpha)\sum_{n=0}^{\infty}\left(-\frac{\alpha}{1-\alpha}\right)^n\Gamma(1-\alpha n)}\right)\right) \pm \csc\left(\sqrt{4\mu\chi-\rho^2}\left(x-\frac{(\alpha-1)ct^{-\alpha}}{B(\alpha)\sum_{n=0}^{\infty}\left(-\frac{\alpha}{1-\alpha}\right)^n\Gamma(1-\alpha n)}\right)\right)\right), \tag{3.54}$$

$$\pounds_{41}(x,t) = -\frac{\sqrt{-(c^2-1)\rho^2}\sqrt{\rho^2-4\mu\chi}}{2\sqrt{2}\sqrt{b}\rho} \tan\left(\frac{1}{4}\sqrt{4\mu\chi-\rho^2}\left(x-\frac{(\alpha-1)ct^{-\alpha}}{B(\alpha)\sum_{n=0}^{\infty}\left(-\frac{\alpha}{1-\alpha}\right)^n\Gamma(1-\alpha n)}\right)\right) \times \left(\cot^2\left(\frac{1}{4}\sqrt{4\mu\chi-\rho^2} \times \left(x-\frac{(\alpha-1)ct^{-\alpha}}{B(\alpha)\sum_{n=0}^{\infty}\left(-\frac{\alpha}{1-\alpha}\right)^n\Gamma(1-\alpha n)}\right)\right)-1\right), \tag{3.55}$$

$$\pounds_{42}(x,t) = \sqrt{-(c^2-1)\rho^2}\left(\sqrt{-(A-B)(A+B)(\rho^2-4\mu\chi)} - A\sqrt{4\mu\chi-\rho^2}\cos\left(\sqrt{4\mu\chi-\rho^2}\left(x - \frac{(\alpha-1)ct^{-\alpha}}{B(\alpha)\sum_{n=0}^{\infty}\left(-\frac{\alpha}{1-\alpha}\right)^n\Gamma(1-\alpha n)}\right)\right)\right)\Big/$$

$$\times\left[\sqrt{2}\sqrt{b}\rho\left(A\sin\left(\sqrt{\rho^2-4\mu\chi}\left(x-\frac{(\alpha-1)ct^{-\alpha}}{B(\alpha)\sum_{n=0}^{\infty}\left(-\frac{\alpha}{1-\alpha}\right)^n\Gamma(1-\alpha n)}\right)\right)+B\right)\right], \tag{3.56}$$

$$\begin{aligned}\pounds_{43}(x,t)=&\sqrt{-(c^2-1)\rho^2}\left(A\sqrt{4\mu\chi-\rho^2}\sin\right.\\&\times\left(\sqrt{4\mu\chi-\rho^2}\left(x-\frac{(\alpha-1)ct^{-\alpha}}{B(\alpha)\sum_{n=0}^{\infty}\left(-\frac{\alpha}{1-\alpha}\right)^n\Gamma(1-\alpha n)}\right)\right)\\&\left.-\sqrt{(B^2-A^2)(\rho^2-4\mu\chi)}\right)\Big/\left[\sqrt{2}\sqrt{b}\rho\left(A\cos\left(\sqrt{4\mu\chi-\rho^2}\left(x\right.\right.\right.\right.\\&\left.\left.\left.\left.-\frac{(\alpha-1)ct^{-\alpha}}{B(\alpha)\sum_{n=0}^{\infty}\left(-\frac{\alpha}{1-\alpha}\right)^n\Gamma(1-\alpha n)}\right)\right)+B\right)\right],\end{aligned} \tag{3.57}$$

where A and B are two nonzero constants satisfying $A^2-B^2>0$.

$$\pounds_{44}(x,t)=\frac{\sqrt{-(c^2-1)\rho^2}\left(1-\dfrac{4\mu\chi}{\rho\sqrt{4\mu\chi-\rho^2}\tan\left(\frac{1}{2}\sqrt{4\mu\chi-\rho^2}\times\left(x-\frac{(\alpha-1)ct^{-\alpha}}{B(\alpha)\sum_{n=0}^{\infty}\left(-\frac{\alpha}{1-\alpha}\right)^n\Gamma(1-\alpha n)}\right)\right)+\rho^2}\right)}{\sqrt{2}\sqrt{b}}, \tag{3.58}$$

$$\begin{aligned}\pounds_{45}(x,t)=&\frac{\sqrt{-(c^2-1)\rho^2}}{\sqrt{2}\sqrt{b}}\left(\left(4\mu\chi\sinh\right.\right.\\&\times\left(\frac{1}{2}\sqrt{4\mu\chi-\rho^2}\left(x-\frac{(\alpha-1)ct^{-\alpha}}{B(\alpha)\sum_{n=0}^{\infty}\left(-\frac{\alpha}{1-\alpha}\right)^n\Gamma(1-\alpha n)}\right)\right)\Bigg)\\&\times\Big/\left(\rho\sqrt{4\mu\chi-\rho^2}\cos\left(\frac{1}{2}\sqrt{4\mu\chi-\rho^2}\right.\right.\\&\times\left.\left(x-\frac{(\alpha-1)ct^{-\alpha}}{B(\alpha)\sum_{n=0}^{\infty}\left(-\frac{\alpha}{1-\alpha}\right)^n\Gamma(1-\alpha n)}\right)\right)\\&-\rho^2\sin\left(\frac{1}{2}\sqrt{4\mu\chi-\rho^2}\right.\\&\times\left.\left.\left.\left(x-\frac{(\alpha-1)ct^{-\alpha}}{B(\alpha)\sum_{n=0}^{\infty}\left(-\frac{\alpha}{1-\alpha}\right)^n\Gamma(1-\alpha n)}\right)\right)\right)+1\right),\end{aligned} \tag{3.59}$$

$$
\begin{aligned}
\pounds_{46}(x,t) =& \frac{\sqrt{-(c^2-1)\rho^2}}{\sqrt{2}\sqrt{b}}\left(1-\left[\left(4\mu\chi\cos\right.\right.\right.\\
&\times\left(\left(\sqrt{4\mu\chi-\rho^2}\left(x-\frac{(\alpha-1)ct^{-\alpha}}{B(\alpha)\sum_{n=0}^{\infty}\left(-\frac{\alpha}{1-\alpha}\right)^n\Gamma(1-\alpha n)}\right)\right)\right)\right)\\
&\Bigg/\left(\rho\sqrt{4\mu\chi-\rho^2}\sin\right.\\
&\times\left(\sqrt{4\mu\chi-\rho^2}\left(x-\frac{(\alpha-1)ct^{-\alpha}}{B(\alpha)\sum_{n=0}^{\infty}\left(-\frac{\alpha}{1-\alpha}\right)^n\Gamma(1-\alpha n)}\right)\right)\\
&+\rho\left(\rho\cos\left(\sqrt{4\mu\chi-\rho^2}\left(x-\frac{(\alpha-1)ct^{-\alpha}}{B(\alpha)\sum_{n=0}^{\infty}\left(-\frac{\alpha}{1-\alpha}\right)^n\Gamma(1-\alpha n)}\right)\right)\right.\\
&\left.\left.\left.\left.\pm\sqrt{4\mu\chi-\rho^2}\right)\right)\right]\right),
\end{aligned}
\tag{3.60}
$$

$$
\begin{aligned}
\pounds_{47}(x,t) =& \frac{\sqrt{-(c^2-1)\rho^2}}{\sqrt{2}\sqrt{b}}\left(1-\left[\left(4\mu\chi\right)\Bigg/\left(\rho^2-\rho\csc\left(\sqrt{4\mu\chi-\rho^2}\left(x\right.\right.\right.\right.\right.\\
&\left.\left.-\frac{(\alpha-1)ct^{-\alpha}}{B(\alpha)\sum_{n=0}^{\infty}\left(-\frac{\alpha}{1-\alpha}\right)^n\Gamma(1-\alpha n)}\right)\right)\\
&\times\left(\sqrt{4\mu\chi-\rho^2}\cos\left(\sqrt{4\mu\chi-\rho^2}\left(x\right.\right.\right.\\
&\left.\left.\left.\left.\left.-\frac{(\alpha-1)ct^{-\alpha}}{B(\alpha)\sum_{n=0}^{\infty}\left(-\frac{\alpha}{1-\alpha}\right)^n\Gamma(1-\alpha n)}\right)\right)\pm\sqrt{4\mu\chi-\rho^2}\right)\right)\right]\right),
\end{aligned}
\tag{3.61}
$$

$$
\begin{aligned}
\pounds_{48}(x,t) =& \left[\sqrt{-(c^2-1)\rho^2}\left(\left(\rho^2-4\mu\chi\right)\sin\right.\right.\\
&\times\left(\frac{1}{2}\sqrt{4\mu\chi-\rho^2}\left(x-\frac{(\alpha-1)ct^{-\alpha}}{B(\alpha)\sum_{n=0}^{\infty}\left(-\frac{\alpha}{1-\alpha}\right)^n\Gamma(1-\alpha n)}\right)\right)\\
&-\rho\sqrt{4\mu\chi-\rho^2}\cos\\
&\left.\left.\times\left(2\sqrt{4\mu\chi-\rho^2}\left(x-\frac{(\alpha-1)ct^{-\alpha}}{B(\alpha)\sum_{n=0}^{\infty}\left(-\frac{\alpha}{1-\alpha}\right)^n\Gamma(1-\alpha n)}\right)\right)\right)\right]
\end{aligned}
$$

$$\Big/\left[\sqrt{2}\sqrt{b}\rho\left(\rho\sin\left(\frac{1}{2}\sqrt{4\mu\chi-\rho^2}\left(x-\frac{(\alpha-1)ct^{-\alpha}}{B(\alpha)\sum_{n=0}^{\infty}\left(-\frac{\alpha}{1-\alpha}\right)^n\Gamma(1-\alpha n)}\right)\right)\right.\right.$$
$$\left.\left.-\sqrt{4\mu\chi-\rho^2}\cos\left(2\sqrt{4\mu\chi-\rho^2}\left(x-\frac{(\alpha-1)ct^{-\alpha}}{B(\alpha)\sum_{n=0}^{\infty}\left(-\frac{\alpha}{1-\alpha}\right)^n\Gamma(1-\alpha n)}\right)\right)\right)\right]. \tag{3.62}$$

For $\left[\mu\chi<0\,\&\,\chi\neq 0\right]$

$$\pounds_{49}(x,t)=-\frac{\sqrt{-(c^2-1)\rho^2}}{\sqrt{2}\sqrt{b}\rho}\left(\sqrt{6}\sqrt{-\mu\chi}\tanh\left(\sqrt{\frac{3}{2}}\sqrt{-\mu\chi}\left(x-\frac{(\alpha-1)ct^{-\alpha}}{B(\alpha)\sum_{n=0}^{\infty}\left(-\frac{\alpha}{1-\alpha}\right)^n\Gamma(1-\alpha n)}\right)\right)\right.$$
$$\left.\pm\left(\sqrt{2}\sqrt{-\mu\chi}\right)-\rho\right), \tag{3.63}$$

$$\pounds_{50}(x,t)=-\frac{\sqrt{-(c^2-1)\rho^2}}{\sqrt{2}\sqrt{b}\rho}\left(\sqrt{6}\sqrt{-\mu\chi}\coth\left(\sqrt{\frac{3}{2}}\sqrt{-\mu\chi}\left(x-\frac{(\alpha-1)ct^{-\alpha}}{B(\alpha)\sum_{n=0}^{\infty}\left(-\frac{\alpha}{1-\alpha}\right)^n\Gamma(1-\alpha n)}\right)\right)\right.$$
$$\left.\pm\left(\sqrt{2}\sqrt{-\mu\chi}\right)-\rho\right), \tag{3.64}$$

$$\pounds_{51}(x,t)=\frac{\sqrt{-(c^2-1)\rho^2}}{\sqrt{2}\sqrt{b}\rho}\left(\sqrt{6}\sqrt{-\mu\chi}\left(-\tanh\left(\sqrt{6}\sqrt{-\mu\chi}\left(x-\frac{(\alpha-1)ct^{-\alpha}}{B(\alpha)\sum_{n=0}^{\infty}\left(-\frac{\alpha}{1-\alpha}\right)^n\Gamma(1-\alpha n)}\right)\right)\right.\right.$$
$$\left.-i\,\text{sech}\left(\sqrt{6}\sqrt{-\mu\chi}\left(x-\frac{(\alpha-1)ct^{-\alpha}}{B(\alpha)\sum_{n=0}^{\infty}\left(-\frac{\alpha}{1-\alpha}\right)^n\Gamma(1-\alpha n)}\right)\right)\right)$$
$$\left.\mp\left(\sqrt{2}\sqrt{-\mu\chi}\right)+\rho\right), \tag{3.65}$$

$$\pounds_{52}(x,t) = -\frac{\sqrt{-(c^2-1)\rho^2}}{\sqrt{2}\sqrt{b}\rho}\Bigg(\sqrt{6}\sqrt{-\mu\chi}\coth \\ \times\left(\sqrt{\frac{3}{2}}\sqrt{-\mu\chi}\left(x-\frac{(\alpha-1)ct^{-\alpha}}{B(\alpha)\sum_{n=0}^{\infty}\left(-\frac{\alpha}{1-\alpha}\right)^n\Gamma(1-\alpha n)}\right)\right) \\ \pm\left(\sqrt{2}\sqrt{-\mu\chi}\right)-\rho\Bigg), \tag{3.66}$$

$$\pounds_{53}(x,t) = \frac{\sqrt{-(c^2-1)\rho^2}}{2\sqrt{2}\sqrt{b}\rho}\Bigg(\sqrt{2}\sqrt{-\mu\chi}\Bigg(-\sqrt{3}\tanh\Bigg(\frac{1}{2}\sqrt{\frac{3}{2}}\sqrt{-\mu\chi}\Bigg(x \\ -\frac{(\alpha-1)ct^{-\alpha}}{B(\alpha)\sum_{n=0}^{\infty}\left(-\frac{\alpha}{1-\alpha}\right)^n\Gamma(1-\alpha n)}\Bigg)\Bigg) \\ -\sqrt{3}\coth\Bigg(\frac{1}{2}\sqrt{\frac{3}{2}}\sqrt{-\mu\chi}\Bigg(x \\ -\frac{(\alpha-1)ct^{-\alpha}}{B(\alpha)\sum_{n=0}^{\infty}\left(-\frac{\alpha}{1-\alpha}\right)^n\Gamma(1-\alpha n)}\Bigg)\Bigg)-\pm 2\Bigg)+2\rho\Bigg), \tag{3.67}$$

$$\pounds_{54}(x,t) = \frac{\sqrt{-(c^2-1)\rho^2}}{\sqrt{2}\sqrt{b}}\Bigg[1+\frac{1}{\rho} \\ \times\left(\frac{\sqrt{6}\left(\sqrt{\mu\chi(-(A^2+B^2))}-A\sqrt{-\mu\chi}\cosh\left(\sqrt{6}\sqrt{-\mu\chi}\left(x-\frac{(\alpha-1)ct^{-\alpha}}{B(\alpha)\sum_{n=0}^{\infty}\left(-\frac{\alpha}{1-\alpha}\right)^n\Gamma(1-\alpha n)}\right)\right)\right)}{A\sinh\left(\sqrt{6}\sqrt{-\mu\chi}\left(x-\frac{(\alpha-1)ct^{-\alpha}}{B(\alpha)\sum_{n=0}^{\infty}\left(-\frac{\alpha}{1-\alpha}\right)^n\Gamma(1-\alpha n)}\right)\right)+B} \\ \mp\left(\sqrt{2}\sqrt{-\mu\chi}\right)\right)\Bigg], \tag{3.68}$$

$$\pounds_{55}(x,t) = \frac{\sqrt{-(c^2-1)\rho^2}}{\sqrt{2}\sqrt{b}}\Bigg[1+\frac{1}{\rho}\Bigg(\mp\left(\sqrt{2}\sqrt{-\mu\chi}\right) \\ -\frac{\sqrt{6}\left(A\sqrt{-\mu\chi}\sinh\left(\sqrt{6}\sqrt{-\mu\chi}\left(x-\frac{(\alpha-1)ct^{-\alpha}}{B(\alpha)\sum_{n=0}^{\infty}\left(-\frac{\alpha}{1-\alpha}\right)^n\Gamma(1-\alpha n)}\right)\right)+\sqrt{\mu\chi(A-B)(A+B)}\right)}{A\cosh\left(\sqrt{6}\sqrt{-\mu\chi}\left(x-\frac{(\alpha-1)ct^{-\alpha}}{B(\alpha)\sum_{n=0}^{\infty}\left(-\frac{\alpha}{1-\alpha}\right)^n\Gamma(1-\alpha n)}\right)\right)+B}\Bigg)\Bigg], \tag{3.69}$$

$$\pounds_{56}(x,t) = \frac{\sqrt{-(c^2-1)\rho^2}}{\sqrt{2}\sqrt{b}}\left[1+(4\mu\chi\cosh\right.$$
$$\times\left(\sqrt{\frac{3}{2}}\sqrt{-\mu\chi}\left(x-\frac{(\alpha-1)ct^{-\alpha}}{B(\alpha)\sum_{n=0}^{\infty}\left(-\frac{\alpha}{1-\alpha}\right)^n\Gamma(1-\alpha n)}\right)\right)\Bigg)$$
$$\Bigg/\left(\rho\left(\sqrt{6}\sqrt{-\mu\chi}\sinh\left(\sqrt{\frac{3}{2}}\sqrt{-\mu\chi}\right.\right.\right.$$
$$\times\left.\left(x-\frac{(\alpha-1)ct^{-\alpha}}{B(\alpha)\sum_{n=0}^{\infty}\left(-\frac{\alpha}{1-\alpha}\right)^n\Gamma(1-\alpha n)}\right)\right) \tag{3.70}$$
$$\mp\sqrt{2}\sqrt{-\mu\chi}\cosh\left(\sqrt{\frac{3}{2}}\sqrt{-\mu\chi}\right.$$
$$\left.\left.\left.\times\left(x-\frac{(\alpha-1)ct^{-\alpha}}{B(\alpha)\sum_{n=0}^{\infty}\left(-\frac{\alpha}{1-\alpha}\right)^n\Gamma(1-\alpha n)}\right)\right)\right)\right)\right],$$

$$\pounds_{57}(x,t) = \frac{\sqrt{-(c^2-1)\rho^2}\left(\dfrac{4\mu\chi}{\sqrt{2}\rho\sqrt{-\mu\chi}\coth\left(\sqrt{\frac{3}{2}}\sqrt{-\mu\chi}\times\left(x-\frac{(\alpha-1)ct^{-\alpha}}{B(\alpha)\sum_{n=0}^{\infty}\left(-\frac{\alpha}{1-\alpha}\right)^n\Gamma(1-\alpha n)}\right)\right)-\rho\left(\pm\left(\sqrt{6}\sqrt{-\mu\chi}\right)\right)}+1\right)}{\sqrt{2}\sqrt{b}}, \tag{3.71}$$

$$\pounds_{58}(x,t) = \frac{\sqrt{-(c^2-1)\rho^2}}{\sqrt{2}\sqrt{b}}\left[1+(4\mu\chi\right.$$
$$\times\cosh\left(\sqrt{6}\sqrt{-\mu\chi}\left(x-\frac{(\alpha-1)ct^{-\alpha}}{B(\alpha)\sum_{n=0}^{\infty}\left(-\frac{\alpha}{1-\alpha}\right)^n\Gamma(1-\alpha n)}\right)\right)\Bigg)$$
$$\Bigg/\left(\rho\left(\left(\sqrt{6}\sqrt{-\mu\chi}\sinh\left(\sqrt{6}\sqrt{-\mu\chi}(x\right.\right.\right.\right.$$
$$\left.\left.-\frac{(\alpha-1)ct^{-\alpha}}{B(\alpha)\sum_{n=0}^{\infty}\left(-\frac{\alpha}{1-\alpha}\right)^n\Gamma(1-\alpha n)}\right)\right)\mp\sqrt{2}\sqrt{-\mu\chi}$$

$$\times\cosh\left(\sqrt{6}\sqrt{-\mu\chi}\left(x-\frac{(\alpha-1)ct^{-\alpha}}{B(\alpha)\sum_{n=0}^{\infty}\left(-\frac{\alpha}{1-\alpha}\right)^{n}\Gamma(1-\alpha n)}\right)\right)\Bigg)$$
$$\pm i\sqrt{6}\sqrt{-\mu\chi}\Bigg)\Bigg)\Bigg], \tag{3.72}$$

$$£_{59}(x,t)=\frac{\sqrt{-(c^2-1)\rho^2}}{\sqrt{2}\sqrt{b}}\Bigg[1+4\mu\chi\Bigg/$$
$$\times\left(\rho\operatorname{csch}\left(\sqrt{6}\sqrt{-\mu\chi}\left(x-\frac{(\alpha-1)ct^{-\alpha}}{B(\alpha)\sum_{n=0}^{\infty}\left(-\frac{\alpha}{1-\alpha}\right)^{n}\Gamma(1-\alpha n)}\right)\right)\right.$$
$$\times\Bigg(\sqrt{6}\sqrt{-\mu\chi}\cosh\Bigg(\sqrt{6}\sqrt{-\mu\chi}\Bigg(x \tag{3.73}$$
$$-\frac{(\alpha-1)ct^{-\alpha}}{B(\alpha)\sum_{n=0}^{\infty}\left(-\frac{\alpha}{1-\alpha}\right)^{n}\Gamma(1-\alpha n)}\Bigg)\Bigg)\pm\sqrt{6}\sqrt{-\mu\chi}\Bigg)$$
$$+\rho\left(\mp\left(\sqrt{6}\sqrt{-\mu\chi}\right)\right)\Bigg)\Bigg],$$

$$£_{60}(x,t)=\frac{\sqrt{-(c^2-1)\rho^2}}{2\sqrt{b}}\Bigg[\sqrt{2}-8\sqrt{-\mu\chi}$$
$$\times\sinh\left(2\sqrt{6}\sqrt{-\mu\chi}\left(x-\frac{(\alpha-1)ct^{-\alpha}}{B(\alpha)\sum_{n=0}^{\infty}\left(-\frac{\alpha}{1-\alpha}\right)^{n}\Gamma(1-\alpha n)}\right)\right)$$
$$\Bigg/\Bigg(\rho\Bigg(2\sqrt{3}\cosh\Bigg(2\sqrt{6}\sqrt{-\mu\chi}$$
$$\times\left(x-\frac{(\alpha-1)ct^{-\alpha}}{B(\alpha)\sum_{n=0}^{\infty}\left(-\frac{\alpha}{1-\alpha}\right)^{n}\Gamma(1-\alpha n)}\right)\Bigg)$$
$$\mp2\sinh\left(\sqrt{\frac{3}{2}}\sqrt{-\mu\chi}\left(x-\frac{(\alpha-1)ct^{-\alpha}}{B(\alpha)\sum_{n=0}^{\infty}\left(-\frac{\alpha}{1-\alpha}\right)^{n}\Gamma(1-\alpha n)}\right)\right)\Bigg)\Bigg)\Bigg], \tag{3.74}$$

$$£_{61}(x,t)=\frac{\sqrt{-(c^2-1)\rho^2}}{\sqrt{2}\sqrt{b}}$$
$$\times\left(\frac{2e_1\mu}{e_2\rho\cosh\left(x-\frac{(\alpha-1)ct^{-\alpha}}{B(\alpha)\sum_{n=0}^{\infty}\left(-\frac{\alpha}{1-\alpha}\right)^{n}\Gamma(1-\alpha n)}\right)+e_3\rho}+1\right), \tag{3.75}$$

$$£_{62}(x,t) = \frac{\sqrt{-(c^2-1)\rho^2}}{\sqrt{2}\sqrt{b}} \times \left(\frac{2e_1\mu}{e_2\rho \sinh\left(x - \frac{(\alpha-1)ct^{-\alpha}}{B(\alpha)\sum_{n=0}^{\infty}\left(-\frac{\alpha}{1-\alpha}\right)^n \Gamma(1-\alpha n)}\right) + e_3\rho} + 1 \right), \tag{3.76}$$

$$\begin{aligned} £_{63}(x,t) = & \frac{\sqrt{-(c^2-1)\rho^2}}{\sqrt{2}\sqrt{b}} \left(4e_1\mu \Big/ \left(e_3\rho \right.\right. \\ & \times \sinh\left(2x - \frac{2(\alpha-1)ct^{-\alpha}}{B(\alpha)\sum_{n=0}^{\infty}\left(-\frac{\alpha}{1-\alpha}\right)^n \Gamma(1-\alpha n)}\right) \\ & + 2e_4\rho \cosh^2 \\ & \left.\left. \times \left(x - \frac{(\alpha-1)ct^{-\alpha}}{B(\alpha)\sum_{n=0}^{\infty}\left(-\frac{\alpha}{1-\alpha}\right)^n \Gamma(1-\alpha n)}\right) + 2e_2\rho\right) + 1\right), \end{aligned} \tag{3.77}$$

$$\begin{aligned} £_{64}(x,t) = & \frac{\sqrt{-(c^2-1)\rho^2}}{\sqrt{2}\sqrt{b}} \left(4e_1\mu \Big/ \left[2e_4\rho \right.\right. \\ & \times \sinh^2\left(x - \frac{(\alpha-1)ct^{-\alpha}}{B(\alpha)\sum_{n=0}^{\infty}\left(-\frac{\alpha}{1-\alpha}\right)^n \Gamma(1-\alpha n)}\right) \\ & + e_3\rho \sinh\left(2x - \frac{2(\alpha-1)ct^{-\alpha}}{B(\alpha)\sum_{n=0}^{\infty}\left(-\frac{\alpha}{1-\alpha}\right)^n \Gamma(1-\alpha n)}\right) \\ & \left.\left. + 2e_2\rho\right] + 1\right], \end{aligned} \tag{3.78}$$

$$£_{65}(x,t) = \frac{\sqrt{-(c^2-1)\rho^2} \times \left(1 - \frac{2e_1e_3\mu}{e_2\rho\left(\exp\left(-e_1\left(x - \frac{(\alpha-1)ct^{-\alpha}}{B(\alpha)\sum_{n=0}^{\infty}\left(-\frac{\alpha}{1-\alpha}\right)^n \Gamma(1-\alpha n)}\right)\right) + c\right)}\right)}{\sqrt{2}\sqrt{b}}, \tag{3.79}$$

$$\pounds_{66}(x,t) = \frac{\sqrt{-(c^2-1)\rho^2}\left(\frac{2e_1\mu\left(\frac{e_3}{\exp\left(e_1\left(x-\frac{(\alpha-1)ct^{-\alpha}}{B(\alpha)\sum_{n=0}^{\infty}\left(-\frac{\alpha}{1-\alpha}\right)^n\Gamma(1-\alpha n)}\right)\right)+e_3}-1\right)}{e_2\rho}+1\right)}{\sqrt{2}\sqrt{b}}, \tag{3.80}$$

$$\pounds_{67}(x,t) = \frac{\sqrt{-(c^2-1)\rho^2}\left(\frac{2e_1\mu}{e_2\rho\cos\left(x-\frac{(\alpha-1)ct^{-\alpha}}{B(\alpha)\sum_{n=0}^{\infty}\left(-\frac{\alpha}{1-\alpha}\right)^n\Gamma(1-\alpha n)}\right)+e_3\rho}+1\right)}{\sqrt{2}\sqrt{b}}, \tag{3.81}$$

$$\pounds_{68}(x,t) = \frac{\sqrt{-(c^2-1)\rho^2}\left(\frac{2e_1\mu}{e_2\rho\sin\left(x-\frac{(\alpha-1)ct^{-\alpha}}{B(\alpha)\sum_{n=0}^{\infty}\left(-\frac{\alpha}{1-\alpha}\right)^n\Gamma(1-\alpha n)}\right)+e_3\rho}+1\right)}{\sqrt{2}\sqrt{b}}, \tag{3.82}$$

$$\pounds_{69}(x,t) = \frac{\sqrt{-(c^2-1)\rho^2}\left(2\mu\phi\left(x-\frac{(\alpha-1)ct^{-\alpha}}{B(\alpha)\sum_{n=0}^{\infty}\left(-\frac{\alpha}{1-\alpha}\right)^n\Gamma(1-\alpha n)}\right)+\rho\right)}{\sqrt{2}\sqrt{b}\rho}, \tag{3.83}$$

$$\begin{aligned}\pounds_{70}(x,t) =& \frac{\sqrt{-(c^2-1)\rho^2}}{\sqrt{2}\sqrt{b}}\\ &\times\left(4e_1\mu\Big/\left[\left(2e_4\rho\sin^2\left(x-\frac{(\alpha-1)ct^{-\alpha}}{B(\alpha)\sum_{n=0}^{\infty}\left(-\frac{\alpha}{1-\alpha}\right)^n\Gamma(1-\alpha n)}\right)\right.\right.\right.\\ &\left.\left.\left.+e_3\rho\sin\left(2x-\frac{2(\alpha-1)ct^{-\alpha}}{B(\alpha)\sum_{n=0}^{\infty}\left(-\frac{\alpha}{1-\alpha}\right)^n\Gamma(1-\alpha n)}\right)+2e_2\rho\right)\right]+1\right),\end{aligned} \tag{3.84}$$

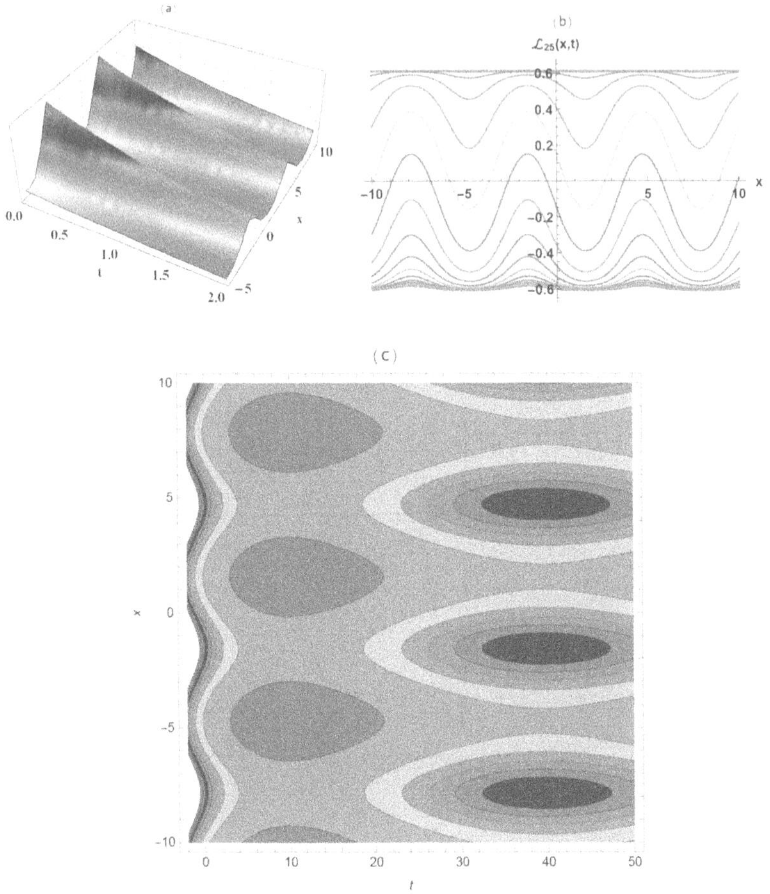

Figure 3.2 Numerical simulation of Eq. (3.39) in three distinct forms.

3.2.3 EXTENDED ($\frac{G'}{G}$) EXPANSION METHOD

Applying the extended Fan expansion method to Eq. (3.10) leads to a general form of solutions according to this technique that takes the following form:

$$\begin{aligned} \pounds(\Xi) &= \sum_{i=1}^{n}\left(\frac{a_i G'(\Xi)}{G(\Xi)} + b_i\left(\frac{G'(\Xi)}{G(\Xi)}\right)^{i-1}\sqrt{\sigma\left(\frac{\left(\frac{G'(\Xi)}{G(\Xi)}\right)^2}{\mu}+1\right)}\right)+a_0 \\ &= \frac{a_1 G'(\Xi)}{G(\Xi)} + a_0 + b_1\sqrt{\sigma\left(\frac{G'(\Xi)^2}{\mu G(\Xi)^2}+1\right)}, \end{aligned} \tag{3.85}$$

where $[a_0, a_1, b_1, \sigma]$ are arbitrary constants. Additionally, $\left(\frac{G'(\Xi)}{G(\Xi)}\right)$ is the solution function of the following auxiliary equation

$$\left(\frac{G''(\xi)}{G(\xi)}\right) = -\mu, \tag{3.86}$$

where $\left[\mu\right]$ is arbitrary constant. Substituting Eq. (3.85) and (3.86) into Eq. (3.10) and collecting all terms with the same power of $\left[\left(\frac{G'(\Xi)}{G(\Xi)}\right)^i, i = 0, 1, 2, ...\right]$ leads to a system of algebraic systems. Using Mathematica 12 to solve this system yields

$$\left[a_0 \to 0, a_1 \to \frac{b_1\sqrt{\sigma}}{\sqrt{\mu}}, c \to \frac{\sqrt{\mu - 2bb_1^2\sigma}}{\sqrt{\mu}}, a \to -bb_1^2, \sigma\right]$$

where $\left[\mu > 2bb_1^2\sigma \neq 0\right]$. Therefore, the explicit wave solutions of the fractional nonlinear KG equation are given under the following conditions with the following formulas:

For $\left[\mu > 0\right]$

$$\begin{aligned}
&\pounds_{71}(x,t) = \\
&b_1\left[\frac{\sqrt{\sigma}}{C_1}\left(\frac{C_1^2+C_2^2}{C_1\tan\left(\sqrt{\mu}x - \frac{(\alpha-1)t^{-\alpha}\sqrt{\mu-2bb_1^2\sigma}}{B(\alpha)\sum_{n=0}^{\infty}\left(-\frac{\alpha}{1-\alpha}\right)^n\Gamma(1-\alpha n)}\right)+C_2} - C_2\right) + \right. \\
&\left. + \sqrt{\frac{\left(C_1^2+C_2^2\right)\sigma}{\left(C_1\sin\left(\sqrt{\mu}x - \frac{(\alpha-1)t^{-\alpha}\sqrt{\mu-2bb_1^2\sigma}}{B(\alpha)\sum_{n=0}^{\infty}\left(-\frac{\alpha}{1-\alpha}\right)^n\Gamma(1-\alpha n)}\right) + C_2\cos\left(\sqrt{\mu}x - \frac{(\alpha-1)t^{-\alpha}\sqrt{\mu-2bb_1^2\sigma}}{B(\alpha)\sum_{n=0}^{\infty}\left(-\frac{\alpha}{1-\alpha}\right)^n\Gamma(1-\alpha n)}\right)\right)^2}}\right].
\end{aligned} \tag{3.87}$$

For $\left[\mu < 0\right]$

$$\begin{aligned}
\pounds_{72}(x,t) = &b_1\left[\mu\sqrt{\sigma}\left(\frac{1}{\mu\tan\left(\sqrt{\mu}x - \frac{(\alpha-1)t^{-\alpha}\sqrt{\mu-2bb_1^2\sigma}}{B(\alpha)\sum_{n=0}^{\infty}\left(-\frac{\alpha}{1-\alpha}\right)^n\Gamma(1-\alpha n)}\right)} - \frac{C_1\sqrt{-\mu^2}}{C_2}\right.\right. \\
&\left. - \frac{C_1}{C_1\mu\cot\left(\sqrt{\mu}x - \frac{(\alpha-1)t^{-\alpha}\sqrt{\mu-2bb_1^2\sigma}}{B(\alpha)\sum_{n=0}^{\infty}\left(-\frac{\alpha}{1-\alpha}\right)^n\Gamma(1-\alpha n)}\right)+C_2\sqrt{-\mu^2}}\right) \\
&+\left(\left(C_1^2-C_2^2\right)\mu\sigma\Big/\left(\left(C_2\sqrt{-\mu}\right.\right.\right.
\end{aligned}$$

$$\times \sin\left(\sqrt{\mu}x - \frac{(\alpha-1)t^{-\alpha}\sqrt{\mu - 2bb_1^2\sigma}}{B(\alpha)\sum_{n=0}^{\infty}\left(-\frac{\alpha}{1-\alpha}\right)^n \Gamma(1-\alpha n)}\right)$$

$$+C_1\sqrt{\mu}\cos\left(\sqrt{\mu}x - \frac{(\alpha-1)t^{-\alpha}\sqrt{\mu - 2bb_1^2\sigma}}{B(\alpha)\sum_{n=0}^{\infty}\left(-\frac{\alpha}{1-\alpha}\right)^n \Gamma(1-\alpha n)}\right)\Bigg)\Bigg)^2\Bigg)\Bigg)^{\frac{1}{2}}\Bigg]. \tag{3.88}$$

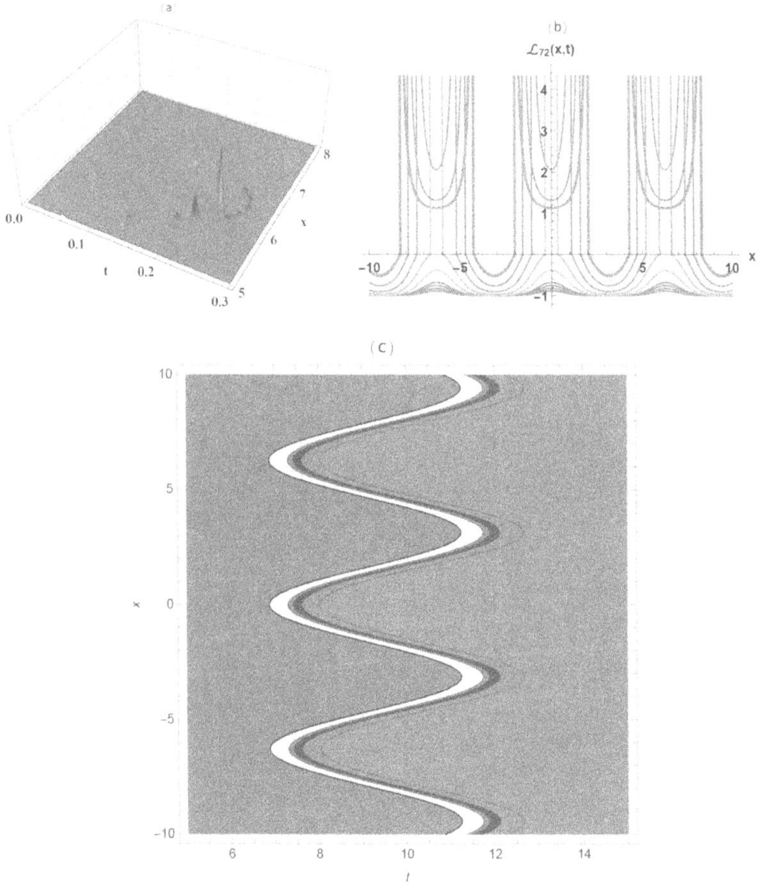

Figure 3.3 Numerical simulation of Eq. (3.88) in three distinct forms.

3.2.4 IMPROVED F-EXPANSION METHOD

Applying the improved F-expansion method to Eq. (3.10) leads to a general form of solutions according to this technique, that takes the following form:

$$\pounds(\Xi) = \sum_{i=-n}^{n} a_i(\mu + \phi(\Xi))^i = \frac{a_{-1}}{\mu + \phi(\Xi)} + a_1(\mu + \phi(\Xi)) + a_0, \tag{3.89}$$

where $\left[a_{-1}, a_0, a_1, \mu\right]$ are arbitrary constants. Additionally, $(\phi(\Xi))$ is the solution function of the following auxiliary equation:

$$\phi'(\Xi) = \phi(\Xi)^2 + r, \tag{3.90}$$

where $\left[r\right]$ is arbitrary constant. Substituting Eq. (3.89) and (3.90) into Eq. (3.10) and collecting all terms with the same power of $\left[\phi^i(\Xi), i = 0, 1, 2, ...\right]$ leads to a system of algebraic systems. Using Mathematica 12 to solve this system yields

$$\left[a_{-1} \to \frac{-a_0\mu^2 - a_0 r}{\mu}, a_1 \to 0, c \to \frac{\sqrt{\mu^2 + r}\sqrt{2\mu^2 - a_0^2 b}}{\sqrt{\mu}\sqrt{2\mu^3 + 2\mu r}}, a \to -\frac{a_0^2 br}{\mu^2}\right],$$

where $\left[\mu \neq 0, r \neq -\mu^2, \mu^2 + r > 0, 2\mu^2 - a_0^2 b > 0, \mu > 0, 2\mu^3 + 2\mu r > 0\right]$. Therefore, the explicit wave solutions of the fractional nonlinear KG equation are given under the following conditions with the following formulas:

For $\left[r < 0\right]$

$$£_{73}(x,t) = a_0\left(1 - \frac{\mu^2 + r}{\mu\left(\sqrt{r}\tan\left(\sqrt{r}\left(\frac{(1-\alpha)\sqrt{\mu^2+r}t^{-\alpha}\sqrt{2\mu^2-a_0^2 b}}{\sqrt{\mu}B(\alpha)\sqrt{2\mu^3+2\mu r}\sum_{n=0}^{\infty}\left(-\frac{\alpha}{1-\alpha}\right)^n\Gamma(1-\alpha n)} + x\right)\right) + \mu\right)}\right), \tag{3.91}$$

$$£_{74}(x,t) = a_0\left(\frac{\mu^2 + r}{\mu\sqrt{r}\cot\left(\sqrt{r}\left(\frac{(1-\alpha)\sqrt{\mu^2+r}t^{-\alpha}\sqrt{2\mu^2-a_0^2 b}}{\sqrt{\mu}B(\alpha)\sqrt{2\mu^3+2\mu r}\sum_{n=0}^{\infty}\left(-\frac{\alpha}{1-\alpha}\right)^n\Gamma(1-\alpha n)} + x\right)\right) - \mu^2} + 1\right). \tag{3.92}$$

For $\left[r < 0\right]$

$$£_{75}(x,t) = a_0\left(1 - \frac{\mu^2 + r}{\mu\left(\sqrt{r}\tan\left(\sqrt{r}\left(\frac{(1-\alpha)\sqrt{\mu^2+r}t^{-\alpha}\sqrt{2\mu^2-a_0^2 b}}{\sqrt{\mu}B(\alpha)\sqrt{2\mu^3+2\mu r}\sum_{n=0}^{\infty}\left(-\frac{\alpha}{1-\alpha}\right)^n\Gamma(1-\alpha n)} + x\right)\right) + \mu\right)}\right), \tag{3.93}$$

$$\pounds_{76}(x,t) = a_0\left(\frac{\mu^2+r}{\mu\sqrt{r}\cot\left(\sqrt{r}\left(\frac{(1-\alpha)\sqrt{\mu^2+r}t^{-\alpha}\sqrt{2\mu^2-a_0^2b}}{\sqrt{\mu}B(\alpha)\sqrt{2\mu^3+2\mu r}\sum_{n=0}^{\infty}\left(-\frac{\alpha}{1-\alpha}\right)^n\Gamma(1-\alpha n)}+x\right)\right)-\mu^2}+1\right). \tag{3.94}$$

For $\left[r=0\right]$

$$\pounds_{77}(x,t) = a_0\left(1-\frac{\mu}{\mu-\frac{1}{x-\frac{(\alpha-1)\sqrt{\mu^2}\sqrt{\mu^3}t^{-\alpha}\sqrt{2\mu^2-a_0^2b}}{\sqrt{2}\mu^{7/2}B(\alpha)\sum_{n=0}^{\infty}\left(-\frac{\alpha}{1-\alpha}\right)^n\Gamma(1-\alpha n)}}}\right). \tag{3.95}$$

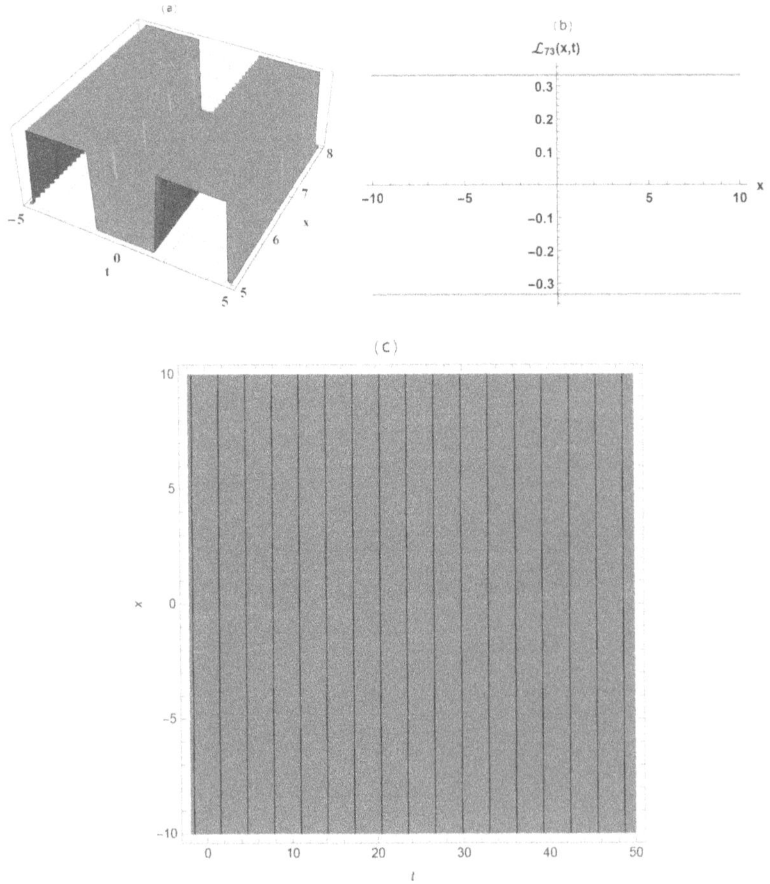

Figure 3.4 Numerical simulation of Eq. (3.91) in three distinct forms.

3.2.5 MODIFIED KHATER METHOD

Applying the modified Khater method to Eq. (3.10) leads to a general form of solutions according to this technique that takes the following form:

$$£(\Xi) = \sum_{i=1}^{n} a_i \mathscr{K}^{i\,\mathscr{F}(\Xi)} + \sum_{i=1}^{n} b_i \mathscr{K}^{-i\,\mathscr{F}(\Xi)} + a_0 = a_1 \mathscr{K}^{\mathscr{F}(\Xi)} + a_0 + b_1 \mathscr{K}^{-\mathscr{F}(\Xi)}, \tag{3.96}$$

where $\left[a_0, a_1, b_1, \mathscr{K}\right]$ are arbitrary constants. Additionally, $\mathscr{F}(\Xi)$ is the solution function of the following auxiliary equation:

$$\mathscr{F}'(\Xi) = \frac{1}{\ln(\mathscr{K})}\left[\delta\,\mathscr{K}^{-\mathscr{F}(\Xi)} + \lambda\,\mathscr{K}^{\mathscr{F}(\Xi)} + \mu\right], \tag{3.97}$$

where $\left[\delta, \lambda, \mu\right]$ are arbitrary constants. Substituting Eq. (3.96) and (3.97) into Eq. (3.10) and collecting all terms with the same power of $\left[\mathscr{K}^{i\,\mathscr{F}(\Xi)}, i = 0, 1, 2, ...\right]$ leads to a system of algebraic systems. Using Mathematica 12 to solve this system yields

Family I

$$\left[a_0 \to \frac{\sqrt{\mu^2 - c^2\mu^2}}{\sqrt{2}\sqrt{b}}, a_1 \to \frac{\sqrt{2}\lambda\sqrt{-\left(c^2-1\right)\mu^2}}{\sqrt{b}\mu}, b_1 \to 0,\right.$$
$$\left. a \to \frac{1}{2}\left(c^2-1\right)\left(4\lambda^2 - \mu^2\right)\right],$$

where $\left[\mu \neq 0, c \neq 1, b < 0\right]$. Therefore, the explicit wave solutions of the fractional nonlinear KG equation are given under the following conditions with the next formulas:

For $\left[\mu^2 - 4\delta\lambda < 0 \,\&\, \lambda \neq 0\right]$

$$£_{78}(x,t) = \frac{\sqrt{-\left(c^2-1\right)\mu^2}\sqrt{4\delta\lambda - \mu^2} \times \tan\left(\frac{1}{2}\sqrt{4\delta\lambda - \mu^2}\left(x - \frac{(\alpha-1)ct^{-\alpha}}{B(\alpha)\sum_{n=0}^{\infty}\left(-\frac{\alpha}{1-\alpha}\right)^n\Gamma(1-\alpha n)}\right)\right)}{\sqrt{2}\sqrt{b}\mu}, \tag{3.98}$$

$$£_{79}(x,t) = \frac{\sqrt{-\left(c^2-1\right)\mu^2}\sqrt{4\delta\lambda - \mu^2}\cot \times \left(\frac{1}{2}\sqrt{4\delta\lambda - \mu^2}\left(x - \frac{(\alpha-1)ct^{-\alpha}}{B(\alpha)\sum_{n=0}^{\infty}\left(-\frac{\alpha}{1-\alpha}\right)^n\Gamma(1-\alpha n)}\right)\right)}{\sqrt{2}\sqrt{b}\mu}. \tag{3.99}$$

For $\left[\mu^2-4\delta\lambda>0\,\&\,\lambda\neq 0\right]$

$$£_{80}(x,t)=-\frac{\sqrt{-\left(c^2-1\right)\mu^2}\sqrt{\mu^2-4\delta\lambda}\times\tanh\left(\frac{1}{2}\sqrt{\mu^2-4\delta\lambda}\left(x-\frac{(\alpha-1)ct^{-\alpha}}{B(\alpha)\sum_{n=0}^{\infty}\left(-\frac{\alpha}{1-\alpha}\right)^n\Gamma(1-\alpha n)}\right)\right)}{\sqrt{2}\sqrt{b}\mu},\tag{3.100}$$

$$£_{81}(x,t)=-\frac{\sqrt{-\left(c^2-1\right)\mu^2}\sqrt{\mu^2-4\delta\lambda}\times\coth\left(\frac{1}{2}\sqrt{\mu^2-4\delta\lambda}\left(x-\frac{(\alpha-1)ct^{-\alpha}}{B(\alpha)\sum_{n=0}^{\infty}\left(-\frac{\alpha}{1-\alpha}\right)^n\Gamma(1-\alpha n)}\right)\right)}{\sqrt{2}\sqrt{b}\mu}.\tag{3.101}$$

For $\left[\mu=\lambda=\kappa\,\&\,\delta=0\right]$

$$£_{82}(x,t)=-\frac{\sqrt{-\left(c^2-1\right)\kappa^2}\coth\left(\frac{1}{2}\kappa\left(x-\frac{(\alpha-1)ct^{-\alpha}}{B(\alpha)\sum_{n=0}^{\infty}\left(-\frac{\alpha}{1-\alpha}\right)^n\Gamma(1-\alpha n)}\right)\right)}{\sqrt{2}\sqrt{b}}.\tag{3.102}$$

For $\left[\delta=0\,\&\,\mu\neq 0\,\&\,\lambda\neq 0\right]$

$$£_{83}(x,t)=\frac{\sqrt{-\left(c^2-1\right)\mu^2}\left(-\frac{4}{\lambda\exp\left(\mu\left(x-\frac{(\alpha-1)ct^{-\alpha}}{B(\alpha)\sum_{n=0}^{\infty}\left(-\frac{\alpha}{1-\alpha}\right)^n\Gamma(1-\alpha n)}\right)\right)-2}-1\right)}{\sqrt{2}\sqrt{b}}.\tag{3.103}$$

Family II

$$\left[a_0\to\frac{\sqrt{\mu^2-c^2\mu^2}}{\sqrt{2}\sqrt{b}},a_1\to 0,b_1\to\frac{\sqrt{2}\lambda\sqrt{-\left(c^2-1\right)\mu^2}}{\sqrt{b}\mu},\right.$$
$$\left.a\to\frac{1}{2}\left(c^2-1\right)\left(4\lambda^2-\mu^2\right),\right]$$

where $\left[\mu\neq 0, c\neq 1, b<0\right]$. Therefore, the explicit wave solutions of the fractional nonlinear KG equation are given under the following conditions with the next formulas:

For $\left[\mu^2-4\delta\lambda<0\,\&\,\lambda\neq 0\right]$

$$£_{84}(x,t)=\frac{\sqrt{-\left(c^2-1\right)\mu^2}\left(1-\frac{4\lambda^2}{\mu^2-\mu\sqrt{4\delta\lambda-\mu^2}\tan\left(\frac{1}{2}\sqrt{4\delta\lambda-\mu^2}\times\left(x-\frac{(\alpha-1)ct^{-\alpha}}{B(\alpha)\sum_{n=0}^{\infty}\left(-\frac{\alpha}{1-\alpha}\right)^n\Gamma(1-\alpha n)}\right)\right)}\right)}{\sqrt{2}\sqrt{b}},\tag{3.104}$$

$$£_{85}(x,t)=\frac{\sqrt{-\left(c^2-1\right)\mu^2}\left(1-\frac{4\lambda^2}{\mu^2-\mu\sqrt{4\delta\lambda-\mu^2}\cot\left(\frac{1}{2}\sqrt{4\delta\lambda-\mu^2}\times\left(x-\frac{(\alpha-1)ct^{-\alpha}}{B(\alpha)\sum_{n=0}^{\infty}\left(-\frac{\alpha}{1-\alpha}\right)^n\Gamma(1-\alpha n)}\right)\right)}\right)}{\sqrt{2}\sqrt{b}}.\tag{3.105}$$

For $\left[\mu^2-4\delta\lambda>0\,\&\,\lambda\neq 0\right]$

$$£_{86}(x,t)=\frac{\sqrt{-\left(c^2-1\right)\mu^2}\left(1-\frac{4\lambda^2}{\mu\sqrt{\mu^2-4\delta\lambda}\tanh\left(\frac{1}{2}\sqrt{\mu^2-4\delta\lambda}\times\left(x-\frac{(\alpha-1)ct^{-\alpha}}{B(\alpha)\sum_{n=0}^{\infty}\left(-\frac{\alpha}{1-\alpha}\right)^n\Gamma(1-\alpha n)}\right)\right)+\mu^2}\right)}{\sqrt{2}\sqrt{b}},\tag{3.106}$$

$$£_{87}(x,t)=\frac{\sqrt{-\left(c^2-1\right)\mu^2}\left(1-\frac{4\lambda^2}{\mu\sqrt{\mu^2-4\delta\lambda}\coth\left(\frac{1}{2}\sqrt{\mu^2-4\delta\lambda}\times\left(x-\frac{(\alpha-1)ct^{-\alpha}}{B(\alpha)\sum_{n=0}^{\infty}\left(-\frac{\alpha}{1-\alpha}\right)^n\Gamma(1-\alpha n)}\right)\right)+\mu^2}\right)}{\sqrt{2}\sqrt{b}}.\tag{3.107}$$

For $\left[\mu = \lambda = \kappa \,\&\, \delta = 0\right]$

$$\pounds_{88}(x,t) = \frac{\sqrt{-(c^2-1)\kappa^2}\left(2\exp\left(\frac{(\alpha-1)c\kappa t^{-\alpha}}{B(\alpha)\sum_{n=0}^{\infty}\left(-\frac{\alpha}{1-\alpha}\right)^n\Gamma(1-\alpha n)} - \kappa x\right) - 1\right)}{\sqrt{2}\sqrt{b}}. \tag{3.108}$$

For $\left[\delta = 0 \,\&\, \mu \neq 0 \,\&\, \lambda \neq 0\right]$

$$\pounds_{89}(x,t) = \frac{\sqrt{-(c^2-1)\mu^2}\left(4\lambda\exp\left(\frac{(\alpha-1)c\mu t^{-\alpha}}{B(\alpha)\sum_{n=0}^{\infty}\left(-\frac{\alpha}{1-\alpha}\right)^n\Gamma(1-\alpha n)} - \mu x\right) - 2\lambda^2 + \mu^2\right)}{\sqrt{2}\sqrt{b}\mu^2}. \tag{3.109}$$

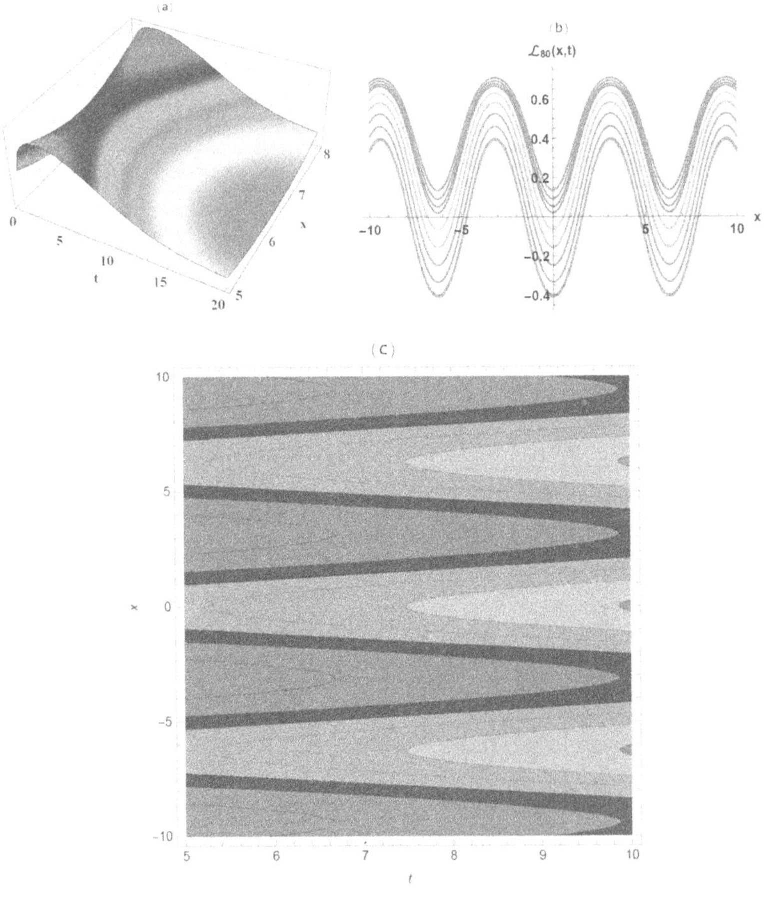

Figure 3.5 Numerical simulation of Eq. (3.100) in three distinct forms.

3.3 STABILITY

This section of the chapter investigates one of the basic properties of any model. It examines the stability property for the fractional nonlinear KG equation by using a Hamiltonian system. The momentum in the Hamiltonian system is given by the following formula:

$$M = \frac{1}{2} \int_{-\mho}^{\mho} \pounds^2(\Xi)\, d\,\Xi, \tag{3.110}$$

where $\mho$ is arbitrary constant. Thus, the condition for stability is given in the next condition:

$$\left.\frac{\partial M}{\partial c}\right|_{c=\varphi} > 0, \tag{3.111}$$

where c, φ are arbitrary constants.
For example, studying the stability of the solution of Eq. (3.6) by using (3.100) with the following values of the constants $\left[b = -2, \delta = 6, \lambda = 1, \mu = 5\right]$ yields:

$$\begin{aligned} M =& \frac{1}{c^2} \Big[-(18. + 4.93038065 \times 10^{-32} i)\,\mathrm{Li}_{2.}((-148.413 - 1.817 \\ & \times 10^{-14} i) \times e^{-1.49071c}) + (18. + 4.93038 \times 10^{-32} i) \\ & \mathrm{Li}_{2.}((-0.0067 + 8.2516052 \times 10^{-19} i) e^{-1.49071c}) + (18. \\ & + 4.930381 \times 10^{-32} i)\,\mathrm{Li}_{2.}((-148.413 - 1.8175370 \\ & \times 10^{-14} i) e^{(0. - 1.49071i)c}) - (18. + 4.9304 \times 10^{-32} i) \\ & \mathrm{Li}_{2.}((-0.00673795 + 8.251605 \times 10^{-19} i) e^{(0. -1.49071 i)c}) \\ & + c((100. + 2.4492936 \times 10^{-15} i)c + (-49.8663 + 218.462i)) \\ & + ((-90. + 56.5487i) - (26.8 - 1.97 \times 10^{-31} i)c) \log(1. \\ & + (0.0067 - 8.2516 \times 10^{-19} i) e^{-1.49c}) + ((26.8328 - 1.97215 \\ & \times 10^{-31} i)c + (-90. + 56.5487i)) \log(1. + (148.413 + 1.81 \\ & \times 10^{-14} i) e^{-1.49c}) + ((90. - 56.5487i) + (1.97 \times 10^{-31} \\ & + 26.8328 i)c) \times \log(1. + (0.00673795 - 8.2516052 \\ & \times 10^{-19} i) e^{(0. -1.49071 i)c}) - (1.972152 \times 10^{-31} \\ & + 26.8328 i)c \log(1. + (148.413 + 1.817537 \times 10^{-14} i) \\ & e^{(0. -1.49071 i)c}) + (90. - 56.5487 i) \times \log(1. + (148.413 \\ & + 1.8175 \times 10^{-14} i) e^{(0. -1.49071 i)c}) - (90. - 56.5487i) \\ & \log(\sin(-0.74c + (1.5708 + 2.5i))) - (90. - 56.5487 i) \end{aligned}$$

$$\begin{aligned}&\times \log(-1.\sin(0.745356c+(1.5708+2.5i)))+(90.-56.5487i)\\&\times \log(\sin((0.+0.745356i)c+(1.5708+2.5i)))+(90.\\&\left.-56.5487i)\times \log(-1.\sin((1.5708+2.5i)-(0.+0.745356i)c))\right].\end{aligned} \tag{3.112}$$

Thus, we obtain

$$\left.\frac{\partial M}{\partial c}\right|_{c=3} = -23.4009+41.8879i < 0. \tag{3.113}$$

This means this solution is unstable, and by applying the same steps to other obtained solutions, the stability property of each one of them can be determined.

3.4 NUMERICAL SOLUTIONS

In this section, we apply the Adomian decomposition method and B-spline scheme to Eq. (3.10). The goals of this section are to investigate the semi-analytical and numerical solutions of the fractional nonlinear KG equation.

3.4.1 SEMI-ANALYTICAL SOLUTIONS

Implement of the Adomian decomposition method enables the rewriting of Eq. (3.10) in the following form:

$$\mathscr{L}\,\pounds(\Xi)+\mathscr{R}\,\pounds(\Xi)+\mathscr{N}\,\pounds(\Xi)=0, \tag{3.114}$$

where $\mathscr{L}$, $\mathscr{R}$, $\mathscr{N}$ represent a differential operator, a linear operator, and a nonlinear term, respectively. Using the inverse operator $\mathscr{L}^{-1}$ on Eq. (3.114), we get

$$\begin{aligned}\sum_{i=0}^{\infty}\pounds_i(\Xi) &= \pounds(0)+\pounds'(0)(\Xi)+\frac{a}{(c^2-1)}\mathscr{L}^{-1}\left(\sum_{i=0}^{\infty}\pounds_i\right)\\&-\frac{b}{(c^2-1)}\mathscr{L}^{-1}\left(\sum_{i=0}^{\infty}A_i\right).\end{aligned} \tag{3.115}$$

Under the following condition $\left[a=-84, b=-2, c=3\right]$ on Eq. (3.100), we get:

$$\pounds_{\text{Exact}} = -\sqrt{2}\tanh\left(\frac{\Xi}{2}\right). \tag{3.116}$$

Thus, we get:

$$\pounds_0(\Xi) = -\frac{\Xi}{\sqrt{2}}, \tag{3.117}$$

$$\pounds_1(\Xi) = \frac{7\,\Xi^3}{4\sqrt{2}} - \frac{\Xi^5}{160\sqrt{2}}, \tag{3.118}$$

$$\pounds_2(\Xi) = \frac{\Xi^{10}}{76800} - \frac{3\,\Xi^8}{512} + \frac{\Xi^7}{640\sqrt{2}} - \frac{147\,\Xi^5}{160\sqrt{2}}. \tag{3.119}$$

$$\begin{aligned}\pounds_3(\Xi) &= -\frac{\Xi^{13}}{10649600\sqrt{2}} - \frac{7\,\Xi^{12}}{6758400} + \frac{21\,\Xi^{11}}{281600\sqrt{2}} \\ &\quad + \frac{7\,\Xi^{10}}{10240} - \frac{83\,\Xi^9}{5120\sqrt{2}} + \frac{157\,\Xi^7}{640\sqrt{2}}.\end{aligned} \tag{3.120}$$

Equations (3.117)–(3.120) lead to the following form of an approximate solution of Eq. (3.10).

$$\begin{aligned}\pounds_{\text{Approximate}} = &- \frac{\Xi^{13}}{10649600\sqrt{2}} - \frac{7\,\Xi^{12}}{6758400} + \frac{21\,\Xi^{11}}{281600\sqrt{2}} + \frac{107\,\Xi^{10}}{153600} \\ &- \frac{83\,\Xi^9}{5120\sqrt{2}} - \frac{3\,\Xi^8}{512} + \frac{79\,\Xi^7}{320\sqrt{2}} - \frac{37\,\Xi^5}{40\sqrt{2}} + \frac{7\,\Xi^3}{4\sqrt{2}} - \frac{\Xi}{\sqrt{2}} + \ldots\end{aligned} \tag{3.121}$$

Table 3.1

Exact and Semi-analytical Values for Different Values Ξ to Explain the Absolute Values of Error Between Them

Values of Ξ	Analytical	Semi-analytical	Absolute Error
0.001	0.000707107	0.000707106	1.17851×10^{-9}
0.002	0.00141421	0.0014142	9.42807×10^{-9}
0.003	0.00212132	0.00212129	3.18196×10^{-8}
0.004	0.00282842	0.00282835	7.5424×10^{-8}
0.005	0.00353553	0.00353538	1.47312×10^{-7}
0.006	0.00424263	0.00424237	2.54553×10^{-7}
0.007	0.00494973	0.00494932	4.04218×10^{-7}
0.008	0.00565682	0.00565622	6.03377×10^{-7}
0.009	0.00636392	0.00636306	8.59096×10^{-7}
0.01	0.00707101	0.00706983	1.17845×10^{-6}

3.4.2 NUMERICAL SOLUTIONS

In this section, the B-spline scheme is applied to the fractional nonlinear KG equation to evaluate its numerical solution and also to show the accuracy of the obtained analytical solutions that were obtained in Section 3.2, by employing the modified Khater method (Section 3.2.5) under the following conditions on Eq. (3.100):

$$\left[b = -2, c = 3, \delta = 6, \lambda = 1, \mu = 5\right].$$

These conditions allow us to apply the B-spline family in the following forms:

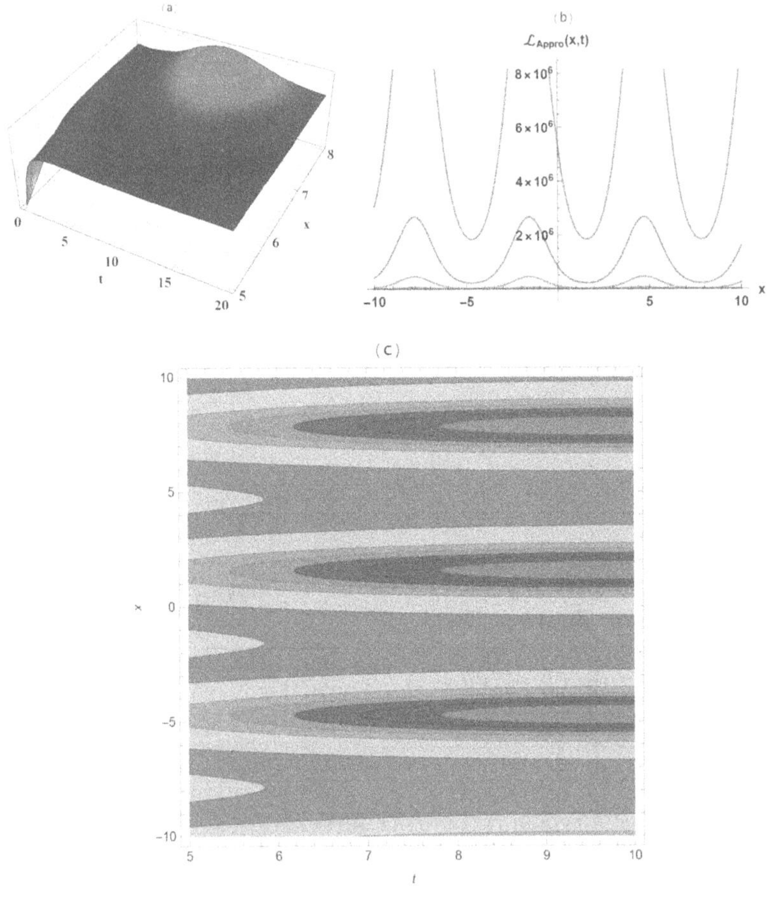

Figure 3.6 Numerical simulation of Eq. (3.121) in three distinct forms.

3.4.2.1 Cubic B-Spline

The cubic B-spline scheme formulates the general solution of Eq. (3.10) in the following form:

$$\pounds(\Xi) = \sum_{\jmath=-1}^{m+1} \mho_\jmath \eth_\jmath, \tag{3.122}$$

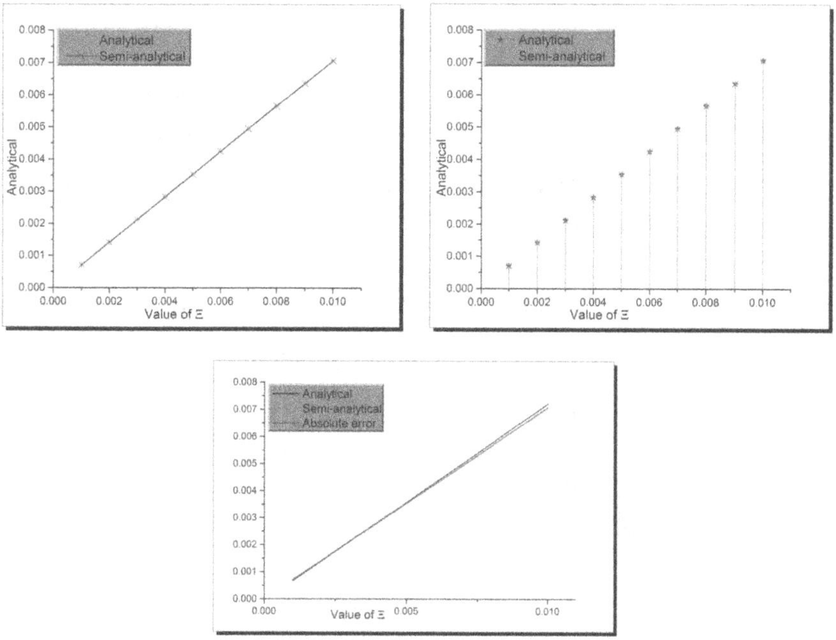

Figure 3.7 Numerical simulation of Eq. (3.121) in three distinct forms.

where $\mho_j$, $\eth_j$ are given in the next mathematical forms, respectively:

$$\mathscr{L}\,\pounds(\Xi) = f(\Xi_j, \pounds(\Xi_j)), \, (j = 0, 1, \ldots, m)$$

and

$$\eth_j(\Xi) = \frac{1}{6\hbar^3}
\begin{cases}
(\Xi - \Xi_{j-2})^3, & \Xi \in [\Xi_{j-2}, \Xi_{j-1}], \\
-3(\Xi - \Xi_{j-1})^3 + 3\hbar(\Xi - \Xi_{j-1})^2 + 3\hbar^2(\Xi - \Xi_{j-1}) + \hbar^3, & \Xi \in [\Xi_{j-1}, \Xi_j], \\
-3(\Xi_{j+1} - \Xi)^3 + 3\hbar(\Xi_{j+1} - \Xi)^2 + 3\hbar^2(\Xi_{i+1} - \Xi) + \hbar^3, & \Xi \in [\Xi_i, \Xi_{i+1}], \\
(\Xi_{i+2} - \Xi)^3, & \Xi \in [\Xi_{i+1}, \Xi_{i+2}], \\
0, & \text{Otherwise},
\end{cases} \tag{3.123}$$

where $j \in [-2, m+2]$. Thus, we obtain

$$\pounds_j(\Xi) = \mho_{j-1} + 4\mho_j + \mho_{j+1}. \tag{3.124}$$

Substituting Eq. (3.124) into Eq. (3.10) yields $(m+3)$ equations. Using Mathematica 12 to solve this system to get the value of $\mho_j$ leads to the following values of analytical, numerical under the different values of Ξ in Table 3.2.

Table 3.2
Computational and Numerical Solutions Values of Eq. (3.10) to Show the Absolute Value of Error Based on the Cubic-B-Spline Scheme

Values of Ξ	Analytical	Approximate	Absolute Error
0.001	−0.000707107	−0.000707223	1.16686×10^{-7}
0.002	−0.00141421	−0.00141444	2.26301×10^{-7}
0.003	−0.00212132	−0.00212164	3.2177×10^{-7}
0.004	−0.00282842	−0.00282882	3.96023×10^{-7}
0.005	−0.00353553	−0.00353597	4.41988×10^{-7}
0.006	−0.00424263	−0.00424308	4.52593×10^{-7}
0.007	−0.00494973	−0.00495015	4.20767×10^{-7}
0.008	−0.00565682	−0.00565716	3.3944×10^{-7}
0.009	−0.00636392	−0.00636412	2.0154×10^{-7}
0.01	0.00707101	−0.00707101	0

3.4.2.2 Quantic–B–spline

The quantic B-spline scheme formulates the general solution of Eq. (3.10) in the following form:

$$\pounds(\Xi) = \sum_{\jmath=-1}^{m+1} \mho_\jmath \eth_\jmath, \tag{3.125}$$

where $\mho_\jmath$, $\eth_\jmath$ are given in the next mathematical forms, respectively:

$$\mathscr{L}\pounds(\Xi) = f(\Xi_\jmath, \pounds(\Xi_\jmath)), \ (\jmath = 0, 1, ..., n)$$

and

$$\eth_\jmath(\Xi) = \frac{1}{\hbar^5}
\begin{cases}
(\Xi - \Xi_{\jmath-3})^5, & \Xi \in [\Xi_{\jmath-3}, \Xi_{\jmath-2}], \\
(\Xi - \Xi_{\jmath-3})^5 - 6(\Xi - \Xi_{\jmath-2})^5, & \Xi \in [\Xi_{\jmath-2}, \Xi_{\jmath-1}], \\
(\Xi - \Xi_{\jmath-3})^5 - 6(\Xi - \Xi_{\jmath-2})^5 + 15(\Xi - \Xi_{\jmath-1})^5, & \Xi \in [\Xi_{\jmath-1}, \Xi_\jmath], \\
(\Xi_{\jmath+3} - \Xi)^5 - 6(\Xi_{\jmath+2} - \Xi)^5 + 15(\Xi_{\jmath+1} - \Xi)^5, & \Xi \in [\Xi_\jmath, \Xi_{\jmath+1}], \\
(\Xi_{\jmath+3} - \Xi)^5 - 6(\Xi_{\jmath+2} - \Xi)^5, & \Xi \in [\Xi_{\jmath+1}, \Xi_{\jmath+2}], \\
(\Xi_{\jmath+3} - \Xi)^5, & x \in [\Xi_{\jmath+2}, \Xi_{\jmath+3}], \\
0, & \text{Otherwise},
\end{cases} \tag{3.126}$$

where $\jmath \in [-2, m+2]$. Thus, we obtain

$$\pounds_\jmath(\Xi) = \mho_{\jmath-2} + 26\mho_{\jmath-1} + 66\mho_\jmath + 26\mho_{\jmath+1} + \mho_{\jmath+2}. \tag{3.127}$$

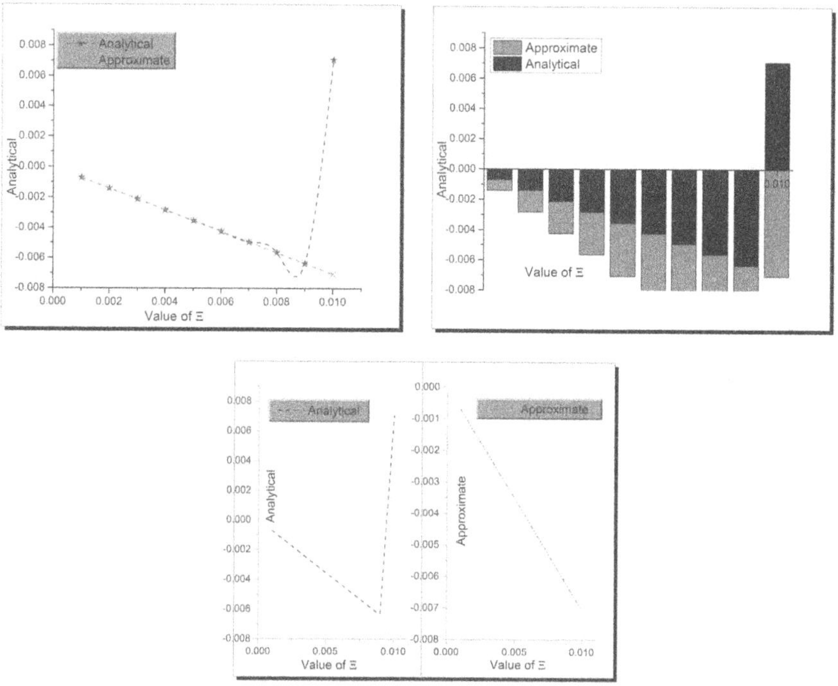

Figure 3.8 Analytical and numerical simulation according to Table 3.2.

Substituting Eq. (3.127) into Eq. (3.10) yields $(m+5)$ equations. Using Mathematica 12 to solve this system to get the value of $\mho_\jmath$ leads to the following values of analytical, numerical under the different values of Ξ in Table 3.3.

3.4.2.3 Septic B-Spline

The septic B-spline scheme formulates the general solution of Eq. (3.10) in the following form:

$$£(\Xi) = \sum_{\jmath=-1}^{n+1} \mho_\jmath \eth_\jmath, \tag{3.128}$$

where $\mho_\jmath$, $\eth_\jmath$ are given in the next mathematical forms, respectively:

$$\mathscr{L}\, £(\Xi) = \mathscr{F}(\Xi_\jmath, £(\Xi_\jmath)), \; (\jmath = 0, 1, ..., m)$$

and

Table 3.3
Computational and Numerical Solutions Values of Eq. (3.10) to Show the Absolute Value of Error Based on Quantic B-Spline Scheme

Values of Ξ	Analytical	Approximate	Absolute Error
0.001	−0.000707107	−0.000707187	8.01534×10^{-8}
0.002	−0.00141421	−0.00141441	1.95938×10^{-7}
0.003	−0.00212132	−0.0021216	2.86129×10^{-7}
0.004	−0.00282842	−0.00282878	3.58202×10^{-7}
0.005	−0.00353553	−0.00353593	4.01043×10^{-7}
0.006	−0.00424263	−0.00424304	4.09201×10^{-7}
0.007	−0.00494973	−0.0049501	3.73166×10^{-7}
0.008	−0.00565682	−0.00565712	2.94004×10^{-7}
0.009	−0.00636392	−0.00636405	2.94004×10^{-7}
0.01	−0.00707101	−0.00707101	8.67362×10^{-19}

$$\mho_j(\Xi) = \frac{1}{h^5}
\begin{cases}
(\Xi - \Xi_{j-4})^7, & \Xi \in [\Xi_{j-4}, \Xi_{j-3}], \\
(\Xi - \Xi_{j-4})^7 - 8(\Xi - \Xi_{j-3})^7, & \Xi \in [\Xi_{j-3}, \Xi_{i-2}], \\
(\Xi - \Xi_{j-4})^7 - 8(\Xi - \Xi_{j-3})^7 + 28(\Xi - \Xi_{j-2})^7, & \Xi \in [\Xi_{j-2}, \Xi_{j-1}], \\
(\Xi - \Xi_{j-4})^7 - 8(\Xi - \Xi_{j-3})^7 + 28(\Xi - \Xi_{j-2})^7 + 56(\Xi - \Xi_{j-1})^7, & \Xi \in [\Xi_{j-1}, \Xi_j], \\
(\Xi_{j+4} - \Xi)^7 - 8(\Xi_{j+3} - \Xi)^7 + 28(\Xi_{j+2} - \Xi)^7 + 56(\Xi_{j+1} - \Xi)^7, & \Xi \in [\Xi_j, \Xi_{j+1}], \\
(\Xi_{j+4} - \Xi)^7 - 8(\Xi_{j+3} - \Xi)^7 + 28(\Xi_{j+2} - \Xi)^7, & \Xi \in [\Xi_{j+1}, \Xi_{j+2}], \\
(\Xi_{j+4} - \Xi)^7 - 8(\Xi_{j+3} - \Xi)^7, & \Xi \in [\Xi_{j+2}, \Xi_{i+3}], \\
(\Xi_{j+4} - \Xi)^7, & \Xi \in [\Xi_{j+3}, \Xi_{i+4}], \\
0, & \text{Otherwise},
\end{cases}
\tag{3.129}$$

where $j \in [-3, m+3]$. Thus, we obtain

$$\begin{aligned}
£_j(\Xi) &= \mho_{j-3} + 120\mho_{j-2} + 1191\mho_{j-1} + 2416\mho_j + 1191\mho_{j+1} \\
&\quad + 120\mho_{j+2} + \mho_{j+3}.
\end{aligned} \tag{3.130}$$

Substituting Eq. (3.130) into Eq. (3.10), yields $(m+7)$ equations. Using Mathematica 12 to solve this system to get the value of $\mho_j$ leads to the following values of analytical, numerical under the different values of Ξ in Table 3.4.

3.5 FIGURES REPRESENTATION

This section explains the physical representation of each of this chapter's figure, to show their dynamical behaviour and to illustrate the spinless relativistic composite particles, such as the pion.

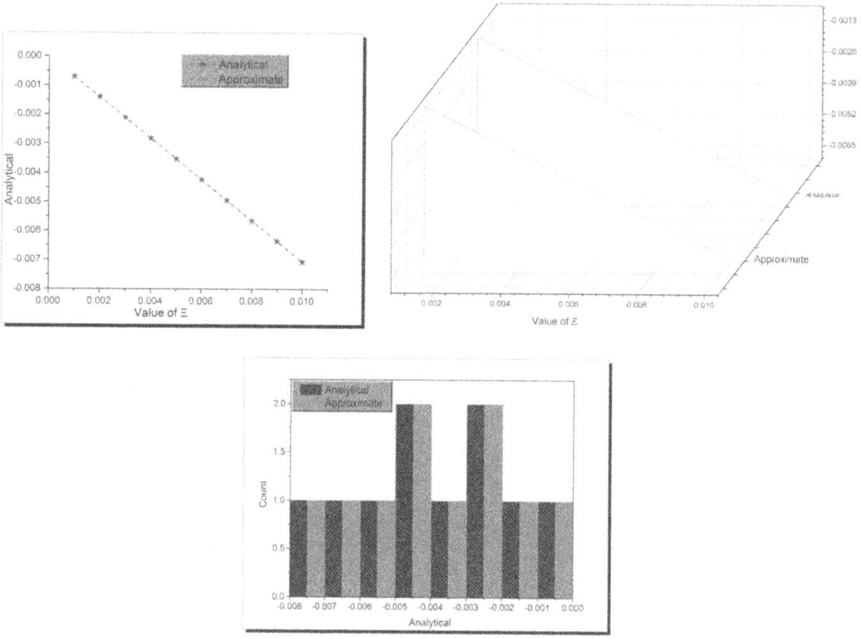

Figure 3.9 Analytical and numerical simulation according to Table 3.3

1. Figure (3.1) shows breath wave solution of Eq. (3.6) according to Eq. (3.13) in three-, two-dimensional, and contour plots with the following parameter values: $\left[\eta=3, b=-4, \beta=5, c=2, \delta=6\right]$.
2. Figure (3.2) shows periodic peak on wave solution of Eq. (3.6) according to Eq. (3.39) in three-, two-dimensional, and contour plots with the following parameter values: $\left[b=-4, \beta=5, c=2, \delta=6, \eta=3\right]$.
3. Figure (3.3) shows singular peak wave solution of Eq. (3.6) according to Eq. (3.88) in three-, two-dimensional, and contour plots with the following parameter values: $\left[b_1=1, b=-2, c=3, C_1=2, C_2=3, \mu=-4, \sigma=-1\right]$.
4. Figure (3.4) shows kink wave solution of Eq. (3.6) according to Eq. (3.91) in three-, two-dimensional, and contour plots with the following parameter values: $\left[a_0=1, b=-2, c=2, \mu=6, r=-4\right]$.
5. Figure (3.5) shows periodic wave solution of Eq. (3.6) according to Eq. (3.100) in three-, two-dimensional, and contour plots with the following parameter values: $\left[b=-2, c=2, \delta=6, \lambda=1, \mu=5\right]$.

Table 3.4
Computational and Numerical Solutions Values of Eq. (3.10) to Show the Absolute Value of Error Based on Septic B-Spline Scheme

Values of Ξ	Analytical	Approximate	Absolute Error
0.001	−0.000707107	−0.000707212	1.04981×10^{-7}
0.002	−0.00141421	−0.00141445	2.38736×10^{-7}
0.003	−0.00212132	−0.00212164	3.23363×10^{-7}
0.004	−0.00282842	−0.00282883	4.03545×10^{-7}
0.005	−0.00353553	0.00353597	4.46132×10^{-7}
0.006	−0.00424263	−0.00424309	4.62128×10^{-7}
0.007	−0.00494973	−0.00495015	4.20868×10^{-7}
0.008	−0.00565682	−0.00565719	3.61988×10^{-7}
0.009	−0.00636392	−0.0063641	1.77367×10^{-7}
0.01	−0.00707101	−0.00707101	8.67362×10^{-19}

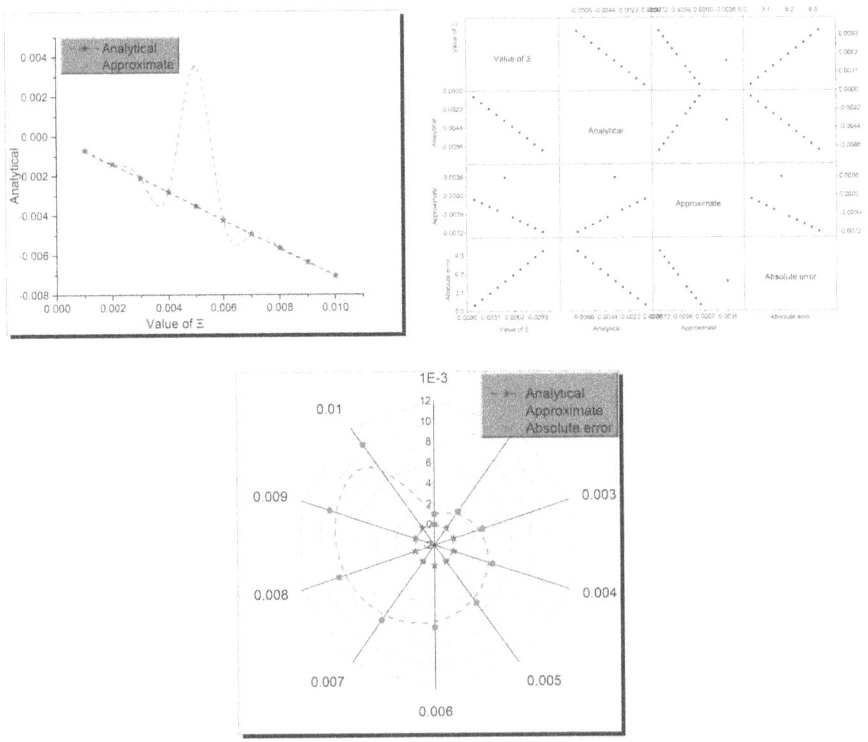

Figure 3.10 Analytical and numerical simulation according to Table 3.4.

6. Figure (3.6) shows soliton wave solution of Eq. (3.6) according to Eq. (3.121) in three-, two-dimensional, and contour plots.
7. Figure (3.7) shows exact and numerical wave solutions with different values Ξ to show the accuracy of obtained solutions of Eq. (3.6) according to Table 3.1 in three different forms.
8. Figure (3.8) shows the accuracy between our obtained analytical and numerical solutions according to the shown values in Table 3.2 in three different sketches with respect to cubic B-spline.
9. Figure (3.9) shows the accuracy between our obtained analytical and numerical solutions according to the shown values in Table 3.3 in three different sketches with respect to quantic B-spline.
10. Figure (3.10) shows the accuracy between our obtained analytical and numerical solutions according to the shown values in Table 3.4 in three different sketches with respect to septic B-spline.

3.6 CONCLUSION

We have successfully applied the eight recent analytical and numerical schemes with a new fractional operator to the fractional nonlinear KG equation, describing spinless relativistic composite particles such as the pion. This new operator is used to avoid the disadvantage of the other fractional operator. Distinct, solitary wave solutions of this equation have been obtained, and for more illustration of the dynamical behaviour of this kind of fluid, some solutions have been sketched (Figures 3.1– 3.10) in three different formulas for each figure (two-, three-dimensional, and contour plots).

REFERENCES

1. M. Eslami, H. Rezazadeh, The first integral method for Wu–Zhang system with conformable time-fractional derivative, Calcolo 53 (3) (2016) 475–485.

2. M. Ekici, M. Mirzazadeh, M. Eslami, Q. Zhou, S. P. Moshokoa, A. Biswas, M. Belic, Optical soliton perturbation with fractional-temporal evolution by first integral method with conformable fractional derivatives, Optik 127 (22) (2016) 10659–10669.

3. H. Aminikhah, A. Refahi Sheikhani, H. Rezazadeh, Sub-equation method for the fractional regularized long-wave equations with conformable fractional derivatives, Scientia Iranica 23 (3) (2016) 1048–1054.

4. O. Ozkan K. Ali, Conformable double laplace transform for fractional partial diferential equations arising in mathematical physics, Mathematical Studies and Applications 4–6 October 2018, 471.

5. A. Atangana, J. Gómez-Aguilar, Numerical approximation of Riemann-Liouville definition of fractional derivative: from Riemann-Liouville to Atangana-Baleanu, Numerical Methods for Partial Differential Equations 34 (5) (2018) 1502–1523.

6. K. M. Owolabi, A. Atangana, Numerical approximation of nonlinear fractional parabolic differential equations with Caputo–Fabrizio derivative in Riemann–Liouville sense, Chaos, Solitons & Fractals 99 (2017) 171–179.

7. S. Vong, P. Lyu, X. Chen, S.-L. Lei, High order finite difference method for time-space fractional differential equations with Caputo and Riemann-Liouville derivatives, Numerical Algorithms 72 (1) (2016) 195–210.

8. J. Hristov, Derivatives with non-singular kernels from the Caputo–Fabrizio definition and beyond: appraising analysis with emphasis on diffusion models, Frontiers in Fractional Calculus 1 (2017) 270–342.

9. H. Yépez-Martínez, J. F. Gómez-Aguilar, A new modified definition of Caputo–Fabrizio fractional-order derivative and their applications to the Multi Step Homotopy Analysis Method (MHAM), Journal of Computational and Applied Mathematics 346 (2019) 247–260.

10. M. A. Dokuyucu, E. Celik, H. Bulut, H. M. Baskonus, Cancer treatment model with the Caputo-Fabrizio fractional derivative, The European Physical Journal Plus 133 (3) (2018) 92.

11. H. Rezazadeh, D. Kumar, A. Neirameh, M. Eslami, M. Mirzazadeh, Applications of three methods for obtaining optical soliton solutions for the Lakshmanan–Porsezian–Daniel model with Kerr law nonlinearity, Pramana 94 (1) (2020) 39.

12. S. M. Mirhosseini-Alizamini, H. Rezazadeh, M. Eslami, M. Mirzazadeh, A. Korkmaz, New extended direct algebraic method for the Tzitzica type evolution equations arising in nonlinear optics, Computational Methods for Differential Equations 8 (1) (2020) 28–53.

13. M. A. Shallal, K. K. Ali, K. R. Raslan, H. Rezazadeh, A. Bekir, Exact solutions of the conformable fractional EW and MEW equations by a new generalized expansion method, Journal of Ocean Engineering and Science 5 (3) (2020) 223–229.

14. R. A. Attia, D. Lu, T. Ak, M. M. Khater, Optical wave solutions of the higher-order nonlinear Schrödinger equation with the non-Kerr nonlinear term via modified Khater method, Modern Physics Letters B (2020) 2050044.

15. M. M. Khater, C. Park, A.-H. Abdel-Aty, R. A. Attia, D. Lu, On new computational and numerical solutions of the modified Zakharov–Kuznetsov equation arising in electrical engineering, Alexandria Engineering Journal 59 (3) (2020).

16. C. Yue, M. M. Khater, R. A. Attia, D. Lu, The plethora of explicit solutions of the fractional KS equation through liquid–gas bubbles mix under the thermodynamic conditions via Atangana–Baleanu derivative operator, Advances in Difference Equations 2020 (1) (2020) 1–12.

17. M. M. Khater, C. Park, D. Lu, R. A. Attia, Analytical, semi-analytical, and numerical solutions for the Cahn–Allen equation, Advances in Difference Equations 2020 (1) (2020) 1–12.

18. H. Qin, M. Khater, R. A. Attia, D. Lu, Approximate simulations for the non-linear long-short wave interaction system, Frontiers in Physics 7 (2020) 230.

19. H. Qin, R. A. Attia, M. Khater, D. Lu, Ample soliton waves for the crystal lattice formation of the conformable time-fractional (n+ 1) Sinh-Gordon equation by the modified Khater method and the Painlevé property, Journal of Intelligent & Fuzzy Systems (preprint) 38 (2020) 2745–2752.

20. M. Al-Raeei, M. S. El-Daher, On: New optical soliton solutions for nonlinear complex fractional Schrödinger equation via new auxiliary equation method and novel (G'/G)-expansion method, Pramana 94 (1) (2020) 9.

21. M. Khater, R. A. Attia, H. Qin, H. Kadry, R. Kharabsheh, D. Lu, On the stable computational, semi-analytical, and numerical solutions of the Langmuir waves in an ionized plasma, Journal of Intelligent & Fuzzy Systems (preprint) 1–13.

22. E. Hosny Mohamad, E. Mostafa Hasan, F. Aly Shoukry, and O. Ahmed El-Dydamoni, Assessment of the microscopic-observation-drug-susceptiblity assay (mods) and gen expert assay for diagnosis of pulmonary tuberculosis, Al-Azhar Medical Journal 49 (1) (2020) 91–102.

23. G. Akram, I. Zainab, Dark Peakon, Kink and periodic solutions of the nonlinear biswas-milovic equation with kerr law nonlinearity, Optik (2020) 164420.

24. R. M. Jena, S. Chakraverty, H. Rezazadeh, D. Domiri Ganji, On the solution of time-fractional dynamical model of Brusselator reaction-diffusion system arising in chemical reactions, Mathematical Methods in the Applied Sciences (2020) DOI: 10.1002/mma.6141.

25. C. Kadapa, W. Dettmer, D. Perić, A fictitious domain/distributed lagrange multiplier based fluid–structure interaction scheme with hierarchical B-spline grids, Computer Methods in Applied Mechanics and Engineering 301 (2016) 1–27.

26. W. Wen, S. Duan, J. Yan, Y. Ma, K. Wei, D. Fang, A quartic B-spline based explicit time integration scheme for structural dynamics with controllable numerical dissipation, Computational Mechanics 59 (3) (2017) 403–418.

27. Q. Bi, J. Huang, Y. Lu, L. Zhu, H. Ding, A general, fast and robust B-spline fitting scheme for micro-line tool path under chord error constraint, Science China Technological Sciences 62 (2) (2019) 321–332.

28. B. P. Moghaddam, J. A. T. Machado, A stable three-level explicit spline finite difference scheme for a class of nonlinear time variable order fractional partial differential equations, Computers & Mathematics with Applications 73 (6) (2017) 1262–1269.

29. S. Sun, H. Lin, L. Zheng, J. Yu, Y. Hu, A real-time and look-ahead interpolation methodology with dynamic B-spline transition scheme for CNC machining of short line segments, The International Journal of Advanced Manufacturing Technology 84(5–8) (2016) 1359–1370.

30. X. Du, J. Huang, L.-M. Zhu, H. Ding, An error-bounded B-spline curve approximation scheme using dominant points for CNC interpolation of micro-line toolpath, Robotics and Computer-Integrated Manufacturing 64 (2020) 101930.

31. H. Bakodah, A. Al Qarni, M. Banaja, Q. Zhou, S. P. Moshokoa, A. Biswas, bright and dark thirring optical solitons with improved Adomian decomposition method, Optik 130 (2017) 1115–1123.

32. K. Nouri, Study on efficiency of the Adomian decomposition method for stochastic differential equations, International Journal of Nonlinear Analysis and Applications 8 (1) (2017) 61–68.

33. H. Rezazadeh, A. Korkmaz, M. M. Khater, M. Eslami, D. Lu, R. A. Attia, New exact traveling wave solutions of biological population model via the extended rational sinh-cosh method and the modified Khater method, Modern Physics Letters B 33 (28) (2019) 1950338.

34. J. Li, Y. Qiu, D. Lu, R. A. Attia, M. Khater, Study on the solitary wave solutions of the ionic currents on microtubules equation by using the modified Khater method, Thermal Science (23) (2019) S2053–S2062.

35. A. Alderremy, R. A. Attia, J. F. Alzaidi, D. Lu, M. Khater, Analytical and semi-analytical wave solutions for longitudinal wave equation via modified auxiliary equation method and Adomian decomposition method, Thermal Science (23) (2019) 1943–1957.

36. M. Khater, R. A. Attia, D. Lu, Modified auxiliary equation method versus three non-linear fractional biological models in present explicit wave solutions, Mathematical and Computational Applications 24 (1) (2019) 1.

37. E. K. Akgül, Solutions of the linear and nonlinear differential equations within the generalized fractional derivatives, Chaos: An Interdisciplinary Journal of Nonlinear Science 29 (2) (2019) 023108.

38. D. Baleanu, A. Fernandez, A. Akgül, On a fractional operator combining proportional and classical differintegrals, Mathematics 8 (3) (2020) 360.

39. A. Akgül, A. Cordero, J. R. Torregrosa, Solutions of fractional gas dynamics equation by a new technique, Mathematical Methods in the Applied Sciences 43 (2020) 1349–1358.

40. A. Atangana, A. Akgül, K. M. Owolabi, Analysis of fractal fractional differential equations, Alexandria Engineering Journal 59 (2020) 1117–1134.

41. A. Atangana, A. Akgül, Can transfer function and Bode diagram be obtained from Sumudu transform?, Alexandria Engineering Journal 59 (2020) 1971–1984.

42. H. Singh, An efficient computational method for the approximate solution of nonlinear Lane–Emden type equations arising in astrophysics, Astrophysics and Space Science 363 (4) (2018) 71.

43. H. Singh, A new numerical algorithm for fractional model of bloch equation in nuclear magnetic resonance, Alexandria Engineering Journal 55 (3) (2016): 2863–2869.

44. H. Singh, H. Srivastava, D. Kumar, A reliable numerical algorithm for the fractional vibration equation, Chaos, Solitons & Fractals 103 (2017) 131–138.

45. H. Singh, D. Kumar, D. Baleanu, Methods of Mathematical Modelling: Fractional Differential Equations, CRC Press, Boca Raton, FL, 2019.

46. H. Singh, Approximate solution of fractional vibration equation using Jacobi polynomials, Applied Mathematics and Computation 317 (2018) 85–100.

47. H. Singh, Operational matrix approach for approximate solution of fractional model of Bloch equation, Journal of King Saud University-Science 29 (2) (2017) 235–240.

48. P. Leal, O. Bertolami, Relativistic dispersion relation and putative metric structure in noncommutative phase-space, Physics Letters B 793 (2019) 240–246.

49. C. Pfeifer, Redshift and lateshift from homogeneous and isotropic modified dispersion relations, Physics Letters B 780 (2018) 246–250.

50. K. M. Fujiwara, Z. A. Geiger, K. Singh, R. Senaratne, S. V. Rajagopal, M. Lipatov, T. Shimasaki, D. M. Weld, Experimental realization of a relativistic harmonic oscillator, New Journal of Physics 20 (6) (2018) 063027.

51. C. Pfeifer, L. Barcaroli, L. K. Brunkhorst, G. Gubitosi, N. Loret, Hamilton geometry: Phase space geometry from modified dispersion relations, in: Fourteenth Marcel Grossmann Meeting-MG14, World Scientific (2018) pp. 3929–3934.

52. K. Hosseini, P. Mayeli, R. Ansari, Modified Kudryashov method for solving the conformable time-fractional Klein–Gordon equations with quadratic and cubic nonlinearities, Optik 130 (2017) 737–742.

53. D. Kumar, J. Singh, D. Baleanu, A hybrid computational approach for Klein–Gordon equations on Cantor sets, Nonlinear Dynamics 87 (1) (2017) 511–517.

54. M. Alaroud, M. Al-Smadi, O. Abu Arqub, R. Rozita Ahmad, M. Abu Hammad, S. Momani, Numerical solutions of linear time-fractional Klein-Gordon equation by using power series approach, in: Proceedings of International Conference on Fractional Differentiation and its Applications (ICFDA), 2018.

55. K. Hosseini, P. Mayeli, R. Ansari, Bright and singular soliton solutions of the conformable time-fractional Klein–Gordon equations with different nonlinearities, Waves in Random and Complex Media 28 (3) (2018) 426–434.

56. X.-J. Yang, D. Baleanu, F. Gao, New analytical solutions for Klein-Gordon and Helmholtz equations in fractal dimensional space, Proceedings of the Romanian Academy. Series A, Mathematics, Physics, Technical Sciences 18 (3) (2017) 231–238.

57. R. J. Le Roy, LEVEL: A computer program for solving the radial Schrödinger equation for bound and quasibound levels, Journal of Quantitative Spectroscopy and Radiative Transfer 186 (2017) 167–178.

58. M. J. Ablowitz, Z. H. Musslimani, Inverse scattering transform for the integrable nonlocal nonlinear Schrödinger equation, Nonlinearity 29 (3) (2016) 915.

59. S.-F. Tian, Initial–boundary value problems for the general coupled nonlinear Schrödinger equation on the interval via the Fokas method, Journal of Differential Equations 262 (1) (2017) 506–558.

60. V. Gerdjikov, A. Saxena, Complete integrability of nonlocal nonlinear Schrödinger equation, Journal of Mathematical Physics 58 (1) (2017) 013502.

61. A. Fokas, Integrable multidimensional versions of the nonlocal nonlinear Schrödinger equation, Nonlinearity 29 (2) (2016) 319.

62. F. Fillion-Gourdeau, S. MacLean, R. Laflamme, Algorithm for the solution of the Dirac equation on digital quantum computers, Physical Review A 95 (4) (2017) 042343.

63. F. Mozaffari, H. Hassanabadi, H. Sobhani, W. Chung, Investigation of the Dirac equation by using the conformable fractional derivative, Journal of the Korean Physical Society 72 (9) (2018) 987–990.

64. W. Bao, Y. Cai, X. Jia, Q. Tang, Numerical methods and comparison for the Dirac equation in the nonrelativistic limit regime, Journal of Scientific Computing 71 (3) (2017) 1094–1134.

65. B. S. T. Alkahtani, Chua's circuit model with Atangana–Baleanu derivative with fractional order, Chaos, Solitons & Fractals 89 (2016) 547–551.

66. A. Atangana, I. Koca, Chaos in a simple nonlinear system with Atangana–Baleanu derivatives with fractional order, Chaos, Solitons & Fractals 89 (2016) 447–454.

4 Applications of Conserved Schemes for Solving Ultra-Relativistic Euler Equations

Mahmoud A.E. Abdelrahman[1,2]
[1]Department of Mathematics, Faculty of Science, Mansoura University, Mansoura, Egypt
[2]Department of Mathematics, College of Science, Taibah University, Al-Madinah Al-Munawarah, Saudi Arabia

CONTENTS

4.1 INTRODUCTION

Relativity has a significant contribution in applied science in such areas as high-energy particle beams, nuclear physics, heavy-ion collisions, and astrophysical phenomena consisting of active galactic nuclei (AG) and gamma-ray bursts (GARBS). A relativistic strategy is necessary to model high-speed flows for characterizing physical operations and to provide the explanation of many physical phenomena. Indeed, as it is so difficult to solve the relativistic equations analytically, numerical schemes are employed. There are some numerical techniques applied to relativistic equations, which yield high-order accuracy in smooth regions of the simulated fluid flow and gain sharp discontinuities for the shock fronts, [1–5]. Except for the kinetic beam technique [6], these methods are dependent on the macroscopic continuum representation. Wernicke, Kunik, and Qamar [7–9] have applied kinetic methods for the relativistic hydrodynamics simulation. These methods are discrete in time but continuous in space.

While there are similarities between the classical and relativistic Euler models, there are significant differences between them. For instance, even the solution for Riemann problem of the classical case may yields vacuum regions, causing analytical investigation to become more complicated see [10]. In contrast, for the relativistic case such behaviour will not occur. A second difference is that, for the numerical methods studied the classical case, one should presume a bound for the eigenvalues, which depend on the initial data to achieve the Courant-Friedrichs-Lewy (CFL) condition. In the relativistic case every speed is bounded by light speed, independent of the initial data. Therefore, the CFL condition for the relativistic case is very straightforward, specifically $\Delta x = 2\Delta t$. Furthermore, the conservation laws of energy and momentum for the ultra-relativistic Euler (URE) equations construct a genuinely nonlinear hyperbolic subsystem, which makes the analysis and numerics for the relativistic case more interesting.

Abdelrahman and Kunik [11] introduced the cone grid and front tracking techniques for solving the URE equations. The cone grid scheme is sturdy and efficacious, and it is easier to implement than the front tracking scheme. Contrarily, the front scheme is better suited for analytical purposes, such as study of shock interactions [12, 13]. The wave tracking method is a vital tool for solving discontinuities in the hyperbolic systems of conservation laws, [14–21]. The well-known conserved schemes such as the Godunov scheme, total variation diminishing (TVD) schemes, finite volume (FV) method, discontinuous Galerkin method, and (W)ENO high-order FV can usually achieve higher-order accuracy with non-oscillatory and sharp shock transmission [22]. Furthermore, the cone grid and front tracking methods are much better than other familiar methods even for initial large density ratios and large pressure ratios [23]. Indeed, there is recent developments in the hyperbolic systems of conservation laws and their applications. See for example [24–27].

In this chapter, the cone grid, front tracking, and Godunov methods are implemented for solving the 1-D URE equations. For validation and comparison, some numerical cases are carried out. We also determine the experimental order of convergence and numerical L^1-stability of the schemes. We also consider one of the important topics of hyperbolic systems of conservation laws, namely the Riemann invariants. These Riemann invariants give a rise to many interesting applications in applied science [28, 29]. Furthermore, in most research works concerning the hyperbolic systems of conservation laws, the measure of wave strength is determined as the change of the Riemann invariants, which is the basic ingredient for many analytical as well as numerical methods [30–35]. We will implement the proposed schemes in this chapter in order to, for the first time, simulate the Riemann invariants corresponding to the URE equations.

4.2 THE URE EQUATIONS

We consider the 1-D URE equations of conservation laws for mass, energy, and momentum given as follows:

$$\begin{aligned} &(n\sqrt{1+u^2})_t + (nu)_x = 0, \\ &(p(3+4u^2))_t + (4pu\sqrt{1+u^2})_x = 0, \\ &(4pu\sqrt{1+u^2})_t + (p(1+4u^2))_x = 0, \end{aligned} \tag{4.1}$$

(e.g. [7, 11, 36]), where $n > 0$, $v = \frac{u}{\sqrt{1+u^2}}$, and $p > 0$ display proper particle density, velocity field, and pressure, respectively, with $|\, v \,| < 1$, $u \in \mathbb{R}$. An extremely important characteristic for URE equations is that the second and third equations construct an independent system for p and u ((p,u) subsystem), which decouples with the relativistic continuity for n.

Equation (4.1) has the characteristic velocities (eigenvalues)

$$\lambda_1 = \frac{2u\sqrt{1+u^2} - \sqrt{3}}{3+2u^2} < \lambda_2 = \frac{u}{\sqrt{1+u^2}} < \lambda_3 = \frac{2u\sqrt{1+u^2} + \sqrt{3}}{3+2u^2}. \tag{4.2}$$

These characteristic velocities may first be achieved in the Lorentz rest frame, where $u = 0$. After that the relativistic additivity law of the velocities yields Eq. (4.2) in the general Lorentz frame. In the Lorentz rest frame, we get the positive speed of sound $\lambda = \frac{1}{\sqrt{3}}$ that is independent of the spatial direction.

The differential equations (4.1) are not sufficient if we consider shock discontinuities. We utilize a weak integral formulation with a piecewise C^1-solution $n, u, p : (0, \infty) \times \mathbb{R} \to \mathbb{R}$, $n, p > 0$ [37]:

$$\begin{aligned} &\oint_{\partial\Omega} n\sqrt{1+u^2}\,dx - nu\,dt = 0, \\ &\oint_{\partial\Omega} 4pu\sqrt{1+u^2}\,dx - p(1+4u^2)\,dt = 0, \\ &\oint_{\partial\Omega} p(3+4u^2)\,dx - 4pu\sqrt{1+u^2}\,dt = 0, \end{aligned} \tag{4.3}$$

where $\Omega \subset \mathbb{R}_0^+ \times \mathbb{R}$ is a bounded and convex region in space-time and with a piecewise smooth, positively oriented boundary.

Over and above this weak solution (4.3) have to obey the entropy inequality

$$\oint_{\partial\Omega} S^0\,dx - S^1\,dt \geq 0, \tag{4.4}$$

$$S^0(p,u,n) = -n\sqrt{1+u^2}\ln\frac{n^4}{p^3}, \quad S^1(p,u,n) = -nu\ln\frac{n^4}{p^3}. \tag{4.5}$$

4.2.1 THE (p,u) SUBSYSTEM

Consider the (p,u) system [12, 13, 34, 38]:

$$\begin{aligned} &(p(3+4u^2))_t + (4pu\sqrt{1+u^2})_x = 0, \\ &(4pu\sqrt{1+u^2})_t + (p(1+4u^2))_x = 0. \end{aligned} \tag{4.6}$$

System Eq. (4.6) has eigenvalues

$$\lambda_1 = \frac{2u\sqrt{1+u^2}-\sqrt{3}}{3+2u^2} < \lambda_3 = \frac{2u\sqrt{1+u^2}+\sqrt{3}}{3+2u^2}. \tag{4.7}$$

The decoupled equation for n in Eq. (4.1) produces contact discontinuities with the characteristic speed λ_2 in Eq. (4.2).

The corresponding eigenvectors are

$$r_1 = (\frac{-4p}{\sqrt{3}\sqrt{1+u^2}}, 1)^T, \quad r_3 = (\frac{4p}{\sqrt{3}\sqrt{1+u^2}}, 1)^T. \tag{4.8}$$

Proposition 4.1: *[38] System Eq. (4.6) is strictly hyperbolic and genuinely nonlinear for $p > 0, u \in \mathbb{R}$.*

The 1- and 3-Riemann invariants of Eq. (4.6) are

$$R = \ln(\sqrt{1+u^2}+u) + \frac{\sqrt{3}}{4}\ln(p) \tag{4.9}$$

and

$$\tilde{R} = \ln(\sqrt{1+u^2}+u) - \frac{\sqrt{3}}{4}\ln(p). \tag{4.10}$$

The function $R = R(p,u)$ is constant across 1-rarefaction waves and $\tilde{R} = \tilde{R}(p,u)$ is constant across 3-rarefaction waves.

Lemma 4.1: *The mapping $(p,u) \mapsto (R,\tilde{R})$ is 1−1 with nonsingular Jacobian for $p > 0$, $u \in \mathbb{R}$.*

Proof. Consider

$$\begin{aligned} \psi &= R+\tilde{R} = 2\ln(\sqrt{1+u^2}+u), \\ \chi &= R-\tilde{R} = \tfrac{\sqrt{3}}{2}\ln(p), \end{aligned} \tag{4.11}$$

Therefore $\psi = \psi(u)$ is a function of u and $\chi = \chi(p)$ is a function of p, where $\psi'(u) = \frac{2}{\sqrt{1+u^2}} > 0$, $\chi'(p) = \frac{\sqrt{3}}{2p} > 0$ for $p > 0$. Thus the mapping $(p,u) \mapsto (\psi,\chi)$ is 1−1.

The corresponding determinant is

$$\begin{vmatrix} \chi'(p) & 0 \\ 0 & \psi'(u) \end{vmatrix} = \chi'(p)\psi'(u) > 0.$$

The mapping $(\psi,\chi) \mapsto (R,\tilde{R})$ is

$$\begin{pmatrix} \psi \\ \chi \end{pmatrix} = \begin{pmatrix} 1 & 1 \\ 1 & -1 \end{pmatrix} \begin{pmatrix} R \\ \tilde{R} \end{pmatrix}. \tag{4.12}$$

This is a nonsingular linear mapping and therefore $(\psi,\chi) \mapsto (R,\tilde{R})$ is 1–1. Thus $(p,u) \mapsto (R,\tilde{R})$ is $1-1$, and determinant for the Jacobian for this mapping is non zero for $p > 0$, $u \in \mathbb{R}$. □

Consider a shock $x = x(t)$ with constant speed $v_s = \dot{x}$, $U_l = (p_l, u_l)$ is the left state and $U_r = (p_r, u_r)$ is the right state with $p_{l,r} > 0$. The Rankine-Hugoniot jump (RHj) conditions are

$$\begin{aligned} v_s\left[p_r(3+4u_r^2) - p_l(3+4u_l^2)\right] &= 4p_r u_r\sqrt{1+u_r^2} - 4p_l u_l\sqrt{1+u_l^2}, \\ v_s\left[4p_r u_r\sqrt{1+u_r^2} - 4p_l u_l\sqrt{1+u_l^2}\right] &= p_r(1+4u_r^2) - p_l(1+4u_l^2). \end{aligned} \tag{4.13}$$

The local form of the entropy inequality in singular points is

$$-v_s(g_r - g_l) + (\psi_r - \psi_l) > 0, \tag{4.14}$$

where

$$g(p,u) = p^{\frac{3}{4}}\sqrt{1+u^2}, \quad \psi(p,u) = p^{\frac{3}{4}}u. \tag{4.15}$$

Inequality (4.14) is equivalent to $u_l > u_r$ see [38].

The following two positive parameters are important to prescribe nonlinear elementary waves

$$\tau := \frac{p_r}{p_l}, \quad \vartheta := \frac{\sqrt{1+u_r^2} - u_r}{\sqrt{1+u_l^2} - u_l}. \tag{4.16}$$

For $\tau > 0$, we define $L_S\colon \mathbb{R}^+ \to \mathbb{R}^+$ by

$$L_S(\tau) = \frac{\sqrt{1+3\tau} + \sqrt{3}\sqrt{3+\tau}}{\sqrt{8}}. \tag{4.17}$$

Based on this function we define the function $K_S\colon \mathbb{R}^+ \to \mathbb{R}^+$ as follows:

$$K_S(\tau) := \frac{L_S(\tau)}{L_S(\frac{1}{\tau})} = \frac{\sqrt{1+3\tau}\sqrt{3+\tau} + \sqrt{3}(\tau-1)}{4\sqrt{\tau}}. \tag{4.18}$$

Lemma 4.2: *[38] The functions L_S and K_S are strictly monotonically increasing.*

Lemma 4.3: *[36, Section 4.4] Given $p_l, p_r > 0$. Define τ and ϑ as in Eq. (4.16).*

1. *Suppose that $\tau > 1$ and $\vartheta = K_S(\tau)$. Then the left state (p_l, u_l) is connected with the right state (p_r, u_r) by a single 1-shock.*
2. *Suppose that $\tau < 1$ and $\vartheta K_S(\tau) = 1$. Then the left state (p_l, u_l) is connected with the right state (p_r, u_r) by a single 3-shock.*

For $\tau > 0$, we define $K_R : \mathbb{R}^+ \to \mathbb{R}^+$ by

$$K_R(\tau) := \tau^{\frac{\sqrt{3}}{4}}. \tag{4.19}$$

Lemma 4.4: *[38] Given $(p_{l,r}, u_{l,r}) \in \mathbb{R}^+ \times \mathbb{R}$. Define τ and β as in Eq. (4.16).*

1. *Suppose that $\tau < 1$ and $\vartheta = K_R(\tau)$. Then the left state (p_l, u_l) is connected with the right state (p_r, u_r) by a single 1-rarefaction wave (1-fan).*
2. *Suppose that $\tau > 1$ and $\vartheta K_R(\tau) = 1$. Then the left state (p_l, u_l) is connected with the right state (p_r, u_r) by a single 3-rarefaction wave (3-fan).*

4.3 THE NUMERICAL SCHEMES

In this section, 1-D cone grid schemes of Abdelrahman and Kunik [11] for solving the 1-D URE equations are derived. For the derivation of the front tracking scheme, see [11] and for the Godunov method, refer to [28].

4.3.1 CONE GRID SCHEME

For the derivation of cone grid scheme in 1-D space [11]. We rewrite the URE equations as

$$W_t + G(W)_x = 0. \tag{4.20}$$

Consider a time step $\Delta t > 0$; and consequently compute the step Δx according to the following CFL condition:

$$\Delta t = \frac{1}{2}\Delta x. \tag{4.21}$$

The Riemann solution of Eq. (4.20) is considered inside the cone, as illustrated in Figure 4.1. We employ the conservation laws with respect to the balance region depicted in Figure 4.1:

$$\oint_{\partial\Omega} W(t,x)dx - G(W(t,x))dt = 0. \tag{4.22}$$

The spatial domain (trapezium) is depicted in Figure 4.2. Given $\alpha, \beta \in \mathbb{R}$ for $\alpha < \beta$, $T > 0$, and $M \in \mathbb{N}$ is the number of grid cells for half of the space interval $[\alpha, \beta]$, that is,

1. The step size given by

$$\Delta x := \frac{\beta - \alpha}{2M}. \tag{4.23}$$

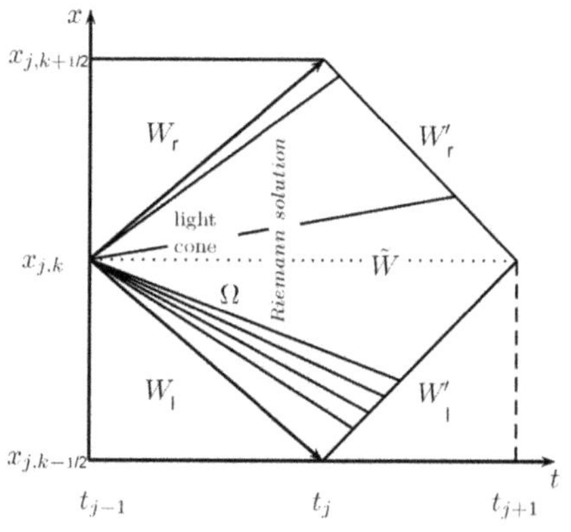

Figure 4.1 Balance regions [11].

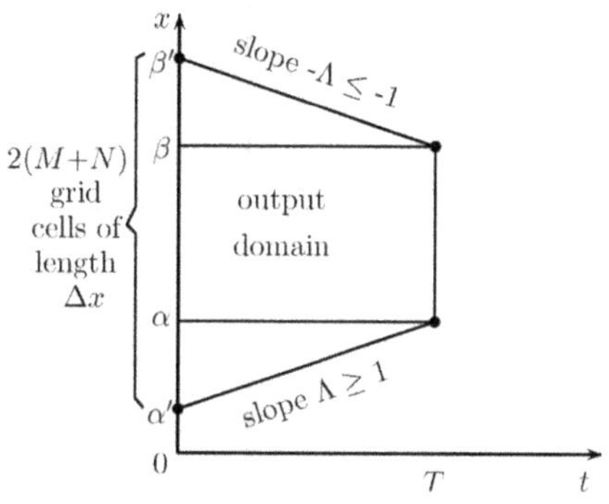

Figure 4.2 Computational domain [11].

2. The number

$$N := \left\lceil \frac{T}{\Delta x} \right\rceil = \left\lceil \frac{2TM}{b-a} \right\rceil \tag{4.24}$$

is the smallest integer, thus $T \leq N \cdot \Delta x$.

3. Put $\Delta t := \frac{T}{2N}$. Then, utilizing 1 and 2 yields

$$\Lambda = \frac{\Delta x}{2\Delta t} = \frac{\Delta x \cdot N}{T} \geq 1, \tag{4.25}$$

thus the CFL condition is obeyed. N is the number of grid cells for half of the time interval $[0,T]$ and Δt is the time step, that is, the interval $[0,T]$ divided into $2N$ subintervals with length Δt.

4. The time steps given as $t_j = j \cdot \Delta t$ for $j \in \{0,\ldots,2N\}$.
5. We define the bounds of the domain as

$$\alpha' = \alpha - T \cdot \Lambda, \quad \beta' = \beta + T \cdot \Lambda, \tag{4.26}$$

to eschew artificial boundary effects in Figure 4.2. The structural intervals $[\alpha', \alpha]$, $[\beta, \beta']$ are divided into N subintervals with length Δx.

6. Within time t_{j-1}, the grid points are defined

$$x_{j,k} = \alpha' + (j-1)\Lambda \cdot \Delta t + (k-1)\Delta x, \quad j = 1,\ldots,2N+1,$$
$$k = 1,\ldots,2(M+N) - j + 2.$$

4.3.2 THE STRUCTURE OF NUMERICAL SOLUTIONS

We consider a constraint of mesh points, which presents the left corner points for every cell in Figures 4.1 and 4.3. We solve the local Riemann problems for each cell. For the corner point $v_l = (t_{j-1}, x_{j,k})$ in Figure 4.3 we present the local Riemann

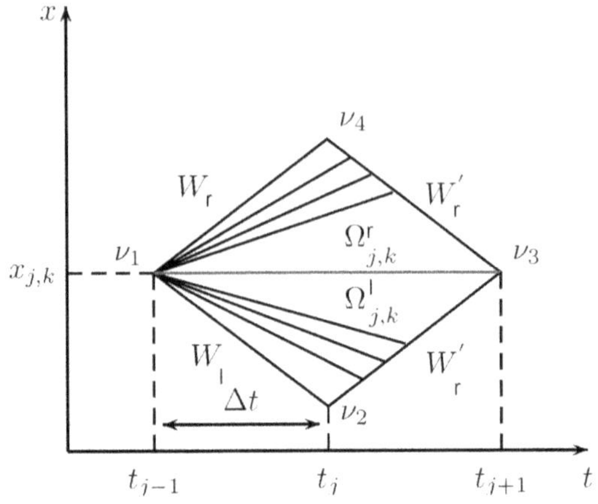

Figure 4.3 Balance cell [11].

solution (4.20) for the initial data

$$W(t_{j-1},x) = \begin{cases} W_l, & x \leq x_{j,k} \\ W_r, & x > x_{j,k} \end{cases}$$

for $t > t_{j-1}$ and assume that $W(t,x)$ is the numerical solution inside the parallelogram.

The fixed value for $W(t,x)$ through the cord $\nu_1\nu_3$ is displayed by $\tilde{W}$. On each cord $\nu_2\nu_3$ and $\nu_3\nu_4$ in Figure 4.3 set the unknown values W_l' and W_r' instead of the numerical solution and require for W and $G(W)$ that

$$\int_{\partial\Omega^{l,r}_{j,k}} W dx - G(W)\, dt = 0. \tag{4.27}$$

Utilizing Eq. (4.25) yields implicit equation for W_r' of the region $\partial\Omega^r_{j,k}$,

$$W_r' - W_r + \frac{1}{\Lambda}\left(G(W_r') + G(W_r) - 2\, G(\tilde{W})\right) = 0. \tag{4.28}$$

In a similar way, an implicit equation of W_l' for region $\partial\Omega^l_{j,k}$ is

$$W_l' - W_l - \frac{1}{\Lambda}\left(G(W_l') + G(W_l) - 2\, G(\tilde{W})\right) = 0. \tag{4.29}$$

The new states $W_{l,r}'$ can be determined uniquely, utilizing Eqs. (4.28), (4.29) as well as the CFL-condition $\Lambda \geq 1$.

Table 4.1
L^1-Error and EOC for Particle Density n

N	Cone-Grid Scheme		Front Tracking Scheme		Godunov Scheme	
	L^1-error	EOC	L^1-error	EOC	L^1-error	EOC
25	0.150303		0.145558		0.017895	
50	0.099396	0.59661	0.016543	0.11333	0.176895	0.48469
100	0.065245	0.60732	0.015429	0.10057	0.118909	0.57303
200	0.043281	0.59213	0.015316	0.01061	0.082585	0.52591
400	0.028982	0.57857	0.013254	0.20861	0.0536501	0.62229
800	0.019458	0.57479	0.0124356	0.09195	0.033375	0.68481

For complete derivation and further details of this scheme, please refer to [11]. Indeed, the cone grid scheme was applied for solving 1-D phonon-Bose model see [39].

4.4 NUMERICAL RESULTS

We introduce numerical applications for the 1-D URE equations to validate the numerical schemes, where the CFL condition is $\Delta x = 2\Delta t$, which egress automatically from the construction of light cones, because each signal velocity is bounded by the speed of light. We implement the proposed schemes to simulate the Riemann invariants.

Example 4.1: *The initial data are*

$$(n,u,p) = \begin{cases} (2.0, 0.0, 10.0), & x \leq 1/2, \\ (1.0, 0.0, 0.3), & x > 1/2. \end{cases}$$

This application depicts the time evolution of an initial discontinuous state of a fluid at rest. The spatial domain is [0, 1] via 400 grid points and final time $t = 0.45$. *The solution composed of a contact bounded by a left transonic rarefaction wave and an right shock wave. Figure 4.4 illustrates particle density n, velocity* $v = \frac{u}{\sqrt{1+u^2}}$ *and pressure p. Figure 4.5 illustrates Riemann invariants for the system* (4.6), *employing the cone grid, front tracking, and Godunov methods, respectively. Table 4.1 depicts* L^1*-errors for the density at different grid points.*

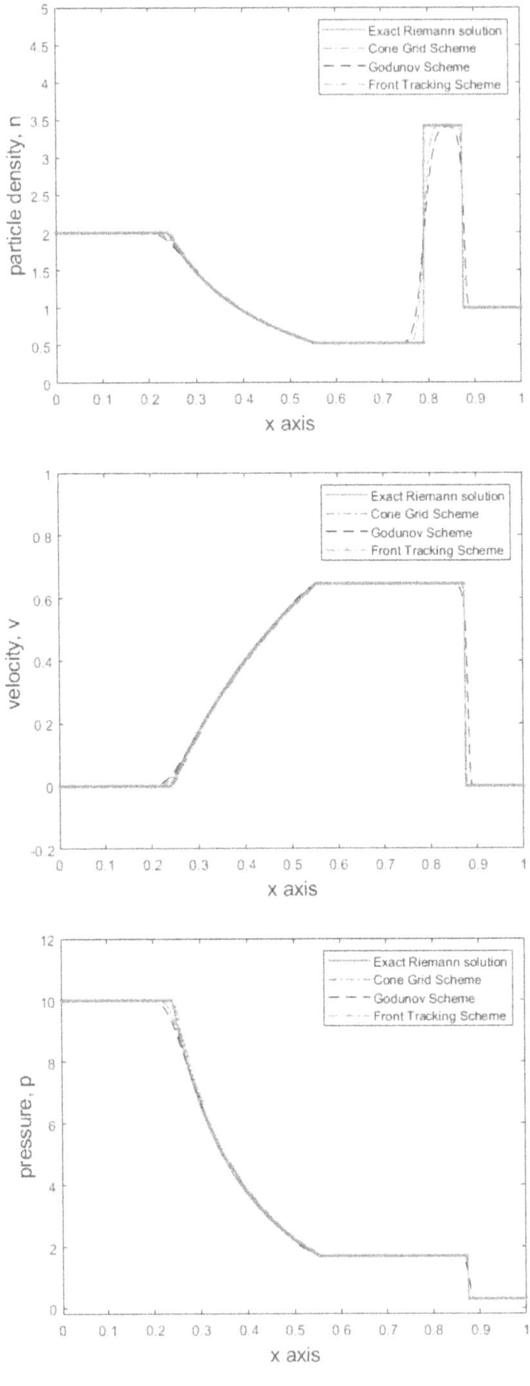

Figure 4.4 Results of Example 4.1 at time $t = 0.45$, utilizing 400 grid points.

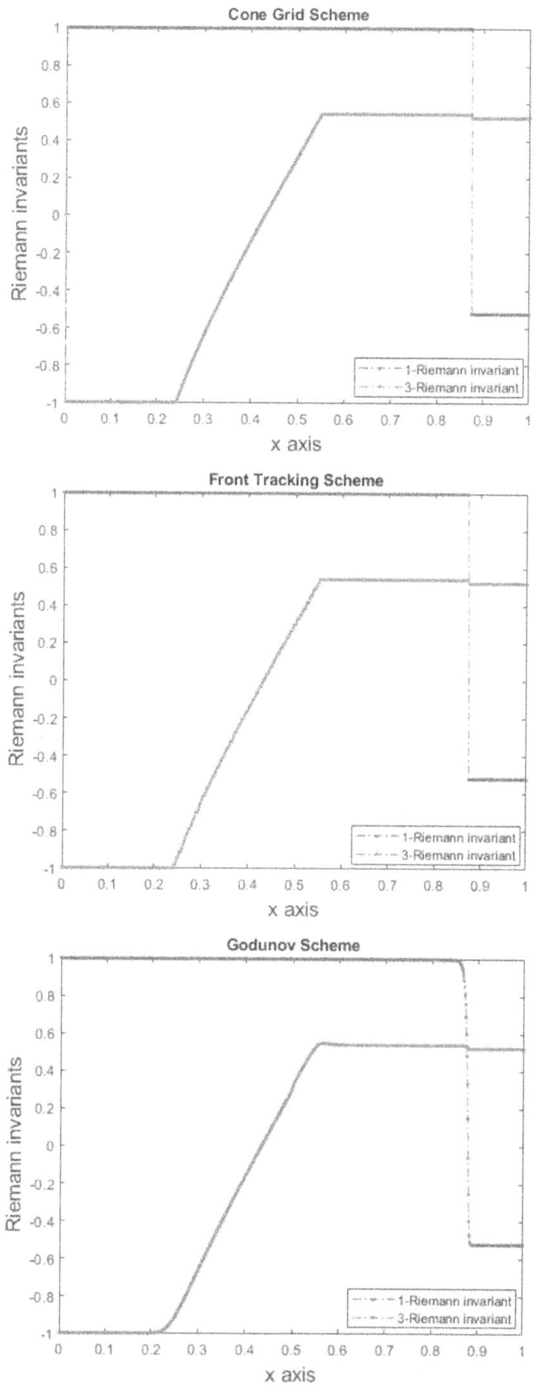

Figure 4.5 Riemann invariants corresponding to Example 4.3.

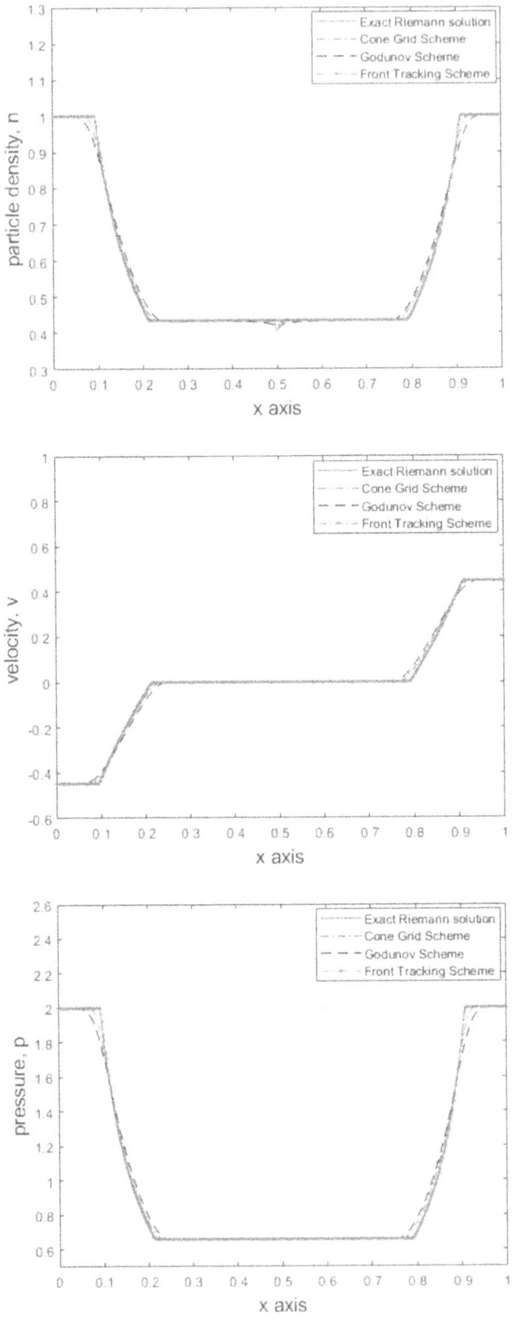

Figure 4.6 Results of Example 4.2 at time $t = 0.5$, utilizing 400 grid points.

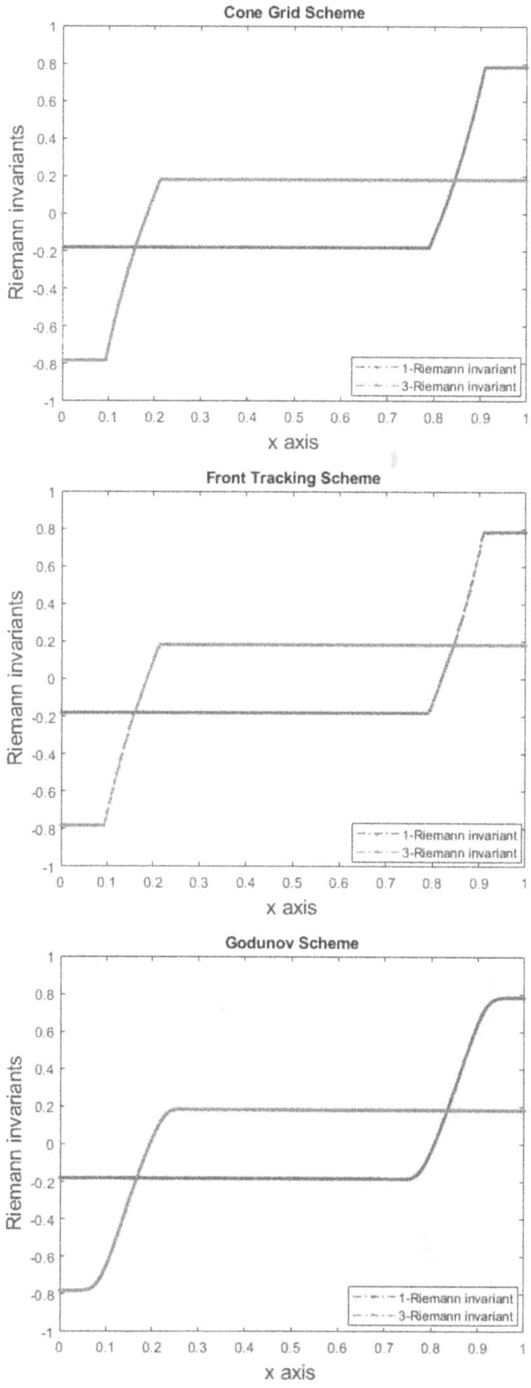

Figure 4.7 Riemann invariants corresponding to Example 4.2.

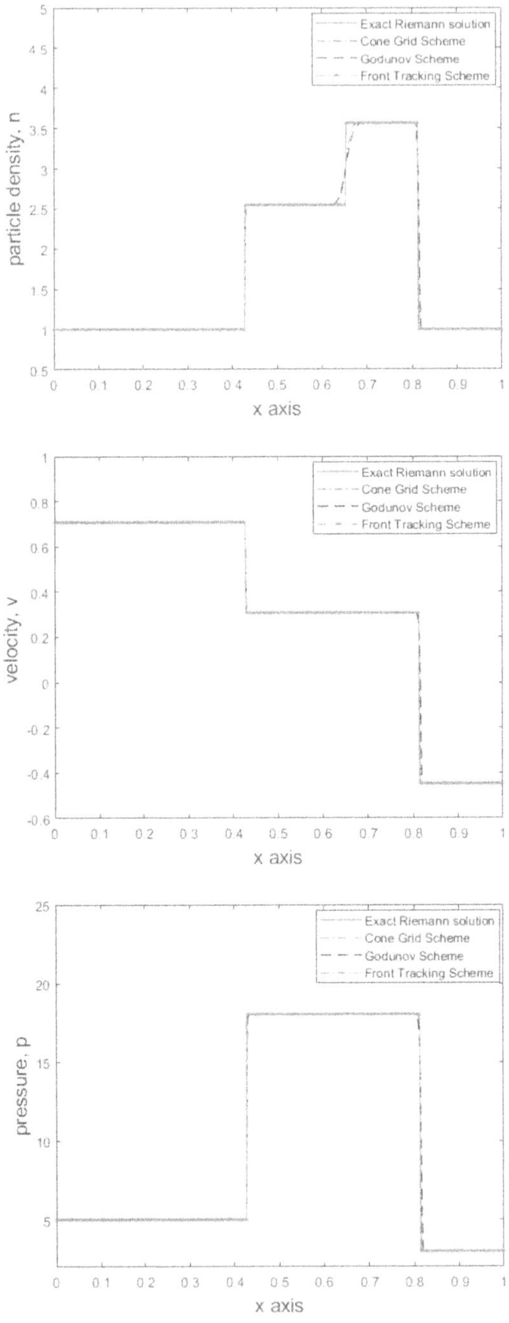

Figure 4.8 Results of Example 4.3 at time $t = 0.5$, utilizing 400 grid points.

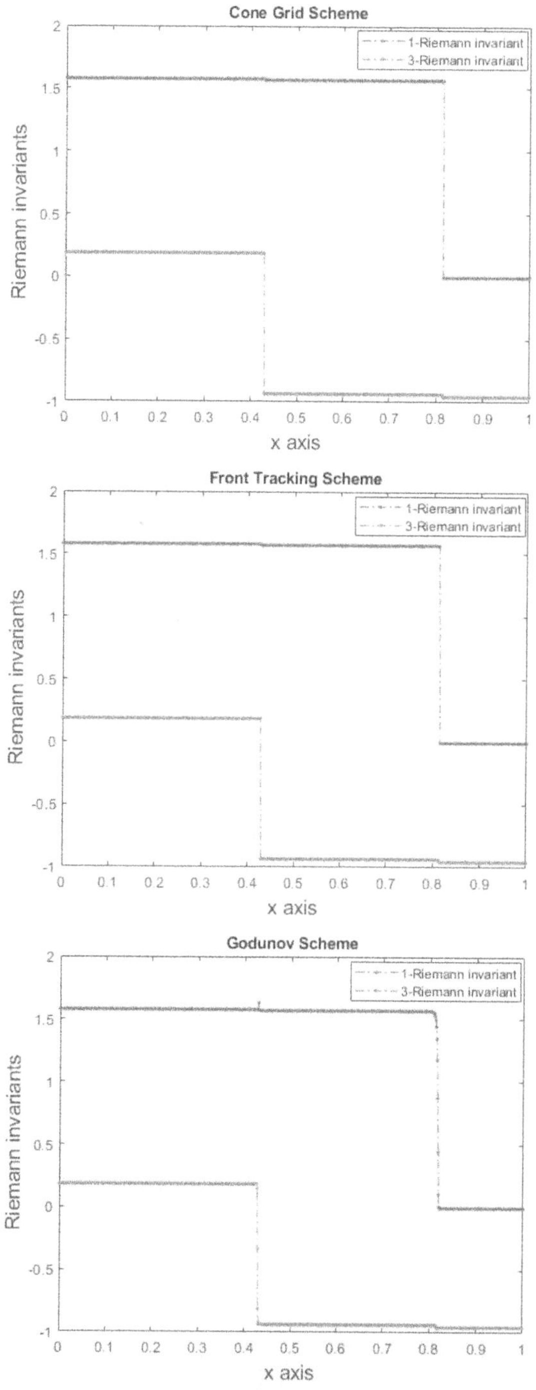

Figure 4.9 Riemann invariants corresponding to Example 4.3.

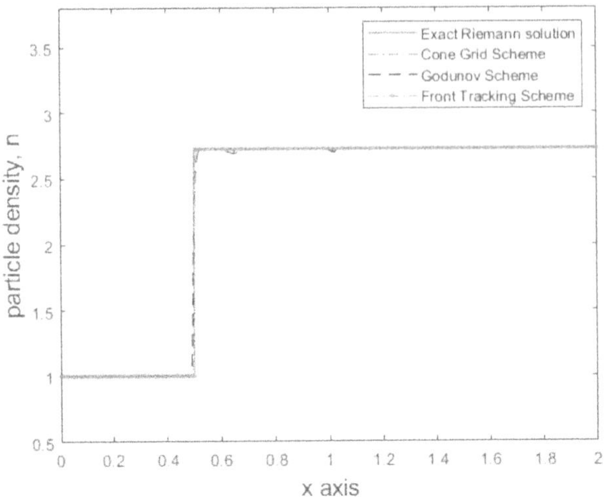

Figure 4.10 A single shock.

Experimental order of convergence

We check the EOC for the proposed schemes. If $\kappa = \Delta x$ is the step size, hence L^1-norm is

$$\| W(.,t) - W_\kappa(.,t) \|_{L^1(\mathbb{R})} = c\kappa^\delta, \tag{4.30}$$

and δ is the order of the L^1-error. W represents the exact solution and W_κ is the numerical solution. The L^1-error is defined as

$$\| W(.,t) - W_\kappa(.,t) \|_{L^1} = \Delta x \sum_{i=1}^{N} | W(x_i,t) - W_\kappa(x_i,t) | .$$

N is the number of grid points. Hence (4.30) yields

$$EOC := \delta = \ln\left(\frac{\| W(.,t) - W_{\frac{\kappa}{2}}(.,t) \|_{L^1}}{\| W(.,t) - W_\kappa(.,t) \|_{L^1}}\right) \Big/ \ln\left(\frac{1}{2}\right). \tag{4.31}$$

Table 4.1 provides L^1-error and EOC for particle density n, utilizing the proposed schemes.

Example 4.2: *The initial data are*

$$(n,u,p) = \begin{cases} (1.0,-0.5,2.0), & x \le 1/2, \\ (1.0,0.5,2.0), & x > 1/2. \end{cases}$$

The solution is composed of two strong rarefaction waves and a trivial stationary contact discontinuity. The computational domain is $[0,1]$ with 400 grid elements and

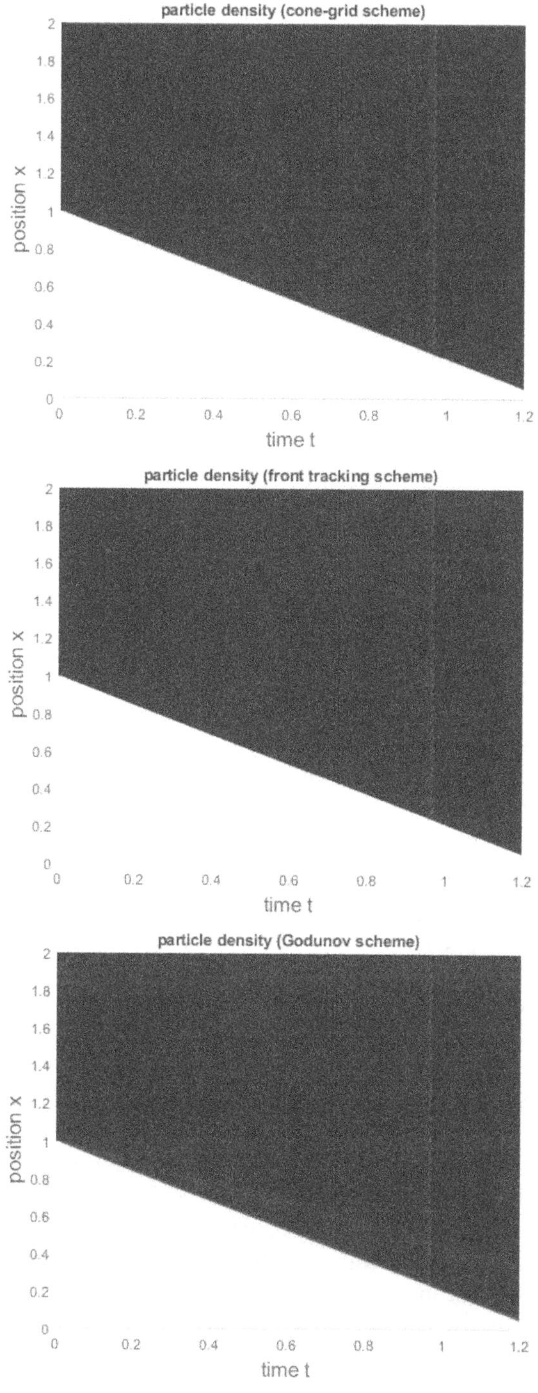

Figure 4.11 A single shock.

final time $t = 1/2$. Figure 4.6 illustrates the solution profiles. Figure 4.7 depicts the Riemann invariants for the system (4.6), using cone grid, front tracking, and Godunov schemes, respectively.

Example 4.3: *The initial data is*

$$(n,u,p) = \begin{cases} (1,1,5), & x \leq 1/2, \\ (1,-0.5,3), & x > 1/2. \end{cases}$$

The solution is composed of a left shock, a contact, and a right shock. The computational domain is $[0,1]$ with 400 grid elements and the final time $t = 1/2$. Figure 4.8 shows the solution profiles. Figure 4.9 depicts the Riemann invariants for the system (4.6), using the proposed schemes.

Example 4.4: *Single shock solution*

The initial data is

$$(n,u,p) = \begin{cases} (1.0,0.0,1.0), & x \leq 1.0, \\ (2.725,-0.6495,4.0), & x > 1.0, \end{cases}$$

This problem was considered by Yang et al. [6]. We provide initial data, from which we know that a single shock solution results from the RHj conditions. We chose the initial data and the space-time range such that the shock exactly reaches the right lower corner at the time axis. Figure 4.10 depicts the shape of particle density for $0 \leq t \leq 1.271$ and $0 \leq x \leq 2$. Figure 4.11 depicts the particle density at the fixed time $t = 0.635$, using cone grid scheme with 400 grid points, front tracking scheme with 400 grid points, and Godunov scheme with 1000 grid points. This example shows that the cone grid and front tracking methods yield a sharp shock resolution and its success indicates that the conservation laws for mass, energy, and momentum, and likewise the entropy inequality, are satisfied.

4.5 CONCLUSIONS

The three conserved schemes were implemented for solving 1-D URE equations. Namely, we compared the cone grid scheme with front tracking and Godunov schemes and the Riemann solution by using some numerical applications. We explained the construction of the solution and its properties for each case. We also checked the experimental order of convergence (EOC) and numerical L^1-error for the proposed schemes. Further, we implemented the proposed schemes in order to simulate the Riemann invariants.

Currently, work is in progress to implement the cone grid and front tracking schemes for 2-D and 3-D URE equations. Moreover, the URE equations with an initial high pressure ratio and with an initial high particle density ratio yield several applications, such as the inverse shock tube problem, interaction of waves, and irreversibility. Finally, it is interesting to implement the cone grid and front tracking schemes for other hyperbolic systems of conservation laws, for example, shallow water system, pressure gradient system, traffic-flow model, blood-flow model, etc.

REFERENCES

1. Del Zanna, L. and Bucciantini, N. (2002). An efficient shock-capturing central-type scheme for multidimensional relativistic flows, I. Hydrodynamics, Astron. Astrophys. 390, 1177–1186.

2. Donat, R. and Marquina, A. (1996). Capturing shock reflections: An improved flux formula, J. Comp. Phys. 125, 42–58.

3. Eulderink, F. and Mellema, G. (1995). General relativistic hydrodynamics with a Roe solver, Astron. Astrophys. Suppl. 110, 587–623.

4. Lucas-Serrano, A., Font, J.A., Ibez, J.M. and Mart, J.M. (2004). Assessment of a high-resolution central scheme for the solution of the relativistic hydrodynamics equations, I. Hydrodynamics, Astron. Astrophys. 428, 703–715.

5. Schneider, V., Katscher, U., Rischke, D.H., Waldhauser, B., Maruhn, J.A. and Munz, C.-D. (1993). New algorithms for ultra-relativistic numerical hydrodynamics, J. Comp. Phys. 105, 92–107.

6. Yang, J.Y., Chen, M.H., Tsai, I.N. and Chang, J.W. (1997). A kinetic beam scheme for relativistic gas dynamics, J. Comput. Phys. 136, 19–40.

7. Kunik, M., Qamar, S. and Warnecke, G. (2003). Kinetic schemes for the ultra-relativistic Euler equations, J. Comput. Phys., 187, 572–596.

8. Kunik, M., Qamar, S. and Warnecke, G. (2004). BGK-type kinetic flux vector splitting schemes for ultra-relativistic Euler equations, SIAM J. Sci. Comput. 26, 196–223.

9. Kunik, M., Qamar, S. and Warnecke, G. (2004). Kinetic schemes for the relativistic gas dynamics, Num. Math. 97, 159–191.

10. Smoller, J. (1994). Shock Waves and Reaction-Diffusion Equations. Springer-Verlag, New York.

11. Abdelrahman, M.A.E and Kunik, M. (2014). A new front tracking scheme for the ultra-relativistic Euler equations, J. Comput. Phys. 275, 213–235.

12. Abdelrahman, M.A.E. and Kunik, M. (2014). The interaction of waves for the ultra-relativistic Euler equations, J. Math. Anal. Appl. 409, 1140–1158.

13. Abdelrahman, M.A.E. (2017). Global solutions for the ultra-relativistic Euler equations, Nonlinear Anal. 155, 140–162.

14. Baiti, P. and Jenssen, H.K. (1998). On the front tracking algorithm, J. Math. Anal. Appl. 217, 395–404.

15. Bressan, A. (1995). Lecture Notes on Conservation Laws, S.I.S.S.A.,Trieste.

16. Bressan, A. (1992). Global solutions to systems of conservation laws by wavefront tracking, J. Math. Anal. Appl. 170, 414–432.

17. Dafermos, C.M. (1972). Polygonal approximations of solutions of the initial value problem for a conservation law, J. Math. Anal. Appl. 38, 33–41.

18. Diperna, R.J. (1976). Global existence of solutions to nonlinear hyperbolic systems of conservation laws, J. Comm. Differ. Equ. 20, 187–212.

19. Risebro, N.H. and Tveito, A. (1992). A front tracking method for conservation laws in one dimension, J. Comput. Phys. 101, 130–139.

20. Risebro, N. H. (1993). A front tracking alternative to the random choice method, Proc. Am. Math. Soc. 117, 1125–1139.

21. Witteveen, J.A.S., Koren, B. and Bakker, P.G. (2007). An improved front tracking method for the Euler equations, J. Comput. Phys. 224, 712–728.

22. Tang, H. and Liu, T. (2006). Note on the conservative schemes for the Euler equations, J. Comput. Phys. 218, 451–459.

23. Abdelrahman, M.A.E. (2018). Conserved schemes with high pressure ratio, high particle density ratio and self-similar method, Eur. Phys. J. Plus. 133, 304. https://doi.org/10.1140/epjp/i2018-12116-9.

24. Singh, H., Sahoo, M.R. and Singh, O.P. (2015). Weak asymptotic solution to non-strictly system of Conservation laws, Electron. J. Diff. Equ., 2015(01), 1–12.

25. Sahoo, M.R. and Singh, H. (2016). Weak asymptotic solution for a non-strictly hyperbolic system of conservation laws-II, Electron. J. Diff. Equ. 2016(94), 1–14.

26. Singh, H., Kumar, D. and Baleanu, D. (2019). Methods of Mathematical Modelling: Fractional Differential Equations, CRC Press/Taylor & Francis Group, Boca Raton, FL.

27. Singh, H. and Srivastava, H.M. (2020). Numerical simulation for fractional-order Bloch equation arising in nuclear magnetic resonance by using the Jacobi polynomials, App. Sci. 10 (8), 2850.

28. Toro, E.F. (2009) Riemann Solvers and Numerical Methods for Fluid Dynamics: A Practical Peer Review Only. Introduction. Springer, Berlin.

29. Abdelrahman, M.A.E. (2017). On the shallow water equations, Z. Naturforsch. 72(9)a, 873–879.

30. Glimm, J. (1995). Solutions in the large for nonlinear hyperbolic systems of equations, Comm. Pure. Appl. Math. 18, 697–715.

31. Chen, J. (1995). Conservation laws for the relativistic p-system, Comm. Partial Diff. Equ. 20, 1602–1646.

32. Chen, G.Q. and Li, Y.C. (2004). Relativistic Euler equations for isentropic fluids: Stability of Riemann solutions with large oscillation, Z. Angew. Math. Phys. 55, 903–926.

33. Pant, V. (1996). Global entropy solutions for isentropic relativistic fluid dynamics, Commu. Partial Diff. Eqs. 21, 1609–1641.

34. Smoller, J. and Temple, B. (1993). Global solutions of the relativistic Euler equations, Commun. Math. Phys, 156, 67–99.

35. Wissman, B.D. (2011) Global solutions to the ultra-relativistic Euler equations, Commun. Math. Phys. 306, 831–851.

36. Kunik, M. (2005). Selected initial and boundary value problems for hyperbolic systems and kinetic equations, Habilitation thesis, Otto-von-Guericke University Magdeburg.

37. Oleinik, O.A. (1957). Discontinuous solutions of nonlinear differential equations, Amer. Math. Soc. Trans. Serv. 26, 95–172.

38. Abdelrahman, M.A.E and Kunik, M. (2015). The ultra-relativistic Euler equations, Math. Meth. Appl. Sci. 38, 1247–1264.

39. Abdelrahman, M.A.E. (2018). Cone-grid scheme for solving hyperbolic systems of conservation laws and one application, Comp. Appl. Math. 37(3), 3503–3513.

5 Notorious Boundary Value Problems: Singularly Perturbed Differential Equations and Their Numerical Treatment

Naresh M. Chadha[1] *and Sunita Kumawat*[2]
[1]Department of Mathematics, DIT University, Dehradun, Uttarakhand, India
[2]Department of Mathematics, Amity University Haryana, India

CONTENTS

5.1 INTRODUCTION

Consider following boundary value problem

$$
\begin{aligned}
&-\varepsilon u''(x) + a(x)u'(x) + r(x)u(x) = f(x) \text{ on } (0,1)\\
&\text{with boundary conditions} \quad u(0) = \alpha, u(1) = \beta,
\end{aligned} \tag{5.1}
$$

where $\varepsilon \ll 1$ denotes a small parameter, $a(x)$ and $r(x)$ denote advection and reaction coefficients, respectively, and $f(x)$ denotes forcing/source term. For numerical solution of Eq. (5.1), we consider a uniform mesh $0 = x_0 < x_2 < \ldots < x_N = 1$, with mesh width $h = 1/N$. The difference operators are defined as

$$
\begin{aligned}
D^- v_j &= \frac{v_j - v_{j-1}}{h}, \quad D^+ v_j = \frac{v_{j+1} - v_j}{h}, \quad D^c = \frac{v_{j+1} - v_{j-1}}{2h},\\
\delta^2 v_j &= \frac{v_{j-1} - 2v_j + v_{j+1}}{h^2}.
\end{aligned} \tag{5.2}
$$

We consider following test problem

$$
-\varepsilon u''(x) + (1+x)u'(x) + 2u(x) = f(x), \qquad u(0) = u(1) = 0, \tag{5.3a}
$$

$$
f(x) = \exp\left(-\frac{3}{2\varepsilon}\right)(3x-1) - 6x - 2. \tag{5.3b}
$$

whose solution is given by

$$
u_{\text{exact}} = 2 - \exp\left(-\frac{3}{2\varepsilon}\right) + (1+x)\left[\exp\left(-\frac{3}{2\varepsilon}\right) + \exp\left(\frac{x^2+2x-3}{2\varepsilon}\right) - 2\right].
$$

The test problem Eq. (5.3) belongs to a class of problems known as *singularly perturbed differential equations*. They are differential equations where the highest order derivative term is multiplied by a parameter $\varepsilon \ll 1$. In general, solutions of these types of differential equations exhibit sharp layers. Inside these layers the solution tends to infinity as $\varepsilon \to 0$, and outside these layers, the solution behaves within a reasonably finite range. Thus, to solve these types of problems numerically, one has to use layer-adapted meshes, which are fine in the layer region and standard outside. These problems arise in many practical applications, such as porous media, crack propagation, mathematical modelling of biochemical reactions and pathways, etc.

We solve our test problem Eq. (5.3) for $\varepsilon = 10^{-2}$ on a uniform mesh, using difference operators defined in Eq. (5.2). Refer to Figure 5.1 for the numerical solution of Eq. (5.3) which clearly demonstrates that uniform mesh is not appropriate to deal with the layer phenomenon depicted in the solution of our test problem Eq. (5.3); to capture the layer requires an excessive number of mesh points, which may not be acceptable for many practical scenarios. The aim of this chapter is to discuss the construction of layer-adapted meshes for singularly perturbed differential equations which require special treatment. The construction of layer-adapted meshes, which include (a) a priori refined meshes, and (b) a posteriori meshes, have been discussed.

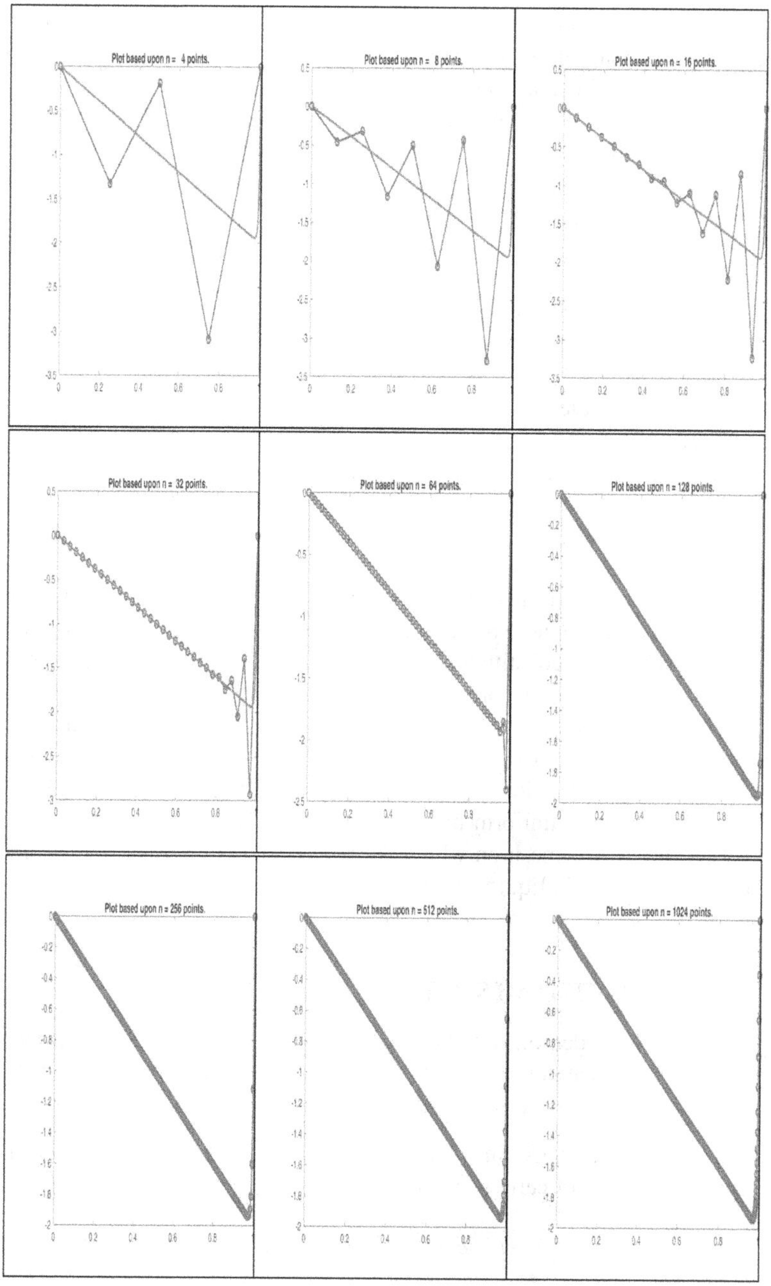

Figure 5.1 Test problem Eq. (5.3) for $\varepsilon = 10^{-2}$ is solved on a uniform mesh. This demonstrates that uniform mesh may not be appropriate to deal with singularly perturbed differential equations. Uniform mesh may either completely fail to depict the layer phenomenon typically exhibited by these type of BVPs or they may need excessive numbers of mesh nodes to capture the layer.

Using these meshes is far superior when compared to using uniform meshes in association with standard finite difference operators. By considering various test problems, both a priori refined meshes and a posteriori refined meshes are compared. The main contribution of the chapter is to highlight the difference between the constructions of different layer-adapted meshes. In particular, it is discussed how to extract monitor functions from given error estimates, and their role in mesh adaptation by employing an algorithm based on the equidistribution principle. Furthermore, for a two-dimensional case, we discuss the extension of these layer-adapted meshes. A priori refined meshes including Bakhvalov mesh and Shishkin mesh can be extended as such in a higher dimension. However, there are limitations of extending the equidistribution principle to higher dimensions. By employing a widely used code BAMG (Bidimensional Anisotropic Mesh Generator), in conjunction with a monitor matrix, we demonstrate how the choice of monitor function may significantly affect the finally adapted mesh.

The chapter is organized as follows: Eq. (5.1) is devoted to demonstrating that uniform mesh is not appropriate to deal with problems, which exhibit layers, a common feature shown by a special class of boundary value problems, commonly referred to as singularly perturbed differential equations. Eq. (5.2.1.1), and Eq. (5.2.1.2) are devoted to discussing the construction of Bakhvalov mesh and Shishkin mesh, respectively. Both these meshes are compared by considering a test problem in Eq. (5.2.1.3). The main constituents of mesh adaptation, and an algorithm based on the equidistribution principle, are discussed in Eq. (5.2.2). In Eq. (5.2.3) we discuss the construction of various probable monitor functions from a priori and a posteriori error estimates; in Eq. (5.2.4), we demonstrate mesh adaptation using algorithm discussed in Eq. (5.2.2) with monitor function constructed in Eq. (5.2.3). The algorithm starting from a uniform mesh redistributes a fixed number of mesh nodes iteratively to produce a mesh on which the computed solution is of a desired accuracy. The chapter ends in Eq. (5.3) with concluding remarks and few possible future extensions.

5.2 LAYER ADAPTED MESHES

In Eq. (5.1), we have demonstrated that uniform mesh may not be appropriate to resolve a layer phenomenon typically shown by singularly perturbed differential equations, and one has to use *layer-adapted meshes* which can be divided into two classes:

1. *A priori refined mesh:* Such a mesh can be constructed by using a priori information in terms of certain available bounds of the type

$$|u^{(k)}(x)| \leq C\left\{1+\varepsilon^{-k}e^{-\beta x/\varepsilon}\right\}, \quad k=0,1,\ldots,q \text{ and } x\in[0,1], \tag{5.4}$$

   where u denotes the exact solution, parameter q depends on the smoothness of the data, and β is some constant. Bakhvalov mesh and Shishkin mesh are the best examples of this class, see Eq. (5.2.1) for details.
2. *A posteriori refined mesh:* The construction of such a mesh is based on a suitable error indicator which can be used to identify the region(s) where the mesh

refinement is most required. An adaptive algorithm (a) employs the error indicator to detect the exact location and the width of a layer, and (b) adaptively increases the accuracy in the desired area(s) by using any one of the following means: (i) by increasing the local grid density of an existing mesh by adding extra nodes in the required region, this approach is referred to as *h-refinement*; or (ii) by employing higher-order schemes to improve the accuracy which is referred to as *p-refinement*; or (iii) keeping the total number of nodes fixed, by reallocating them strategically over the whole domain this approach is referred to as *r-refinement*.

5.2.1 A PRIORI REFINED MESHES

Let Ω_p be the physical domain where the actual problem is posed, and let Ω_c denote the computational domain, solely and purposely chosen for mesh generation; x and ξ denote the variables in Ω_p and Ω_c, respectively. Without loss of generality, the domains Ω_p, and Ω_c are assumed to be $[0,1]$. Then, an adaptive mesh for Ω_p can be generated by using the following one-to-one map:

$$x = x(\xi), \quad \xi \in [0,1], \qquad \text{with} \quad x(0) = 0, \quad x(1) = 1. \tag{5.5}$$

A priori refined meshes use an explicit relation $x(\xi)$, between x and ξ; this relation is commonly referred to as a *mesh generating function*. They usually involve a transition point which distinguishes the grid density of the so-called smooth region with the grid density of the layer region. The key here is to decide the location of the transition point(s) for which one can use a priori information (if available) regarding the solution and its derivatives.

5.2.1.1 Bakhvalov-Type Meshes

The following mesh generating function was introduced by Bakhvalov [1] for dealing with exponential boundary layer at $x = 0$:

$$x(\xi) = \begin{cases} x_1(\xi) := & -\frac{\sigma\varepsilon}{\beta}\ln\left(1-\frac{\xi}{q}\right) & \text{for} \quad \xi \in [0,\tau], \\ x_2(\xi) := & x_1(\tau) + x_1'(\tau)(\xi-\tau) & \text{for} \quad \xi \in [\tau,1]. \end{cases} \tag{5.6}$$

Here τ denotes a transition point which separates fine and coarse parts of the mesh; $q \in (0,1)$ and $\sigma > 0$ are user-defined parameters. The transition point τ is a solution of the following non-linear equation:

$$x_1(\tau) + x_1'(\tau)(1-\tau) = 1. \tag{5.7}$$

Geometrically, Eq. (5.7) implies that the line $x_2(\xi)$ passes through the point (1,1). Note that Eq. (5.7) is nonlinear and cannot be solved explicitly. However, for all practical purposes, the following formulation can be used:

$$x(\xi) = \begin{cases} x_{\text{layer}}(\xi) := & \varepsilon\lambda \log\frac{b}{b-\xi} & \text{for} \quad \xi \in [0,\tau], \\ x_{\text{smth}}(\xi) := & x_{\text{layer}}(\tau) + \frac{1-x_{\text{layer}}(\tau)}{1-\tau}(\xi-\tau) & \text{for} \quad \xi \in [\tau,1]. \end{cases} \tag{5.8}$$

In the above representation of mesh generating function. $\lambda = \sigma/\beta$ is used to simplify the notation. The MATLAB code for generating Bakhvalov mesh is given as follows.

```
% function x = BakhvalovMesh(N,epsilon)
N = 16; epsilon = 10^(-2);
jhi = linspace(0,1,N+1)'; % uniform mesh
lambda = 3; b = 0.5;
theta = b-epsilon*lambda; % transition point on jhi axis
x_theta = epsilon*lambda*log(b/(lambda*epsilon)); % transition point for mesh
plot(x_theta, theta, 'o', 'LineWidth',1.5,'MarkerEdgeColor','k','MarkerFaceColor','g','MarkerSize',4);
hold on
if epsilon > b / lambda
    x=jhi;
else
    % coding for mesh generating function
    % for layer region now
    layindx = (jhi <= theta); jhilay = jhi(layindx); xlayer = epsilon * lambda * log( b./ (b - jhilay ));
    % for smooth region now
    jhismth = jhi(length(jhilay)+1: N+1); xlayerattheta = epsilon*lambda*log(b/(b-theta));
    xsmth = xlayerattheta + ( (1- xlayerattheta)/(1-theta) )*(jhismth - theta);
    % concatenate both to yield the desired mesh
    x = [xlayer; xsmth];
end
plot(x,jhi, 'k','linewidth', 2); xlabel('\bf{Bakhvalov mesh}'); ylabel('\bf{Uniform grid points}')
gtext('\bf{Mesh generating function}'); set(gca,'xtick',x,'ytick',jhi,'GridLineStyle','-')
grid on; set(gca,'xticklabel','','yticklabel','')
```

The output of the code is shown in Figure 5.2. Also refer to, for example, [2, 3] for a modified Bakhvalov mesh proposed by Vulvanovic.

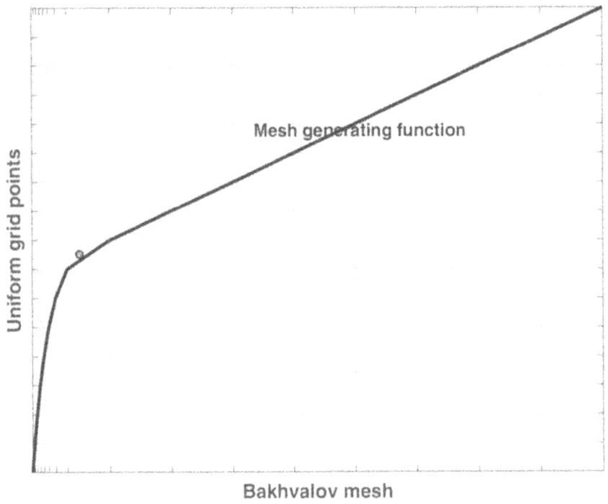

Figure 5.2 Bakhvalov mesh generated by using the code; green circle is used to show the transition point.

5.2.1.2 Shishkin-Type Meshes

Another widely used mesh for the purpose is the so-called Shishkin mesh [4], and refer to [4–6] for its applications. For a model convection–diffusion problem

$$-\varepsilon u''(x) - b(x)u'(x) + c(x)u(x) = f(x) \quad \text{in} \quad (0,1), \qquad u(0) = u(1) = 0, \quad (5.9)$$

where $b(x) \geq \beta > 0$ and $c(x) \geq 0$ on $(0,1)$. The boundary value problem given by Eq. (5.9) exhibits layer at $x = 0$ which behaves like the exponential function $\exp(-\beta x/\varepsilon)$. Let q and $\sigma > 0$ be two parameters. The transition point is defined as

$$\tau = \min\left\{q, \frac{\sigma\varepsilon}{\beta}\ln N\right\}. \quad (5.10)$$

Then, the regions $[0,\tau]$ and $[\tau,1]$, separated by τ, are divided into qN and $(1-q)N$ subintervals of equal length, respectively. The parameters β and σ are usually chosen so that $\exp(-\beta x/\varepsilon) \leq N^{-\sigma}$ on the smooth part of the mesh. Here, the parameter σ usually depends on the order of the discretization scheme. The parameter q sets the number of the mesh points used to depict the layer, similar to this parameter in the Bakhvalov mesh. The Shishkin mesh corresponds to the following mesh generating function:

$$x(\xi) = \begin{cases} x_1(\xi) := & \frac{\sigma\varepsilon}{\beta}\ln N\frac{\xi}{q} & \text{for} \quad \xi \in [0,\tau], \\ x_2(\xi) := & 1 - \left(1 - \frac{\sigma\varepsilon}{\beta}\ln N\right)\frac{1-\xi}{1-q} & \text{for} \quad \xi \in [\tau,1]. \end{cases} \quad (5.11)$$

Refer to Figure 5.3 for a MATLAB code and corresponding Shishkin mesh.

```
%function shismesh = xshis(N,epsilon)
N = 16; epsilon = 10^(-2);
tau = min(1/2, epsilon*log(N));
xlayer = linspace(0,tau,N/2+1);
xsmth = linspace(tau,1,N/2+1);
x = union(xlayer, xsmth);
t=(0:N)'/N;
plot(x,t, 'k','linewidth', 2)
xlabel('\bf{Shishkin mesh}')
ylabel('\bf{Uniform grid points}')
gtext('\bf{Mesh generating function}')
set(gca,'xtick',x,'ytick',t,
'GridLineStyle','-')
grid on
set(gca,'xticklabel','','yticklabel','')
```

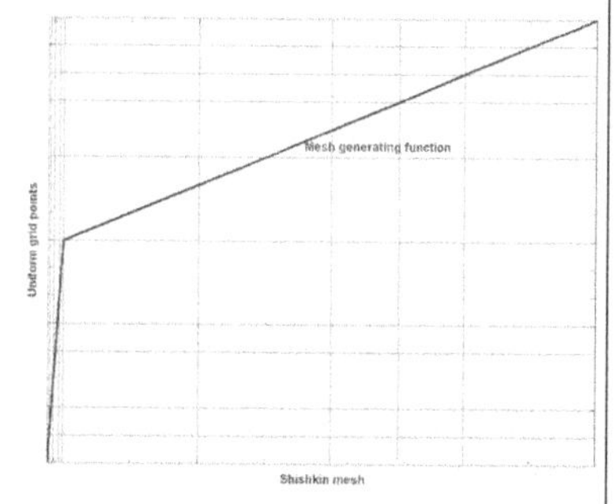

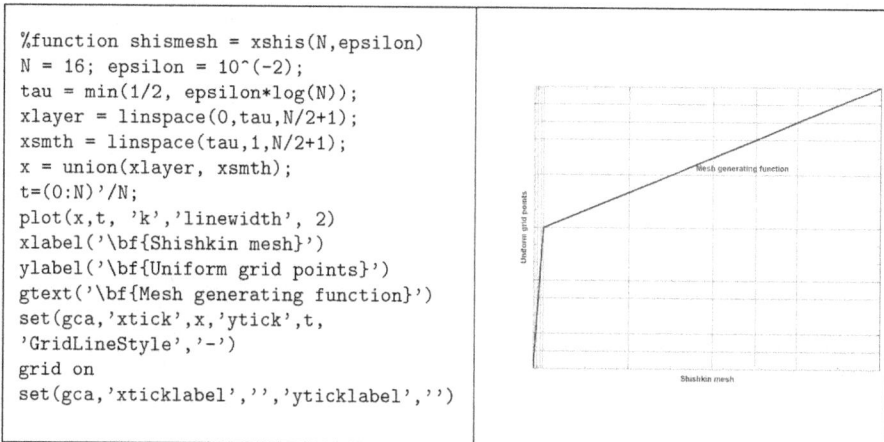

Figure 5.3 Code to generate Shishkin mesh and its output.

5.2.1.3 Comparison Between Bakhvalov Mesh and Shishkin Mesh

Shishkin-type meshes have simpler structure; two uniform meshes are simply glued together. Thus, analyzing numerical schemes may be comparatively easier for these meshes. In contrast, Bakhvalov-type meshes offer much better numerical accuracy [7]. This is not surprising as Bakhvalov-type meshes are generated using a smooth function, which gradually increases the mesh size from the beginning of the layer to the smooth region. While Shishkin-type meshes have only one transition point, often too many mesh nodes are unnecessarily placed in the layer region. Moreover, the mesh width of the subintervals on both sides of the transition point is very different; refer to Figure 5.4 and 5.5 for a visualization of the consequence. To

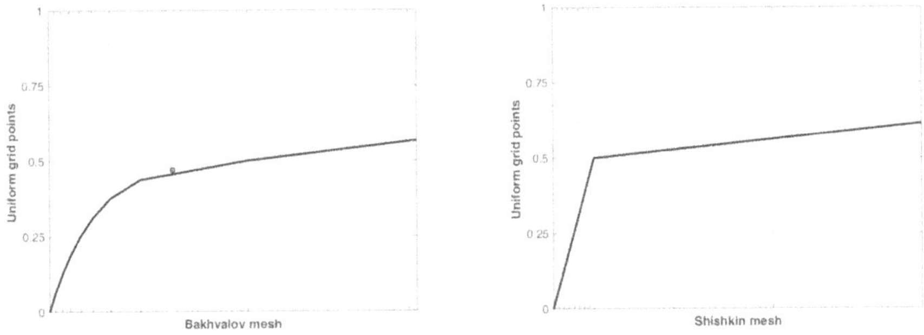

Figure 5.4 Layer-adapted meshes zoomed near the transition point: Bakhvalov Mesh (left), Shishkin Mesh (right). There is a gradual decrease in the density of the mesh nodes in Bakhvalov mesh. In contrast, in the Shishkin mesh two uniform meshes (separated by the transition point) are glued together, and there is a big jump near the transition point.

appreciate the structural difference between these meshes, and consequently its effect on the computed solution, let us take another test problem given as follows:

$$\varepsilon u''(x) + u'(x) = 0, \quad u(0) = 1, \; u(1) = 0. \tag{5.12a}$$

The exact solution is given by
$$u_{\text{exact}} = \frac{\exp(-x/\varepsilon) - \exp(-1/\varepsilon)}{1 - \exp(-1/\varepsilon)}. \tag{5.12b}$$

Next, we consider arbitrary mesh $\{x_j\}_{i=0}^{N}$ with $0 = x_0 < x_1 < \ldots < x_N = 1$, where N is a positive integer with local mesh size $h_j := x_j - x_{j-1}$ for $i = 1 \ldots N$ in the interval $[0,\ 1]$. The finite difference operators are as follows:

$$\begin{aligned} D^- v_j &= \frac{v_j - v_{j-1}}{h_j}, \quad D^+ v_j = \frac{v_{j+1} - v_j}{h_{j+1}}, \quad D^c = \frac{v_{j+1} - v_{j-1}}{x_{j+1} - x_{j-1}}, \\ \delta^2 v_j &= \frac{2}{h_j + h_{j+1}} \left(\frac{v_{j+1} - v_j}{h_{j+1}} - \frac{v_j - v_{j-1}}{h_j} \right), \end{aligned} \tag{5.13}$$

where $h_j = x_j - x_{j-1}$ and $h_{j+1} = x_{j+1} - x_j$.

For our model test problem Eq. (5.1), in view of the operators Eq. (5.13), we have the following difference scheme:

$$-\varepsilon\frac{2}{h_j+h_{j+1}}\left(\frac{v_{j+1}-v_j}{h_{j+1}} - \frac{v_j-v_{j-1}}{h_j}\right) + a(x_j)\frac{v_{j+1}-v_{j-1}}{h_j+h_{j+1}} + r(x_j)v_j = r(x_j). \tag{5.14}$$

The finite difference scheme Eq. (5.14) can be rewritten as

$$\left[\frac{1}{h_j+h_{j+1}}\left(-\frac{2\varepsilon}{h_j} - a(x_j)\right)\right]v_{j-1} + \left[\frac{2\varepsilon}{h_jh_{j+1}} + r(x_j)\right]v_j$$
$$+\left[\frac{1}{h_j+h_{j+1}}\left(-\frac{2\varepsilon}{h_j} + a(x_j)\right)\right]v_{j+1} = f(x_j),$$

which leads to solving a tridiagonal matrix of the form

$$\alpha v_{j-1} + \beta v_j + \gamma v_{j+1} = f(x_j),$$

which can easily be coded in MATLAB. Here, we present a MATLAB code, where matrices corresponding to the operators Eq. (5.13) have been constructed and used to design the following solver.

```
function U = CentralDiffSPPSolver(epsilon, a, r, f, alpha, beta, x)
%%% The function file implements FDM for the problem
%%% -epsilon u''(x) + a(x)u'(x) + r(x) u(x) = f(x) on (0,1)
%%% u(0) = alpha, u(1) = beta.
%%% on a non-uniform mesh
%%% Using central differences to discretize u'(x)

N = length(x)-1; j = 2:N;
hj = x(j) - x(j-1); hj1 = x(j+1) - x(j);
%% following is delta2 operator
D = - 2./(hj.*hj1); LD =  2./( hj.*(hj + hj1) ); UD =  2./( hj1.*(hj + hj1) );
Delta2Mat = sparse([j, j, j], [j-1, j, j+1], [LD, D, UD], N+1, N+1);
clear D UD LD
%% D+ operator
D = -(1./hj1); UD = 1./hj1; LD = zeros(N-1, 1);
Dplus = sparse([j, j, j], [j-1, j, j+1], [LD, D, UD], N+1, N+1);
clear D UD LD
%% D- operator
D = -(1./hj); LD =  -1./hj; UD = zeros(N-1, 1);
Dminus = sparse([j, j, j], [j-1, j, j+1], [LD, D, UD], N+1, N+1);
clear D UD LD
%% Dc operator
LD =  -1./(hj+ hj1); UD =  1./(hj+ hj1); D = zeros(N-1, 1);
DcMat = sparse([j, j, j], [j-1, j, j+1], [LD, D, UD], N+1, N+1);
clear D UD LD
% for variable coefficient case: for a(x)*y'(x) term
```

```
for i = 1:N+1
    DcMat(i,:) = a(x(i)).*DcMat(i,:);
end
% r(x)*y term
RecMat = sparse(j, j, r(x(2:N)), N+1, N+1);
A = -epsilon*Delta2Mat + DcMat + RecMat;
A(1, 1:2) = [1,0];
A(N+1, N:N+1) = [0, 1];
%% right hand vector
b = f(x);
b(1) = alpha;  b(N+1) = beta;     % Dirichlet Bnd conditions
%% Solve the linear system
U = A\b;
```

Refer to Figure 5.5, for a comparison of the computed solution of our test problem given by Eq. (5.12). Uniform mesh completely fails and produces spurious oscillations. Both layer-adapted meshes produce acceptable solutions. Note the behaviour of the computed solution at the transition point due to the difference in the structure of these meshes; the scale on x-axis is changed from $[0, 1]$ to $[0, .3]$ to appreciate the behaviour of the computed solution near the transition point in both the meshes. A few modifications in these meshes are available in the literature; for example, Vulanovic [8] proposed dividing the layer region into more equidistant parts in the Shishkin mesh, and Linß [9] proposed using Bakhvalov-type mesh in the layer re-

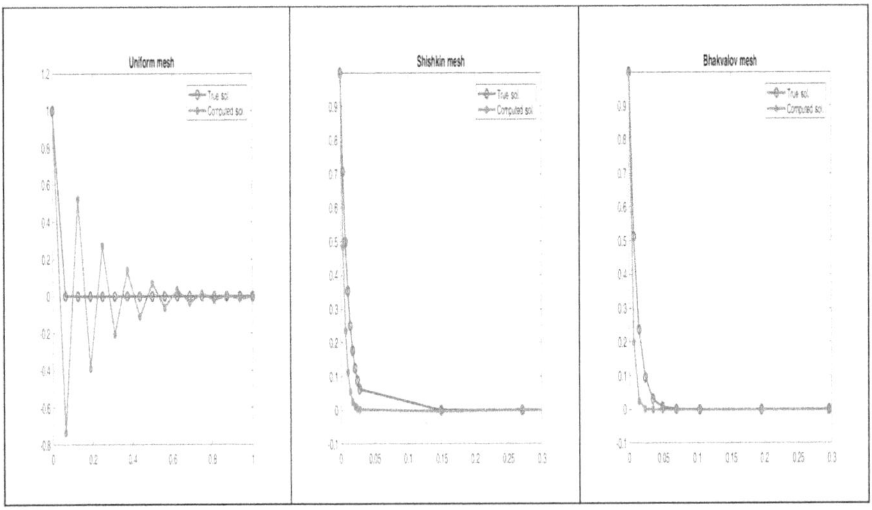

Figure 5.5 Computed solution compared to exact solution on different meshes for a model problem $\varepsilon u'' + u' = 0$ with $u(0) = 1$, and $u(1) = 0$, whose exact solution is uexact $= \frac{\exp(-x/\varepsilon)-\exp(-1/\varepsilon)}{1-\exp(-1/\varepsilon)}$: (a) uniform mesh, (b) Shishkin mesh, and (c) Bakhvalov mesh. To highlight the behaviour of computed solution near the transition point in both the meshes, the scale is changed from $[0, 1]$ to $[0, .3]$.

gion while retaining the original structure of a Shishkin-type mesh away from the layer. These type of meshes are known as Bakhvalov–Shishkin-type meshes.

As the name suggests, a priori refined meshes that we discussed above require some a priori information regarding the exact solution and its derivatives, which is rarely available in practice. Therefore, for most of the practical applications, one has to use some adaptive algorithm which by employing an error indicator redistributes the mesh nodes on a uniform mesh iteratively to yield a mesh on which the computed solution has desirable accuracy. These iteratively generated meshes are commonly referred to as *a posteriori refined meshes*, discussed in the next section.

5.2.2 A POSTERIORI REFINED MESHES

We are mainly interested in the construction of a posteriori refined meshes using a r-adaptive method. Recall that a r-adaptive method, keeping the total number of mesh nodes fixed, reallocates them using certain error indicators. The main components of a r-adaptive method for constructing a suitable nonuniform mesh are as follows:

1. use a uniformly stable discretizing scheme such as upwinding for a convection-diffusion problem;
2. construct/choose a suitable error indicator, commonly referred to as *monitor function* which should preferably be based on a posteriori error estimates; and
3. reallocate the mesh nodes in the regions suggested by the indicator.

This whole procedure canbe formally defined within the framework of the equidistribution principle, see, for example [9–14] and the references therein. The underlying idea of the equidistribution principle is to adjust the mesh points in such a way that the error of the numerical solution is uniformly distributed throughout the domain. Thus, the most natural way to formulate the equidistribution principle is to formulate it in terms of a current mesh and a monitor function.

On an arbitrary mesh $\{x_i\}_{i=0}^N$ with local mesh width $h_i := x_i - x_{i-1}$ for $i = 1,\dots,N$ in $[0,1]$, consider an *abstract stable numerical method* (possibly involving $\varepsilon \ll 1$) given by

$$T^N u_i^N = 0 \qquad \text{for} \quad i = 0,\dots,N. \tag{5.15}$$

Equidistribution Problem: Find $\{(x_i, u_i^N)\}$, with the $\{u_i^N\}$ computed on an arbitrary mesh $\{x_i\}_{i=0}^N$ by means of Eq. (5.15), such that

$$h_i\tilde{M}_i = \frac{1}{N+1}\sum_{j=1}^{N+1} h_j\tilde{M}_j, \qquad \text{for} \quad i = 1,2,\dots,N+1. \tag{5.16}$$

Here, $\tilde{M}_i$ denotes a discrete counterpart of the monitor function $M(x)$. Note that in formulation Eq. (5.16) both the the mesh $\{x_i\}$ and the computed solution $\{u_i^N\}$ are a priori unknown.

Note that the equidistribution problem warrants simultaneous solution of Eqs. 5.15 and 5.16. The following algorithm can be used to iteratively update the mesh and the computed solution, and at the termination, the algorithm provides the desired accuracy.

Given an arbitrary mesh, and keeping the number of mesh nodes N fixed, the algorithm aims to solve the equidistibution principle. The algorithm is given below [15].

1. Initialize mesh: Start with a uniform mesh with mesh width $h = 1/N$.
2. For $k = 0, 1, \ldots$, given the mesh $\{x_i^{(k)}\}$, obtain the computed solution $\{u_i^{(k)}\}$ satisfying an abstract numerical scheme given by

$$T^{(k)}u^{(k)} = 0 \quad \text{on}\{x_i^{(k)}\}, \qquad \text{with} \quad u_0^{(k)} = u_N^{(k)} = 0. \tag{5.17}$$

   Set $h_i^{(k)} = x_i^{(k)} - x_{i-1}^{(k)}$ for each i. Let $M^{(k)}(x)$ be a piecewise-constant monitor function Then, evaluate the total integral given by

$$I^{(k)} := \int_0^1 M^{(k)}(x)dx = \sum_{i=1}^{N} M_i^{(k)} h_i^{(k)}.$$

3. Test mesh: For a user-defined constant $C_0 > 1$, if

$$\max_{1 \le i \le N} \{M_i^{(k)} h_i^{(k)}\} \le C_0 \frac{I^{(k)}}{N}, \tag{5.18}$$

   then go to step 5, or else continue to step 4.
4. Generate a new mesh $0 = x_0^{(k+1)} < x_1^{(k+1)} < \ldots < x_N^{(k+1)} = 1$ such that

$$\int_{x_{i-1}^{(k+1)}}^{x_i^{(k+1)}} M^{(k)}(x)dx = I^{(k)}/N, \qquad i = 0, \ldots, N. \tag{5.19}$$

   Return to Step 2.
5. Set $\{x_0^*, x_1^*, \ldots, x_N^*\} = \{x_i^{(k)}\}$ and $\{u_i^*\} = \{u_i^{(k)}\}$ then stop.

Remark 5.1: *The choice of $C_0 > 1$ in Eq. (5.18) implies that rather than solving the equidistribution problem Eqs 5.15 and 5.16, the algorithm solves the following quasi-equidistribution problem given by Find $\{(x_i, u_i^N)\}$, with the $\{u_i^N\}$ computed on arbitrary mesh $\{x_i\}$ by means of Eq. (5.15), such that*

$$M_i^N h_i \le C_0 \frac{1}{N} \sum_{j=1}^{N} M_j^N h_j \quad \textit{for} \quad i = 1, 2, \ldots, N.$$

□

Step 4 in the algorithm can be interpreted as follows: For an arbitrary input mesh $\{x_i^{(k)}\}$, and piecewise constant monitor function

$$M^{(k)}(x) = M_i^{(k)}, \quad x \in (x_{i-1}^{(k)}, x_i^{(k)}), \quad \text{for} \quad i = 1, \ldots, N. \tag{5.20}$$

The equidistribution problem can be formulated as follows:

$$\int_{x_{i-1}^{(k+1)}}^{x_i^{(k+1)}} M^{(k)}(x)dx = \frac{1}{N}\int_0^1 M^{(k)}(x)dx. \quad (5.21)$$

For redistributing the mesh points, see Eq. (5.19), consider the points $\{I_i^{(k)}\}$ on the I-axis corresponding to the mesh points $\{x_i^{(k)}\}$, given by, (see Figure 5.6)

$$I_i^{(k)} = \sum_{j=1}^{i} h_j^{(k)} M_j^{(k)}, \quad \text{for} \quad i = 1,\ldots,N, \quad (5.22)$$

where we set $I_0^{(k)} = 0$. Joining points $(I_i^{(k)}, x_i^{(k)})$, for $i = 1,\ldots,N$, using linear

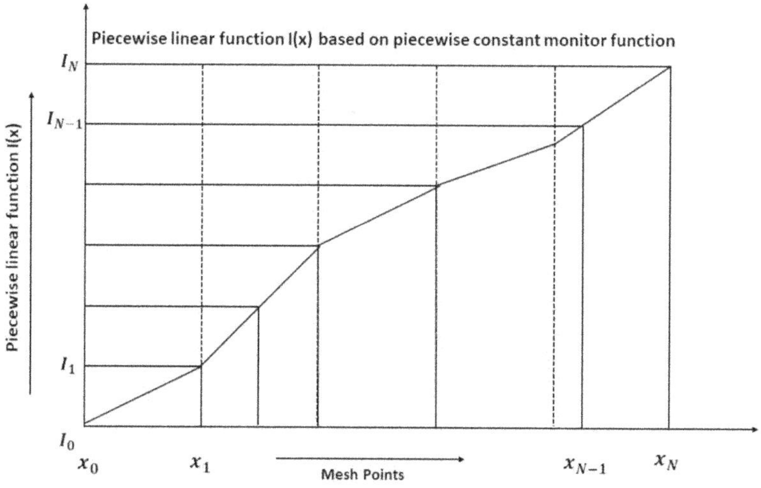

Figure 5.6 Geometrical representation of redistribution of mesh points using Step 4 in the algorithm. It may be interesting to compare this with graphs of mesh generating functions of Bakhvalov mesh and Shishkin mesh.

interpolation would give piecewise linear function $I^{(k)}(x)$,

$$I^{(k)}(x) = I(x_{i-1}^{(k)}) + (x - x_{i-1}^{(k)})M_i, \quad x \in (x_{i-1}^{(k)}, x_i^{(k)}), \quad \text{for} \quad i = 1,\ldots,N. \quad (5.23)$$

We divide $[0, I_N^{(k)}]$ in N equal parts on the I-axis and denote these points by $\tilde{I}_i^{(k)}$,

$$\tilde{I}_i^{(k)} = \frac{i}{N} I_N^{(k)}, \qquad \text{for} \quad i = 0, 1, \ldots, N. \quad (5.24)$$

Project $\tilde{I}_i^{(k)}$ for $i = 0, 1, \ldots, N$ on the I-axis, onto piecewise linear function $I^{(k)}(x)$. The abscissa of the points after projection will generate a new mesh $\{x_i^{(k+1)}\}$.

One of the key constituents of mesh adaptation is to extract or design an appropriate monitor function. It is important to note that the success of any mesh adaptation strategy strongly hinges on the choice of monitor function. In the next section, we discuss how to construct a monitor function from a given error estimate.

5.2.3 ERROR ESTIMATES AND THE CONSTRUCTION OF A MONITOR FUNCTION

Given an error estimate a monitor function can be extracted by representing the right hand of the estimate in terms of a product of h_i, the local mesh width; and a function M_i which depends on either on the derivatives of the exact solution (in the case of a priori error estimates) or on its discrete derivatives (in the case of a posteriori error estimates), that is,

$$\|\text{error}\| \approx \bar{C} \max_i |h_i M_i|^p, \tag{5.25}$$

where h_i is the local mesh spacing, and p is some positive number (usually the order of the numerical method). This function M_i provides certain candidate monitor functions which may be used for mesh adaptation. Here, we want to emphasize that a monitor function constructed from an error estimate may not be unique. One has to use one's intuition and experience to choose a monitor function before using it in an adaptive algorithm.

We put forward our point by taking two examples of constructing monitor function(s), (a) from a priori error estimates, and (b) from a posteriori error estimates.

5.2.3.1 Constructing a Monitor Function from a Priori Error Estimates

Consider reaction diffusion problem $\quad -\varepsilon^2 u''(x) + p(x)u(x) = f(x) \quad$ for $x \in (0,1), \quad u(0) = u(1) = 0,$

with appropriate conditions on p and f. We now state the error estimate(s) derived in Natalia et al. [16].

Lemma 5.1: *[16, Theorem 3.1] Let $e_i = u_i^N - u(x_i)$ for $i = 0, \ldots, N$, denote the error at x_i in the computed solution. Then there exists a constant C such that*

$$\|e\|_\infty \le C \left\{ \max_i \max_{x \in [x_{i-1}, x_i]} \varepsilon h_i^2 |u'''(x)| + \max_i \max_{x \in [x_{i-1/2}, x_{i+1/2}]} \hbar_i^2 \left[1 + |(bu)''(x)|\right] \right\}. \tag{5.26}$$

□

One can rewrite Eq. (5.26) in terms of lower-order derivatives as follows:

Lemma 5.2: *[16, Corollary 3.1] There exists a constant C such that*

$$\|e\|_\infty \le C \left\{ \max_i \max_{x \in [x_{i-1}, x_i]} h_i^2 [1 + |u''(x)|] + \max_i \max_{x \in [x_{i-1/2}, x_{i+1/2}]} \hbar_i^2 [1 + |(bu)''(x)|] \right\}. \tag{5.27}$$

□

Given a priori error bounds as in Eqs. 5.26 and 5.27, it is reasonable to consider following discrete analogues of Eqs. 5.26 and 5.27 given by

$$\|e\|_\infty \leq C \max_i \left[\hbar_i^2 (1 + |\delta^2 (b_i u_i^N)|) + h_i^2 \varepsilon |\delta^3 u_i^N|\right], \tag{5.28}$$

and

$$\|e\|_\infty \leq C \max_i \hbar_i^2 \left[1 + \frac{|\delta^2 u_i^N| + |\delta^2 (b_i u_i^N)|}{2}\right]. \tag{5.29}$$

To extract a monitor function from Eq. (5.28) or Eq. (5.29) one has to rearrange the error bounds such that they are in the following form

$$\|e\|_\infty \leq [h_i M_i]^k, \quad k \in N^* = \{1, 2, \ldots\}, \tag{5.30}$$

for this representation, M_i will be a candidate monitor function. For instance, consider the expression given in Eq. (5.29). To get rid of $\hbar_i$ in Eq. (5.29), one may consider an equivalent form of the error bound Eq. (5.29) given by

$$\|e\|_\infty \leq h_i^2 \left[1 + \frac{|\delta^2 u_{i-1}^N|^{1/2} + |\delta^2 (b_{i-1} u_{i-1}^N)|^{1/2} + |\delta^2 u_i^N|^{1/2} + |\delta^2 (b_i u_i^N)|^{1/2}}{4}\right]^2. \tag{5.31}$$

This leads to a monitor function

$$M_i = (\tilde{M}_{i-1} + \tilde{M}_i)/2; \quad \tilde{M}_i = 1 + \frac{|\delta^2 u_i^N|^{1/2} + |\delta^2 (b u^N)_i|^{1/2}}{2}. \tag{5.32}$$

An analogue monitor function from error estimate Eq. (5.28) takes the following form

$$M_i = (\tilde{M}_{i-1} + \tilde{M}_i)/2; \quad \tilde{M}_i = 1 + \frac{\varepsilon^{1/2} |\delta^3 u_i^N|^{1/2} + |\delta^2 (b u^N)_i|^{1/2}}{2}, \tag{5.33}$$

One can get rid of ε in Eq. (5.33), by using the relations $\varepsilon^{1/2} |u'''(x)|^{1/2} \leq C[1 + |u'''(x)|^{1/3}]$, see Corollary 3.2 and Lemma 2.1 in [16], to obtain

$$\tilde{M}_i = 1 + \frac{|\delta^3 u_i^N|^{1/3} + |\delta^2 (b u^N)_i|^{1/2}}{2}. \tag{5.34}$$

in the corresponding term in Eq. (5.33). This clearly shows that, from a given error estimate, various monitor functions can be extracted.

5.2.3.2 Constructing a Monitor Function from a Posteriori Error Estimates

To demonstrate constructing a monitor function using a posteriori error estimates, we consider the following singularly perturbed semilinear reaction-diffusion problem

$$-\varepsilon^2 u''(x) + b(x, u) = 0 \quad \text{for} \quad x \in (0, 1), \quad u(0) = u(1) = 0, \tag{5.35}$$

where $\varepsilon \ll 1$ is a parameter, and $b(x, u)$ is sufficiently smooth and satisfies other desired conditions in order to possess a unique solution refer to [15] for details.

The difference scheme for Eq. (5.35) on an arbitrary mesh is given by

$$-\varepsilon^2\delta^2 u_i^N + b(x_i,u_i^N) = 0 \quad \text{for} \quad i=1,\ldots,N-1, \quad u_0^N = u_N^N = 0, \tag{5.36a}$$

where
$$\delta^2 u_i^N := \frac{2}{h_i+h_{i+1}}\left(\frac{u_{i+1}^N-u_i^N}{h_{i+1}} - \frac{u_i^N-u_{i-1}^N}{h_i}\right),$$

$i = 1,\ldots,N-1.$ (5.36b)

For the construction of probable monitor functions, we invoke a posteriori error estimates in [17] given below.

Theorem 5.1: *[17, Theorem 3.3] Let u be a solution of Eq. (5.35), $\{u_i^N\}$ be a solution of Eq. (5.36), and $u^N(x)$ be its linear interpolant. Then*

$$\|u^N(\cdot)-u(\cdot)\|_\infty \le C \max_{1\le i\le N}\left\{h_i^2\left[|\delta^2 u_i^N| + \varepsilon|\delta^3 u_i^N| + (|D^- u_i^N|^2+1)\right]\right\}. \tag{5.37}$$

Here, the discrete derivatives $\delta^2 u_i^N$ are formally extended to the mesh nodes $i = 0,N$ as $\delta^2 u_0^N := \varepsilon^{-2} b(x_0,u_0^N) = b(0,0)$, $\delta^2 u_N^N := \varepsilon^{-2} b(x_N,u_N^N) = b(1,0)$. □

We perform the same exercise to extract a monitor function from a posteriori error estimates which we did in the case of a priori error estimates; that is, representing the right-hand side of the estimate Eq. (5.37) as $[h_i M_i]^k$, $k \in N^* = \{1,2\ldots\}$.

We combine Eq. (5.37) with following lemma:

Lemma 5.3: *[17, Lemma 3.5] Under the conditions of Eq. (5.24), we have*

$$h_i^2\varepsilon|\delta^3 u_i^N| + h_i^2|D^- u_i^N|^2 \le \max_{0\le i\le N}\left\{h_i^2(|\delta^2 u_{i-1}^N| + |\delta^2 u_i^N| + 1)\right\}, \quad i=1,\ldots,N. \tag{5.38}$$

□

To obtain a neat error bound, we have

$$\|u^N(\cdot)-u(\cdot)\|_\infty \le \max_{0\le i\le N}\left\{h_i^2(|\delta^2 u_{i-1}^N| + |\delta^2 u_i^N| + 1)\right\}, \tag{5.39}$$

which readily provides a monitor function

$$M_i = (|\delta^2 u_{i-1}^N| + |\delta^2 u_i^N| + 1)^{1/2}. \tag{5.40}$$

One can replace the quantity $|\delta^2 u_i^N|$ by $|\delta^2 u_{i-1}^N|$, or $\min\{|\delta^2 u_{i-1}^N|, |\delta^2 u_i^N|\}$, or $(|\delta^2 u_{i-1}^N| + |\delta^2 u_i^N|)/2$ in error bounds Eqs. 5.37 and 5.39. This provides more monitor functions which may be considered. For example, a monitor function

$$M_i = \left[\min\{|\delta^2 u_{i-1}^N|, |\delta^2 u_i^N|\}^{1/2} + (\varepsilon|\delta^3 u_i^N|)^{1/2}\right]/2+1. \tag{5.41}$$

follows from Eqs. 5.37 and 5.38. If one replaces the quantity $|\delta^2 u_i^N|$ by $|\delta^2 u_{i-1}^N|$, or $\min\{|\delta^2 u_{i-1}^N|, |\delta^2 u_i^N|\}$, or $(|\delta^2 u_{i-1}^N| + |\delta^2 u_i^N|)/2$ in error bounds Eqs. 5.37 and 5.39, there will be more monitor functions available, for example

$$M_i = \frac{|\delta^2 u_{i-1}^N|^{1/2} + |\delta^2 u_i^N|^{1/2}}{2} + 1, \quad \text{and} \quad M_i = \min\{|\delta^2 u_{i-1}^N|, |\delta^2 u_i^N|\}^{1/2} + 1. \tag{5.42}$$

We conclude this section with following remarks, which may be helpful for choosing a monitor function as per the requirement.

Remark 5.2: *It is desirable that a monitor function satisfies the following two conditions:*

$$M(x) \geq C_1, \quad \text{for} \quad x \in (0,1), \tag{5.43}$$

$$\int_0^1 M(x)dx \leq C_2, \quad \text{for} \quad x \in (0,1), \tag{5.44}$$

where C_1 and C_2 are some positive constants [12, Remark 4.15]. The condition given by Eq. (5.43) is useful to avoid 'mesh starvation', when a monitor function is employed by some adaptive algorithm. The condition given by Eq. (5.44) puts a check on the total error bound. Furthermore, combining this bound with the conditions given by Eq. (5.43), Eq. (5.44), we get $\|error\| \leq CN^{-p}$ in view of Eq. (5.25). □

Remark 5.3: *Picking the most appropriate monitor function out of several possible choices for a particular problem is a skill rather than a science.* □

5.2.4 NUMERICAL EXPERIMENTS FOR MESH ADAPTATION ON A TEST PROBLEM

For comparison purposes, we take the following test problem:

$$\varepsilon^2 u'' + p(x)u = f(x), \quad u(0) = 2, u(1) = -1; \quad p(x) = 4(1+x)^{-4}[1 + \varepsilon(1+x)]. \tag{5.45}$$

The right-hand side f is chosen so that the exact solution is satisfied. The exact solution is given by

$$u(x) = -\cos(2\pi t) + 3(e^{-t/\varepsilon} - e^{-1/\varepsilon})[1 - e^{-1/\varepsilon}]^{-1}, \qquad t := 2x/(x+1). \tag{5.46}$$

The problem is from [15–18]. We use two monitor functions, namely, $M_{1,i}^N = \min\{|\delta^2 u_{i-1}^N|, |\delta^2 u_{2,i}^N|\}^{1/2} + 1$ and $M_2^N = \max\{|\delta^2 u_{i-1}^N|, |\delta^2 u_i^N|\}^{1/2} + 1$ for mesh adaptation using the algorithm. Figure 5.7 and 5.8 show the intermediate computed solution at each iteration while using $M_{1,i}^N = \min\{|\delta^2 u_{i-1}^N|, |\delta^2 u_{2,i}^N|\}^{1/2} + 1$, and $M_{2,i}^N = \max\{|\delta^2 u_{i-1}^N|, |\delta^2 u_{2,i}^N|\}^{1/2} + 1$, respectively. In Figure 5.8, only three intermediate iterations, namely iterations, 2, 3, and 4 are shown to highlight the phenomenon

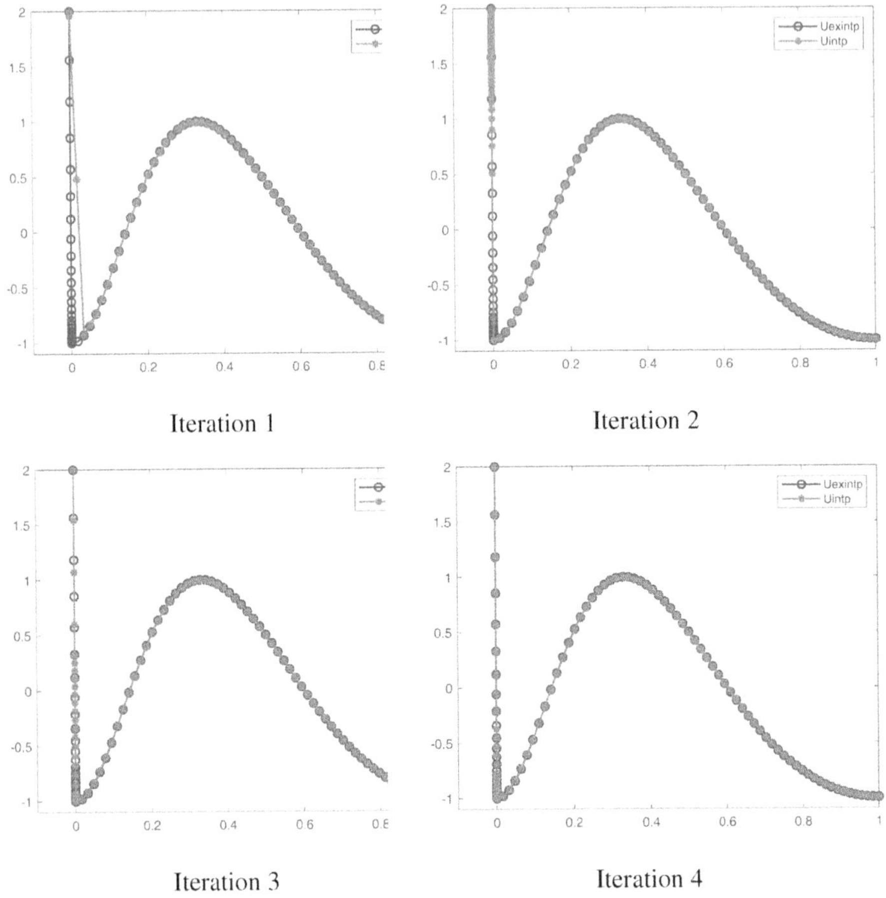

Figure 5.7 Intermediate computed solution shown for $N = 32$; $\varepsilon = 10^{-5}$ when the monitor function $M^N_{1,i} = \min\{|\delta^2 u^N_{i-1}|, |\delta^2 u^N_{2,i}|\}^{1/2} + 1$ is used in the algorithm for solving Eq. (5.44). The algorithm, starting from a uniform mesh, rearrange the mesh nodes and place them and appropriate place(s), and at the termination yields the computed solution with desired accuracy.

of mesh starvation, which indicates that, for a given number of mesh nodes, most of the mesh nodes are concentrated in certain region. This may often be undesirable; Figure 5.7 shows that in iteration 3 itself this mesh starvation is rectified, thus leading to a faster convergence to the desired solution. The mesh node distributions at each iteration while using both the monitor functions in the algorithm are shown in Figure 5.9.

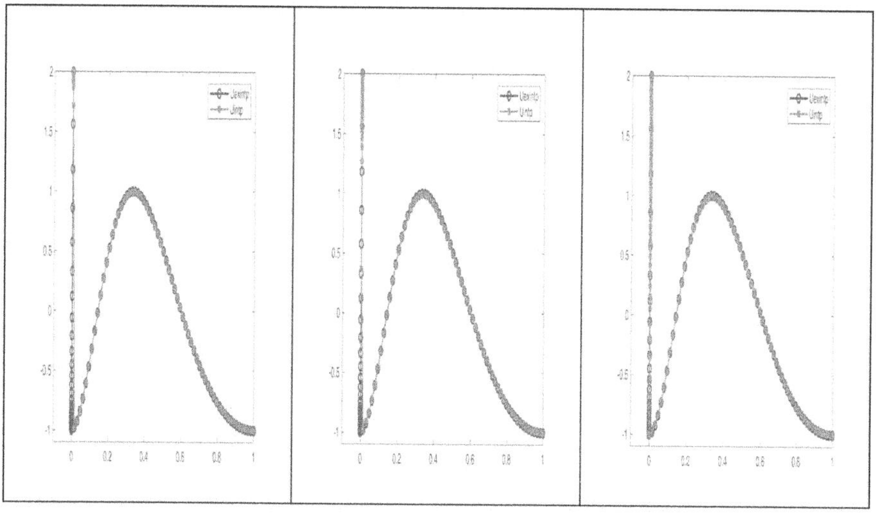

Figure 5.8 Intermediate computed solution shown for $N = 32$; $\varepsilon = 10^{-5}$ when the monitor function $M_{2,i}^{N} = \max\{|\delta^2 u_{i-1}^{N}|, |\delta^2 u_{2,i}^{N}|\}^{1/2} + 1$ is used in the algorithm for solving Eq. (5.45). Only three intermediate iterations: iteration, 2, 3, and 4 to show the phenomenon of mesh starvation, where most of the mesh nodes are concentrated in a certain region, often an undesirable phenomenon.

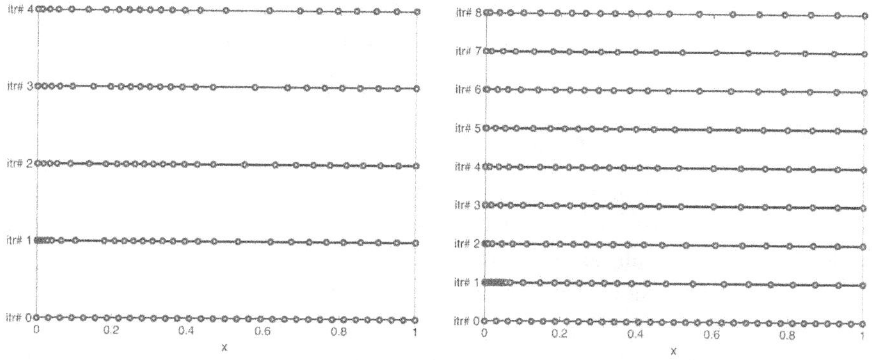

Figure 5.9 Mesh node distribution after every iteration of the algorithm while solving Eq. (5.45) using an algorithm with the monitor functions: $M_{1,i}^{N} = \min\{|\delta^2 u_{i-1}^{N}|, |\delta^2 u_{2,i}^{N}|\}^{1/2} + 1$ is used for left figure; $M_{2,i}^{N} = \max\{|\delta^2 u_{i-1}^{N}|, |\delta^2 u_{i}^{N}|\}^{1/2} + 1$ is used for right figure.

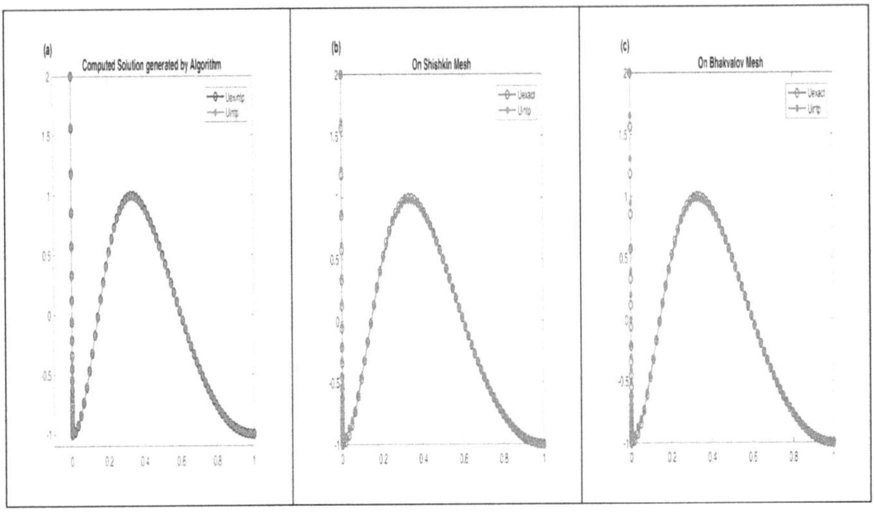

Figure 5.10 Comparison of computed solution. (a) Computed solution using algorithm using monitor function $M^N_{1,i} = \min\{|\delta^2 u^N_{i-1}|, |\delta^2 u^N_{2,i}|\}^{1/2} + 1$, (b) On Shishkin mesh, (c) On Bakhvalov mesh. Note that while using Shishkin mesh or Bakhvalov mesh, it is essential to have some a priori information regarding the solution to set the transition point. No such information is required for generating adapted meshes using an algorithm in conjunction with an appropriately chosen monitor function.

In Fig. 5.10, we demonstrate that the algorithm involving the monitor function given by $M^N_{1,i} = \min\{|\delta^2 u^N_{i-1}|, |\delta^2 u^N_{2,i}|\}^{1/2} + 1$ yields computed sloution which is comparable with the solution obtained by using layer adapted meshes. However, it is important to note that certain a priori information is required to design the layer adapted meshes which is not the case with adapted meshes generated by using the algorithm in conjunction wth an appropriately chosen monitor function.

5.3 CONCLUDING REMARKS

A priori refined meshes discussed in Eq. (5.2.1) can be extended to higher dimensions without any difficulty see, e.g. [5, 19, 20] and references therein. Note that these meshes are by default tensor product meshes, refer to Figure 5.11 for the Shishkin mesh in two dimensions, here τ_x and τ_y are the transition points in x and y directions, respectively.

To demonstrate the construction of a posteriori refined meshes, we invoke a posteriori error estimates in maximum norm constructed for a semilinear reaction-diffusion problem posed in the unit square [19, Theorem 2.1] to construct certain metric tensors. Furthermore, we will compare the meshes generated by using these metric tensors with those generated by using metric tensors based on the Hessian matrix. It is important to mention that the equidistribution principle cannot be extended as such for mesh adaptation in higher dimensions. Nonetheless, most of the mesh

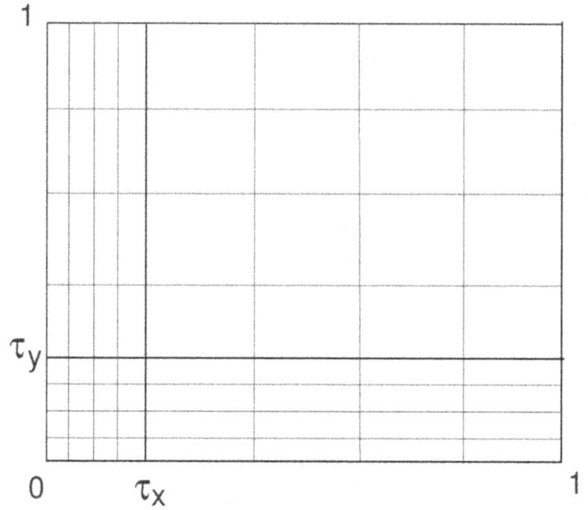

Figure 5.11 Shishkin mesh with $N = 8$, τ_x and τ_y denote the transition points in x and y directions, respectively.

adaptation strategies/softwares are based on the equidistribution principle; refer to a meshing software survey by Owen [21] for various algorithms for mesh adaptation.

In higher dimensions, monitor function is replaced by a monitor matrix or a metric tensor. They both are equivalent and play the same role in mesh adaptation. A continuous form of the metric tensor corresponding to the estimate [19, Theorem 2.1] can be written as

$$\mathscr{M} = \text{diag}(M_{11}, M_{22}), \tag{5.47a}$$

where $M_{11} = \ln(2+\varepsilon/\kappa)\left|\frac{\partial^2 u}{\partial x^2}\right| + \varepsilon\left|\frac{\partial^3 u}{\partial x^3}\right| + \left|\frac{\partial u}{\partial x}\right|^2 + 1;$

$$M_{22} = \ln(2+\varepsilon/\kappa)\left|\frac{\partial^2 u}{\partial y^2}\right| + \varepsilon\left|\frac{\partial^3 u}{\partial y^3}\right| + \left|\frac{\partial u}{\partial y}\right|^2 + 1, \tag{5.47b}$$

with $\kappa = \min\{\min_i\{h_i\}, \min_j\{k_j\}\}$, h_i and k_j are local mesh widths in the x and y directions, respectively. Another metric tensor can be obtained by simply dropping first- and third-order terms in Eq. (5.47), as given below

$$\tilde{\mathscr{M}} = \text{diag}(\tilde{M}_{11}, \tilde{M}_{22}), \tag{5.48a}$$

where $\tilde{M}_{11} = \ln(2+\varepsilon/\kappa)\left|\frac{\partial^2 u}{\partial x^2}\right| + 1; \quad \tilde{M}_{22} = \ln(2+\varepsilon/\kappa)\left|\frac{\partial^2 u}{\partial y^2}\right| + 1$

(5.48b)

for mesh adaptation. Numerical results in [19] also support this argument of dropping the first- and third-order terms in Eq. (5.48).

For the construction of adapted meshes using the metric tensors given by Eq. (5.47) and Eq. (5.48), we employ BAMG (Bidimensional Anisotropic Mesh Generator) [22] which is a part of freely available software FreeFEM++ [23, 24]. BAMG is based on Delaunay-type triangulation in which a metric tensor which is defined on an initial or a background mesh can be supplied by the user. Once the metric is given, BAMG aims to construct a mesh which is isotropic with respect to the given metric. In the process to complete the mesh, it employs certain local operations to optimize the mesh (see [25] for more details). For instance, in the case of an isotropic mesh adaptation, let h_i be the desired mesh size near the vertex i on an initial mesh. Then, a metric can be defined as $\mathcal{M}_i = h_i^{-2}I$, I is the identity matrix. For an anisotropic case, let the metric at the vertex i be defined as

$$\mathcal{M}_i = \begin{pmatrix} a_{11_i} & a_{12_i} \\ a_{12_i} & a_{22_i} \end{pmatrix}.$$

This controls the desired edge length h in a vicinity of the vertex i. For instance, in direction $\mathbf{v} \in \mathbb{R}^2$, h is equal to $\frac{|\mathbf{v}|}{\sqrt{\mathbf{v}..\mathcal{M}_i\mathbf{v}}}$, where $|\cdot|$ is the classical norm. In a case where a function is given, the software uses its Hessian for mesh adaptation.

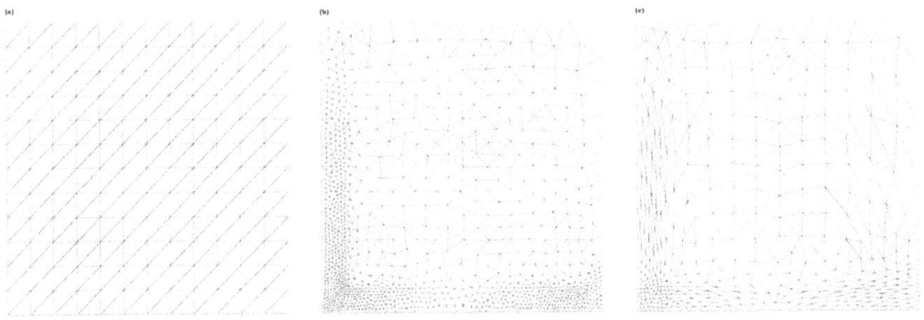

Figure 5.12 Isotropic and anisotropic mesh adaptation based on the Hessian, using BAMG; (a) initial mesh, number of vertices = 169, number of triangles = 288; (b) isotropic mesh, number of vertices = 1232, number of triangles = 2308; (c) anisotropic mesh, number of vertices = 458, number of triangles = 830.

We demonstrate mesh adaptation using various metrics for a test function

$$f(x,y) = \left(\cos(\pi x) - \frac{e^{-x/\varepsilon} - e^{-1/\varepsilon}}{1 - e^{-1/\varepsilon}}\right)\left(1 - y - \frac{e^{-y/\varepsilon} - e^{-1/\varepsilon}}{1 - e^{-1/\varepsilon}}\right), \quad \varepsilon = 10^{-4}.$$

See Figure 5.12 for isotropic and anisotropic mesh adaptation; the mesh adaptation is performed using the Hessian matrix. It is apparent from Figure 5.13 (b), (c) that the mesh generated by metrics Eq. (5.47) and Eq. (5.48) are more or less the same with a slight change in the number of triangles used. Moreover, it is clear from Figures 5.12 and 5.13 that the meshes obtained by using metric tensors based on an a posteriori

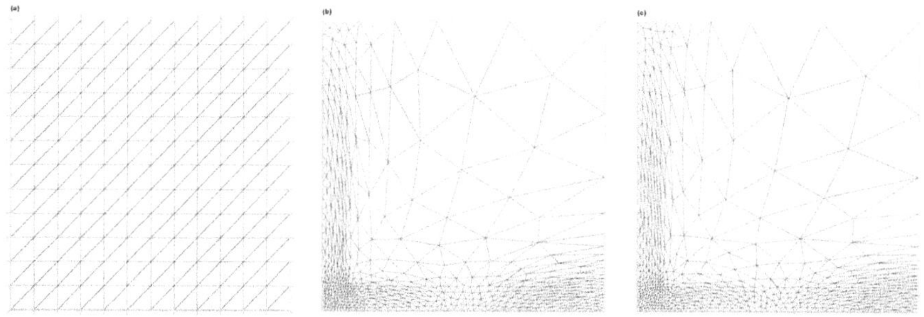

Figure 5.13 Anisotropic mesh adaptation based on metrics Eq. (5.47), Eq. (5.48), using BAMG; (a) initial mesh with nb vertices = 169, nb triangles = 288; (b) anisotropic mesh using Eq. (5.47), nb of vertices = 554, nb of triangles = 1001; (c) anisotropic mesh using Eq. (5.48) anisotropic mesh with nb of vertices = 569, nb of triangles = 1031.

error estimate [19, Theorem 2.1], are more concentrated in the layer region in comparison to the mesh(es) obtained by using the Hessian. Though the overall quality of a mesh is usually judged considering various aspects, such as skewness, isotropy, equidistribution of error, the meshes obtained by metrics based on a posteriori error estimates seem to be equally good if not better than the meshes obtained by the Hessian. Given the choice, it would be wise to use a metric based on a posteriori error estimates, for obvious reasons.

The code for generating Figures 5.12 and 5.13 using BAMG is given below.

```
// Calculate desired derivatives of u symbolically
// Mesh adaptation is performed using epsilon = .0001
// isotropic adaptation using Hessian
mesh HessIso = square(12,12); // initial mesh
plot(HessIso,wait = 1, ps = "HessInitial.eps");// plotting initial mesh
HessIso = adaptmesh(HessIso,u, iso = 1); // adapting mesh using Hessian metric
plot(HessIso, wait = 1, ps = "HessianIsoAdapted.eps");
// anisotropic adaptation using Hessian
mesh HessAniso = square(12,12);
//plot(HessAniso , wait = 1);
HessAniso = adaptmesh(HessAniso,u); // anisotropic mesh adaptation (by deault)
plot(HessAniso,wait = 1, ps = "HessAnisoAdapted.eps");
// second order derivatives
// this is claimed to be the main metric
mesh ErrMatIIder = square(12,12);
//plot(ErrMatIIder, wait = 1);
func m11IIder = uxx*(log(2+k/12))+1;
func m22IIder = uyy*(log(2+k/12))+1;
ErrMatIIder = adaptmesh(ErrMatIIder, [m11IIder,0,m22IIder]);
plot(ErrMatIIder, wait = 1, ps = "ErrMatIIderAdapted.eps");
// full metric as such is used
mesh ErrMatfull = square(12,12);
//plot(ErrMatfull, wait = 1);
func m11full = 1 + abs(ux)*abs(ux) + k*abs(uxxx) + abs(uxx)*(log(2+k/12));
```

```
func m22full = 1 + abs(uy)*abs(uy) + k*abs(uyyy) + abs(uyy)*(log(2+k/12));
ErrMatfull = adaptmesh(ErrMatfull, [m11full, 0, m22full]);
plot(ErrMatfull, wait = 1, ps = "ErrMatfullAdapted.eps");
```

5.3.1 FUTURE DIRECTIONS

The success of any mesh adaptation strategy strongly depends on the choice of monitor function. Thus, it is recommended to extract monitor functions from sharp a posteriori error estimates in maximum norm for the problem under consideration, so that large variations in the solution (such as layers) can be captured. Another important aspect is to design efficient adaptive algorithms which take these monitor functions (or monitor metrics in higher dimension) into account, and yield a mesh at the end of the iteration on which the computed solution is sufficiently accurate. Note that most of the algorithms use certain criteria to locally optimize the mesh, for example skewness, isotropy, and equidistribution of error. Furthermore, the extension of the equidistribution principle in higher dimensions is not straightforward. It has been reported in the literature that the equidistribution principle is not sufficient in higher dimension. Thus, future extension of the work would be in the direction of constructing sharp monitor metrics based on sharp error estimates, and to develop efficient algorithms for mesh adaptation. Recently, a lot of interest has been shown in fractional differential boundary value problems, due to their wide physical applications in various fields, which include electromagnetics, acoustics, viscoelasticity, electrochemistry, and material science (see e.g [26–28] and references therein). The interested reader may also refer to [29–32] for various other applications of fractional differential equations. It may be interesting to explore and extend the mesh adaptation strategies discussed here to deal with fractional differential equations when they exhibit the layer phenomenon.

REFERENCES

1. N. Bakhvalov, Towards optimization of methods for solving boundary value problems in the presence of boundary layers. Zhurnal Vychislitel'noi Matematiki i Matematicheskoi Fiziki, 9 (4) (1969) 841–859.

2. R. Vulanovic, Mesh construction for discretization of singularly perturbed boundary value problems, Ph. D. thesis, University of Novi Sad (1986).

3. R. Vulanović, A uniform numerical method for quasilinear singular perturbation problems without turning points, Computing 41 (1–2) (1989) 97–106.

4. G. Shishkin, Grid approximation of singularly perturbed elliptic and parabolic equations Russian Journal of Numerical Analysis and Mathematical Modelling 7 (2) (1992) 145–167.

5. M. Stynes, Steady-state convection-diffusion problems, Acta Numerica 14 (2005) 445–508.

6. G. Sun, M. Stynes, An almost fourth order uniformly convergent difference scheme for a semilinear singularly perturbed reaction-diffusion problem, Numerische Mathematik 70 (4) (1995) 487–500.

7. R. Vulanović, Fourth order algorithms for a semilinear singular perturbation problem, Numerical Algorithms 16 (2) (1997) 117–128.

8. R. Vulanović, A priori meshes for singularly perturbed quasilinear two-point boundary value problems, IMA Journal of Numerical Analysis 21 (1) (2001) 349–366.

9. T. Linß, Uniform pointwise convergence of finite difference schemes using grid equidistribution, Computing 66 (1) (2001) 27–39.

10. G. Beckett, J. Mackenzie, Convergence analysis of finite difference approximations on equidistributed grids to a singularly perturbed boundary value problem, Applied Numerical Mathematics 35 (2) (2000) 87–109.

11. N. Kopteva, Maximum norm a posteriori error estimates for a one-dimensional convection-diffusion problem, SIAM Journal on Numerical Analysis 39 (2) (2001) 423–441.

12. N. Kopteva, Adaptive Computations: Theory and Algorithms, Convergence Theory of Moving Grid Methods, Beijing: Science Press (2007) 159–210.

13. N. Kopteva, M. Stynes, A robust adaptive method for a quasi-linear one-dimensional convection-diffusion problem, SIAM Journal on Numerical Analysis 39 (4) (2001) 1446–1467.

14. Y. Qiu, D. Sloan, Analysis of difference approximations to a singularly perturbed two-point boundary value problem on an adaptively generated grid, Journal of Computational and Applied Mathematics 101 (1–2) (1999) 1–25.

15. N. M. Chadha, N. Kopteva, A robust grid equidistribution method for a one-dimensional singularly perturbed semilinear reaction–diffusion problem, IMA Journal of Numerical Analysis 31 (1) (2011) 188–211.

16. N. Kopteva, N. Madden, M. Stynes, Grid equidistribution for reaction–diffusion problems in one dimension, Numerical Algorithms 40 (3) (2005) 305–322.

17. N. Kopteva, Maximum norm a posteriori error estimates for a 1d singularly perturbed semilinear reaction-diffusion problem, IMA Journal of Numerical Analysis 27 (3) (2007) 576–592.

18. E. O'Riordan, M. Stynes, A uniformly accurate finite-element method for a singularly perturbed one-dimensional reaction-diffusion problem, Mathematics of Computation 47 (176) (1986) 555–570.

19. N. Kopteva, Maximum norm error analysis of a 2d singularly perturbed semilinear reaction-diffusion problem, Mathematics of Computation 76 (258) (2007) 631–646.

20. T. Linß, Layer-adapted meshes for convection–diffusion problems, Computer Methods in Applied Mechanics and Engineering 192 (9–10) (2003) 1061–1105.

21. S. Owen, Meshing software survey, Structured Grid Generation Software, web page: http://www.andrew.cmu.edu/user/sowen/software/structured.html.

22. F. Hecht, Bamg: bidimensional anisotropic mesh generator, User Guide. INRIA, Rocquencourt.

23. F. Hecht, New development in freefem++, Journal of Numerical Mathematics 20 (3–4) (2012) 251–266.

24. F. Hecht, O. Pironneau, A. Le Hyaric, K. Ohtsuka, Freefem++ Manual (2005).

25. M. Castro-Díaz, F. Hecht, B. Mohammadi, O. Pironneau, Anisotropic unstructured mesh adaption for flow simulations, International Journal for Numerical Methods in Fluids 25 (4) (1997) 475–491.

26. H. Singh, D. Kumar, D. Baleanu, Methods of Mathematical Modelling: Fractional Differential Equations, Boca Raton, FL: CRC Press, (2019).

27. R. P. Agarwal, M. Benchohra, S. Hamani, Boundary value problems for fractional differential equations, Georgian Mathematical Journal 16 (3) (2009) 401–411.

28. S. Zhang, Positive solutions for boundary-value problems of nonlinear fractional differential equations, Electronic Journal of Differential Equations (EJDE) [electronic only] 2006 (36) (2006) 1–12.

29. H. Weitzner, G. Zaslavsky, Some applications of fractional equations, Communications in Nonlinear Science and Numerical Simulation 8 (3–4) (2003) 273–281.

30. R. Gorenflo, F. Mainardi, Fractals and fractional calculus in continuum mechanics, Fractional calculus: Integral and differential equations of fractional order. Wien, New York: Springer-Verlag (1997) 223–276.

31. H. Singh, M. R. Sahoo, O. P. Singh, Numerical method based on galerkin approximation for the fractional advection-dispersion equation, International Journal of Applied and Computational Mathematics 3 (3) (2017) 2171–2187.

32. H. Singh, R. K. Pandey, H. M. Srivastava, Solving non-linear fractional variational problems using Jacobi polynomials, Mathematics 7 (3) (2019) 224.

6 Review on Non-Standard Finite Difference (NSFD) Schemes for Solving Linear and Non-linear Differential Equations

Kushal Sharma,[1] *Seema Swami,*[1] *Vimal Kumar Joshi,*[2] *S.B. Bhardwaj*[3]

[1]Department of Mathematics, MNIT Jaipur, India

[2]KCC Institute of Technology and Management, Greater Noida, India

[3]Department of Physics, Pt. C. L. S. Govt. College, Karnal, India

CONTENTS

6.1 INTRODUCTION

In real-life applications such as the dynamics of HIV transmission [1], linear inhomogeneous time-fractional equations, continuous-time predator-prey systems, linear space-fractional telegraph equations, linear inhomogeneous fractional Burgers equations, fractional wave equations, etc. researchers are encouraged to efficiently provide highly accurate solutions. Predator-prey systems are among the most interesting topics in mathematical biology. Because of their universal existence and importance [2–5], their models continue to attract the attention of researchers from both applied mathematics and ecology. Biological systems including predator-prey systems often are described by ordinary differential equations (ODEs) or partial differential equations (PDEs). Actually, the conversion of continuous systems to discrete systems preserving the features of the original continuous systems is of great importance. Some continuous models constituted as real-life applications [9] are discussed fully in a descriptive way and their discrete counterparts were also investigated. There are many ways for converting continuous models into their discrete counterparts. The most popular is to use standard difference methods such as Euler and Runge-Kutta schemes, etc. However, in many non-linear problems, the standard difference schemes have a major consequence known as numerical instability [6–11]; there are many examples which analyze the numerical instabilities when standard difference schemes are under consideration.

When the discretization parameter is sufficiently small, in general, standard difference schemes preserve the properties of the modelled system in the form of differential equations. Therefore, when studying dynamical models in big time intervals, the option of small-time steps will need very large computational effort. Hence, these discrete models are inefficient. Besides, for some special dynamical problems, standard difference schemes cannot preserve the properties of the problems for any step sizes. The net objective of the analysis was that there existed solutions to the finite difference equations that do not correspond to any solution to the differential equations, and these solutions lead to numerical instabilities. In order to defeat the numerical instabilities phenomena, Mickens [12] developed the concept of non-standard finite difference (NSFD) schemes and after that explained the NSFD scheme for many problems in his research works. According to Mickens, NSFD schemes obey the set of five basic rules: positivity, periodicity, monotonicity, stability, and some other invariants including energy and shapes defined geometrically. Nevertheless, in contrast to the usual methods for constructing discrete models of ODEs and PDEs [8, 13], for the purpose of numerical integration, these schemes generally require non-local representation and algorithms for non-linear terms.

An NSFD model is a free and more accurate translation of the differential equations. The behaviour or stability of the NSFD scheme will be seen by plotting graphs using MATLAB [14]. NSFD schemes are a general set of techniques in the numerical analysis that gives numerical solutions to differential equations by constructing a discrete model [15, 16]. Also, an NSFD scheme was constructed by Mickens and Gumel [17]; and the associated properties were studied by Sachdev [18] for the following Burgers-Fisher PDE as an application

$$\frac{\partial u}{\partial t} + au\frac{\partial u}{\partial x} = D\frac{\partial^2 u}{\partial x^2} + \lambda u(1-u)$$

Here, a, D, and λ are non-negative parameters. Actually, it was noted [19] to model turbulence in one dimension. Also, the above equation correlates with the Burgers equation, having a non-linear reaction. The logistic reaction in sound waves for a viscous medium can be modelled by this equation in an interesting way. An equivalent of this equation is to consider it as a modified Fisher equation [20]. An NSFD scheme was also demonstrated by Mickens [21] for the modified Fisher PDE.

Numerical studies indicate that the derived non-standard scheme provides excellent numerical solutions. For the above equation, it was also observed [17] that this effectively explicit NSFD scheme satisfies both the positivity and boundedness conditions and is obvious to implement. A new kind of finite difference scheme is presented [22] for special second-order non-linear two-point boundary value problems subject to some boundary conditions. Much research work has been done by researchers, on the numerical integration of non-linear boundary value problems. Usually, the adopted integration methods are the finite difference method [23, 24], the shooting method [25, 26], etc. A variety of quality recent works [27–32] have also been published to demonstrate applications of fractional differential equations along with the solutions of a different kind. Also, an NSFD scheme was developed by Moaddya [33] to solve the linear PDEs with time and space fractional derivatives, and the Grunwald Letnikov method was used to approximate the fractional derivatives.

Recently, Dang and Hoang [34] transformed a continuous-time predator-prey system into a discrete-time model by NSFD scheme and proved their results by numerical simulations and also discussed that the standard finite difference schemes such as Euler, second-order Runge-Kutta, and the classical fourth-order Runge-Kutta schemes cannot conserve the features of the continuous model for big step sizes, whereas the NSFD scheme conserves the qualitative properties positivity and stability for any finite step size, which is essential for the continuous system. This report introduces some of the features and uses of the NSFD scheme. Here, we start with a review of the elements of finite differences with a variety of problems, and how discrete differences are introduced in place of continuous derivatives. Then the non-standard ideas are presented. The scheme has many benefits over classical techniques and provides an efficient numerical solution. The laws of nature are usually best described by ODEs and PDEs, which usually apply to continuous materials pertained to continuous solutions. From the calculus point of view, the analytic results give classic solutions for a limited set of problems.

The main objective of the finite difference literature survey is to minimize the hazards that occur when continuous functions are replaced at finite intervals. An important application of finite differences is in numerical analysis, especially in numerical differential equations, which aim at the numerical solution of ODEs and PDEs, respectively. The idea is to replace the derivatives appearing in the differential equation by finite differences that approximate them. The resulting methods are called finite difference methods. In this chapter we review and discuss the widely used NSFD scheme for the ODEs and PDEs with time derivatives, and its efficiency is shown by graphical representations.

Exact Schemes: Consider a first-order ODE

$$\frac{dv}{dx} = g(v, x, \lambda), \; v(x_0) = v_0 \tag{6.1}$$

where λ is a parameter and $g(v,x,\lambda)$ is such that the equation has a unique solution over the interval (x_0, X). Now express the solution for (6.1) as

$$v(x) = \varphi(\lambda, v_0, x_0, x) \text{ with initial condition } \varphi(\lambda, v_0, x_0, x_0) = v_0. \quad (6.2)$$

A discrete model can be visualized if Eq. (6.1) can be written as

$$v_{i+1} = g(\lambda, h, v_i, x_i), \quad (6.3)$$

where $h = \Delta x$ and $x_i = hi$. Hence, the solution of the discrete equation takes the form $v_i = \varphi(\lambda, h, v_0, x_0, x_i)$ with $\varphi(\lambda, h, v_0, x_0, x_0) = v_0$. The preceding system, defined by Eq. (6.1) and the corresponding discrete Eq. (6.3), is said to have the same general solution iff $v_i = v(x_i)$. Also, a scheme is called an exact finite difference scheme if the difference equation and differential equation have the same general solution [9].

Consider that the ODE

$$\frac{dv}{dx} = f(v,x,\lambda), \; v(x_0) = v_0 \quad (6.4)$$

has an exact finite difference scheme $v_{k+1} = \varphi[\lambda, \; h, \; v_k, \; t_k, \; t_{k+1}]$, where φ is given by Eq. (6.2). Here, Eq. (6.4) gives the solution

$$v(x+h) = \varphi[\lambda, \; v(x), \; x, \; x+h]. \quad (6.5)$$

We can assume the identifications $x \to x_k, \; v(x) \to v_k$, in Eq. (6.5) gives

$$v_{k+1} = \varphi[\lambda, h, v_k, x_k, x_{k+1}]. \quad (6.6)$$

Here, the general solution for a difference Eq. (6.6) is the same to the general solution of the system defined in Eq. (6.4). Further, Eq. (6.6) is said to be an exact scheme for the initial value problem (4).

Some properties of the NSFD scheme

1. A major consequence of the theorem is that the solution of the difference scheme is exactly equal to the solution of the ODE on the computational grid for fixed but arbitrary step size h.
2. If an ODE has a solution, then an exact difference scheme exists, and if the solution is known, the scheme can be demonstrated.
3. The theorem gives no guidance as to how to construct the exact scheme for a general ODE for which the solution is not known.
4. The discrete model is explicit and a difference scheme replaces the derivatives.

Let $\{v^{(j)}(x)\}; j = 1,2,...,N$; be the set of linearly independent functions. We can construct an Nth-order linear ODE that has the corresponding discrete functions, $\{v_i^{(j)} \equiv v^{(j)}(x_i)\}$, as solution. Further, the required equation can be constructed by the following determinant [35]

$$\begin{vmatrix} v_i & v_i^{(1)} & \cdots & v_i^{(N)} \\ v_{i+1} & v_{i+1}^{(1)} & \cdots & v_{i+1}^{(N)} \\ \vdots & \vdots & \ddots & \vdots \\ v_{i+N} & v_{i+N}^{(1)} & \cdots & v_{i+N}^{(N)} \end{vmatrix} = 0.$$

Problem 6.1: The ODE

$$\frac{dy}{dx} = y\cos x,\ y(0) = 0,\ \left.\frac{dy}{dx}\right|_{x=0} = 1;$$

by the use of forward Euler scheme, can be written as $\frac{y_{i+1} - y_i}{h} = y_i \cos(x_i)$.

Table 6.1

Comparison of Solution Values by Exact Method and Solution Values by Finite Difference (FD) Method of $y_{i+1} = (1 + h\cos(x_i))y_i$, Corresponding to x_i Values When $h = 0.1$. A comparison between exact solution and FD scheme is given in Table 6.1.

S.No.	Value of x	Exact Solution	FD Solution
1	0.0	1.0000	1.1000
2	0.3	1.3438	1.4549
3	0.5	1.6151	1.7283
4	1.0	2.3197	2.4119
5	1.3	2.6210	2.6825
6	1.5	2.7115	2.7474
7	2.0	2.4826	2.4508
8	3.0	1.1515	1.0674
9	4.0	0.4691	0.4000
10	5.0	0.3833	0.3615

Problem 6.2: Consider the decay equation

$$\frac{dv}{dx} = -\lambda v,\quad v(0) = v_0.$$

It is obvious that the solution of this equation is $v(x) = v_0 e^{-\lambda x}$. Therefore,

$$\begin{vmatrix} v_i & e^{-\lambda x_i} \\ v_{i+1} & e^{-\lambda x_{i+1}} \end{vmatrix} = e^{-\lambda x_i} \begin{vmatrix} v_i & 1 \\ v_{i+1} & e^{-\lambda h} \end{vmatrix} = 0,$$

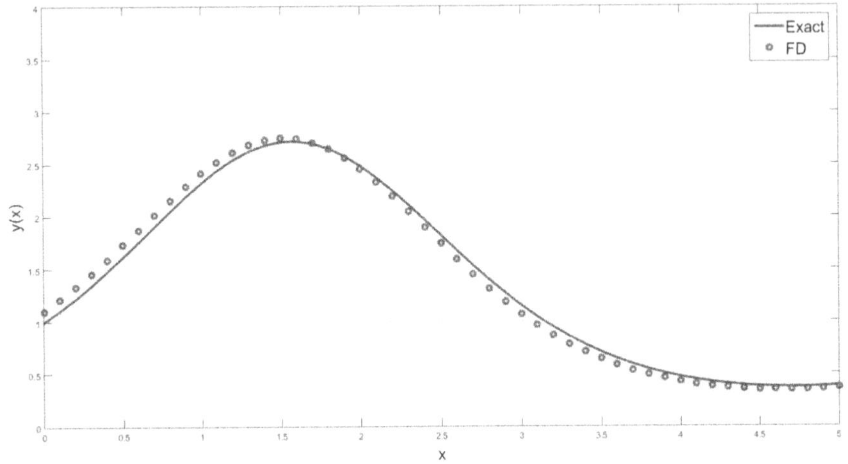

Figure 6.1 Graph of the solution $y' = y\cos(x)$ with the initial conditions $y(0) = 0,\ y'(0) = 1$, and $h = 0.1$. Numerical results are shown graphically in Figure 6.1 to 6.3.

and the exact scheme for this equation is

$$v_{i+1} = e^{-\lambda h} v_i \text{ and } \frac{v_{i+1} - v_i}{\left(\frac{1 - e^{-\lambda h}}{\lambda}\right)} = -\lambda v_i.$$

The discrete first-order derivative for this equation is $\frac{dv}{dx} \to \frac{v_{i+1} - v_i}{\phi(h)}$, where $\phi(h) = \frac{1 - e^{-\lambda h}}{\lambda} = \phi(\lambda,\ h) = h + O(\lambda h^2)$ $\left(\text{since } e^{-\lambda h} = \sum_{m=0}^{\infty} \frac{(-\lambda h)^m}{m!}\right)$ and big O represents the order of accuracy.

Problem 6.3: The general solution of the second-order linear differential equation

$$\frac{d^2 v}{dx^2} = \lambda \frac{dv}{dx},$$

is the linear combination of the two functions $v^{(1)}(x) = 1$ and $v^{(2)}(x) = e^{(\lambda x)}$, that is, $c_1 v^{(1)}(x) + c_2 v^{(2)}(x)$. The exact finite difference scheme

$$\frac{v_{i+1} - 2v_i + v_{i-1}}{\frac{(e^{\lambda h} - 1)h}{\lambda}} = \lambda \left(\frac{v_i - v_{i-1}}{h}\right)$$

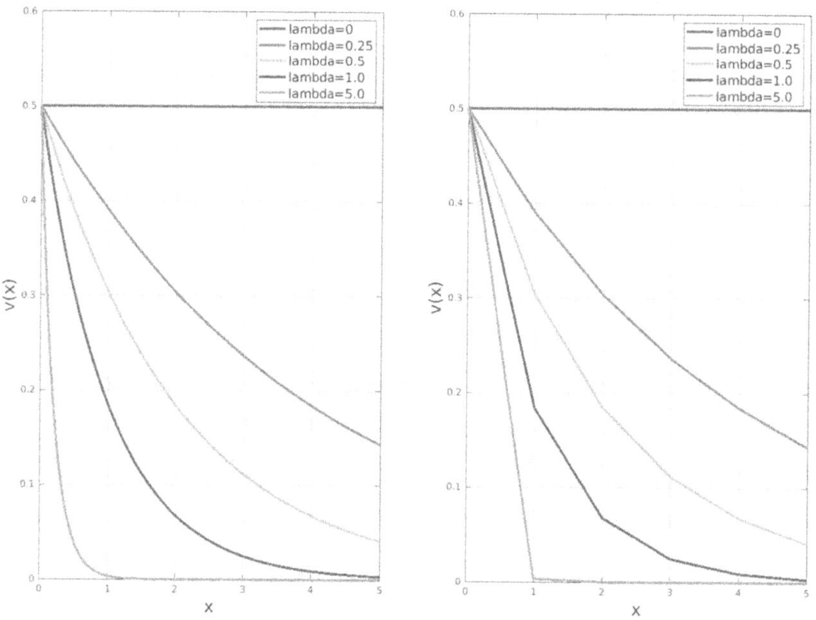

(a) Numerical approximation of the solution of equation $\frac{dv}{dx} = -\lambda v$, for $h = 0.01$

(b) Numerical approximation of the solution of equation $\frac{dv}{dx} = -\lambda v$, for $h = 1$

Figure 6.2 Graph of the solution of decay equation $\frac{dv}{dx} = -\lambda v$, using exact scheme to the IVP with different step sizes h and different values of parameter λ that is, $\lambda = 0, 0.25, 0.5, 1, 5$.

is discovered by the use of the determinant

$$\begin{vmatrix} v_i & 1 & e^{\lambda hi} \\ v_{i+1} & 1 & e^{\lambda h(i+1)} \\ v_{i+2} & 1 & e^{\lambda h(i+2)} \end{vmatrix} = 0.$$

Also, system of ODEs and PDEs can be descretized with finite differences [36]. Further, there is an occurrence of non-linear equations in various techniques applied for the solutions of system of differential equations. Here, we are presenting the syntax used in MATLAB for a particular example to find zeroes of a system of non-linear equations. Here, we have to start our search for a solution at some initial point $x0$.

$$e^{-(x_1)} = x_2(1+x_1^2), \quad x_1 \sin x_2 + x_2 \cos x_1 = 2.$$

Table 6.2
Exact Finite Difference Schemes for ODEs, Compared with Standard Finite Difference schemes. A comparison between exact FD and FD scheme is given in Table 6.2.

System of ODEs	FD Scheme	Exact FD Scheme
$\frac{dv}{dt} = \omega$	$\frac{v_{i+1} - v_i}{h} = \omega_i$	$\frac{v_{i+1} - \cos(\omega h) v_i}{\left(\frac{\sin(\omega h)}{\omega}\right)} = \omega_i$
$\frac{d\omega}{dt} = \omega^2 v$	$\frac{\omega_{i+1} - \omega_i}{h} = -\omega^2 v_i$	$\frac{\omega_{i+1} - \cos(\omega h)\omega_i}{\left(\frac{\sin(\omega h)}{\omega}\right)} = -\omega^2 v_i$

MATLAB Syntax: First, write a function that computes f, the values of the equations at x, and then create the initial point $x0 = [0\ 0]$;
$f = @(x)[exp(-(x(1))-x(2)*(1+x(1)^2);\ x(1)*\sin(x(2))+x(2)*\cos(x(1))-2]$;
$x0 = [0\ 0]$
$[x, fval] = fsolve(f, x0)$.
The *fval* output gives the function value, which should be zero at a solution.

6.2 NON-STANDARD FINITE DIFFERENCE (NSFD) SCHEMES

The NSFD was given by Mickens [37]. A great benefit of having exact finite difference schemes over the solution of differential equations is that various doubts related to the issue of convergence of the scheme [6, 38–40] do not arise. Also, the uncertainty related to stability analysis and the consistency of the scheme should not be encountered. The NSFD scheme has the following benefits which make this method preferable and interesting.

- The NSFD schemes are a general set of methods in numerical analysis that give numerical solution to differential equations by constructing a discrete model.
- An NSFD model is free and more accurate at solving differential equations.
- The NSFD scheme is more consistent and stable than the standard finite difference methods in solving differential equations.

There are some rules to provide guidance for the construction of NSFD models of differential equations. The orders of the discrete derivatives should not differ from the orders of the corresponding derivatives of differential equations. In general, for the discrete derivatives denominator functions must be expressed in terms of more complex functions of the step sizes h, and non-local discrete representations should be given in place of non-linear terms. All those conditions that hold for the solutions of the given differential equations must also hold for the solutions of the constructed finite difference scheme, and the unauthentic/spurious solution should be excluded if it has appeared in the scheme.

Here are some important points for the above rules that we can keep in mind.

1. If we have higher order of the discrete derivatives than there order of the given differential equation, then we obtain unauthentic solutions [38]. The discrete form (use the central difference approximation) for decay equation $du/dt = -u$ is $(u_{i+1} - u_{i-1})/2h = -u_i$. The solution of this discrete equation can be taken as $u_i = A(r_1)^i + B(r_2)^i$, $r_1 = -h + \sqrt{1+h^2}$, $r_2 = -h - \sqrt{1+h^2}$, where A and B are arbitrary constants. In the solution, the second term of the right-hand side oscillates with an amplitude, which increases exponentially with i, whereas the first term decreases monotonically to zero. The actual solution behaviour of the decay equation is modelled by the first term, whereas the second term in the solution is an extra solution, which is constructed by the second-order discrete approximation to the first-order derivative in the given differential equation.
2. The discrete first derivative takes the form $\dfrac{du}{dt} \to \dfrac{u_{i+1} - \varphi u_i}{\phi(h)}$, where φ and ϕ have the properties given by $\varphi(h) = 1 + O(h^2)$, $\phi(h) = h + O(h^2)$.
3. The non-linear term u^2 is replaced by $u_{i+1}u_i$ and other general forms may be used, $u^2 \to 2(u_i)^2 - u_{i+1}u_i$. If the discrete equations do not satisfy the given conditions, then numerical instabilities can be observed. If the solution of discrete equations becomes negative, instability in numerical computations will arise.

6.2.1 COMPARISON BETWEEN STANDARD AND NON-STANDARD FINITE DIFFERENCE METHODS

The exact solution of the first-order ODE $y' = y\sin(x)$, $y(0) = 1$, $y'(0) = 0$. is

$$y(x) = \frac{\exp(-\cos(x))}{\cosh(1) - \sinh(1)}.$$

Standard method: $y_{i+1} = (1 + h\sin(x_i))y_i$.
Non-standard method: $y_{i+1} = (1 + (e^h - 1)\sin(x_i))y_i$.

Here, Figure 6.3 represents the comparative behaviour of solution by standard and non-standard finite difference schemes. It is shown that the solution obtained by the NSFD scheme is closed to the exact solution that is more stable than the standard method.

6.2.2 APPLICATIONS OF NSFD SCHEME

There are so many applications of the NSFD scheme to real-life problems. Some of these are now discussed.

Table 6.3

Comparison Between Exact, Standard, and Non-standard Solutions When $h = 0.2$. A comparison between exact solution, standard and non-standard scheme is given in Table 6.3.

S.No.	x_i	y_i (Exact Method)	y_i (Standard Method)	y_i (Non-standard Method)
1	0	1	1	1
2	0.4	1.082138	1.1207	1.1340
3	0.8	1.354312	1.4262	1.4784
4	1.4	2.293394	2.3665	2.5773
5	1.8	3.411671	3.3926	3.8264
6	2.4	5.682514	5.2872	6.2301
7	2.8	6.974233	6.2231	7.4559
8	3.0	7.315476	6.3987	7.6889
9	4.0	5.225987	4.0738	4.6481
10	5.0	2.046923	1.4123	1.4179

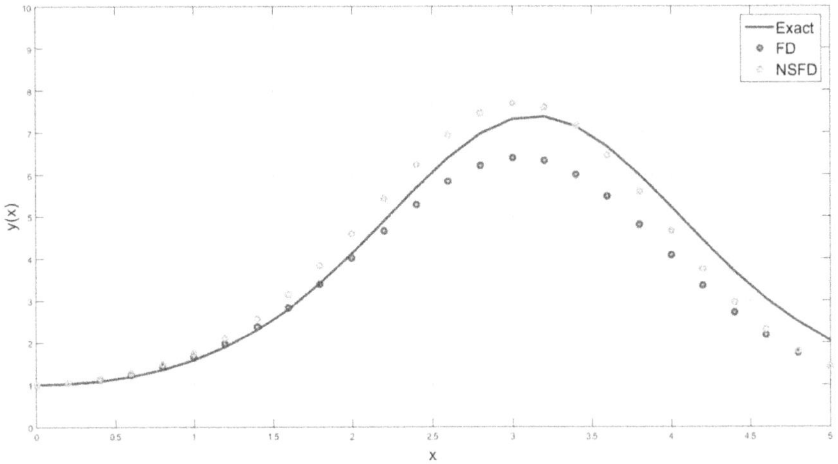

Figure 6.3 Graph of numerical solution of FD and NSFD methods, and exact solution, of equation $y' = y\sin(x)$ with $h = 0.2$.

6.2.2.1 Applications to Modelled ODEs

1. The fixed points $(\bar{u})$ are the constant solutions of the first-order equation $\dfrac{du}{dt} = g(u)$, and also are solutions to the equation

$$g(\bar{u}) = 0. \tag{6.7}$$

Now we construct a non-standard scheme $\dfrac{du}{dt} \rightarrow \dfrac{u_{i+1} - u_i}{\varphi(h)}$ for this equation. Let the given model have m number of real solutions. Define A_i to be

$$A_i = \frac{df}{du}|_{\bar{u}^{(i)}},$$

where $\{\bar{u}^{(i)}; i = 1,2,...,m\}$ is the set of fixed points, and define $A^\star$ as the maximum of $|A_i|$.

A special φ is $\varphi(h,A^\star) = \dfrac{1-e^{-A^\star h}}{A^\star}$, which has the properties

$$\varphi(h,A^\star) = h + O(A^\star h^2).$$

Problem 6.4: Consider a first-order non-linear ODE

$$\frac{du}{dt} = u^2(1-u), \quad u(0) = u_0 > 0, \tag{6.8}$$

which represents the combustion model.

Here $g(u) = u^2(1-u)$ and three fixed points $\bar{u}(1) = \bar{u}(2) = 0$, $\bar{u}(3) = 1$, exist, and here $A_1 = 0 = A_2$, $A_3 = 1$. Also, the denominator function $\varphi(h)$ is $(1-e^{-h})$, and the discrete derivative is

$$\frac{du}{dt} \to \frac{u_{i+1}-u_i}{1-e^{-h}}.$$

Non-negativity is required as $u(t) \geq 0$, for which the corresponding condition for u_i is $u_i \geq 0 \Rightarrow u_{i+1} \geq 0$. As $u_0 > 0$, all the solutions to Eq. (6.8) reach to the value $\bar{u}(3) = 1$ monotonically. Now u^2 and u^3 are replaced by non-local description $2(u_i)^2 - u_{i+1}u_i$ and $u^3 \to u_{i+1}(u_i)^2$, respectively. Therefore,

$$\frac{u_{i+1}-u_i}{\varphi(h)} = 2(u_i)^2 - u_{i+1}u_i - u_{i+1}(u_i)^2.$$

Clearly, this equation is linear in u_{i+1}, so

$$u_{i+1} = \frac{(1+2\varphi u_i)u_i}{1+\varphi[u_i+(u_i)^2]}. \tag{6.9}$$

Since $h > 0$, $0 < \varphi(h) < 1$; Eq. (6.9) has the fixed points $\bar{u}(1) = 0 = \bar{u}(2)$, and $\bar{u}(3) = 1$. For $u_0 > 0, h > 0$; all the solution u_i of Eq. (6.9) monotonically approaches $\bar{u}(3) = 1$ as $i \to \infty$.

2. Consider Duffing oscillator second-order ODE

$$\frac{d^2u}{dt^2} = g(u), \quad u(0) = u_0, \quad \frac{du(0)}{dt} = u_0',$$

where $g(u)$ is defined as a non-linear polynomial. The discrete equation for this model is

$$\frac{u_{i+1}-2u_i+u_{i-1}}{h^2\varphi} = g(u_i), \tag{6.10}$$

where

$$\varphi(z) = \frac{\cos\sqrt{-z}-1}{z/2}, \quad z = h^2\frac{dg}{du}|_{u=u_i}.$$

Equation (6.10) can be written as

$$g(u_i) = \frac{-u_{i+1} + 2u_i - u_{i-1}}{(4h^2/z)\sin^2(\sqrt{-z})/2)}.$$

The Duffing equation is

$$\frac{d^2u}{dt^2} = -u - \varepsilon u^3, \tag{6.11}$$

which is also discussed by Micken [9], where ε is some real parameter. So, comparing to (6.10)

$$g(u) = -u - \varepsilon u^3, \quad \frac{dg}{du} = -1 - 3\varepsilon u^2,$$

which implies $z = -h^2(1 + 3\varepsilon u_i{}^2)$, and

$$\frac{u_{i+1} - 2u_i + u_{i-1}}{(4/(1+3\varepsilon u_i{}^2))\sin^2(h/2\sqrt{1+3\varepsilon u_i{}^2})} + u_i + \varepsilon u_i{}^3 = 0.$$

Now u^3 can be replaced by the non-local representation $u_i{}^2\left(\frac{u_{i+1} + u_{i-1}}{2}\right)$. Putting all these in the above equation, we get NSFD scheme as

$$\frac{u_{i+1} - 2u_i + u_{i-1}}{(4/(1+3\varepsilon u_i{}^2))\sin^2(h/2\sqrt{1+3\varepsilon u_i{}^2})} + u_i + u_i{}^2\left(\frac{u_{i+1} + u_{i-1}}{2}\right) = 0.$$

3. The advection diffusion equation is

$$u_t + u_x = \alpha u_{xx}, \ t > 0, \ x > 0, \ \alpha > 0. \tag{6.12}$$

This equation has advection and diffusion properties. The advection diffusion equation has the three sub-parts as

$$u_t + u_x = 0, \ u_x = \alpha u_{xx}, \ u_t = \alpha u_{xx}.$$

The first and second equations have exact finite difference scheme as

$$\frac{u_i^{n+1} - u_i^n}{\Delta t} + \frac{u_i^n - u_{i-1}^n}{\Delta x} = 0, \ \text{with} \ \Delta t = \Delta x \ \text{and}$$

$$\frac{u_i - u_{i-1}}{\Delta x} = \alpha\left(\frac{u_{i+1} - 2u_i + u_{i-1}}{\alpha\Delta x(\exp(\Delta x/\alpha) - 1)}\right).$$

The scheme's finite difference has both of the sub-equation properties, as

$$\frac{u_i^{n+1} - u_i^n}{\Delta t} + \frac{u_i^n - u_{i-1}^n}{\Delta x} = \alpha\left(\frac{u_{i+1} - 2u_i + u_{i-1}}{\alpha\Delta x(\exp(\Delta x/\alpha) - 1)}\right). \tag{6.13}$$

On rearranging the terms in Eq. (6.13), we get the NSFD scheme [9, 41] which is

$$u_i^{n+1} = \beta_1 u_{i+1}^n + (1-\alpha_1-2\beta_1)u_i^n + (\alpha_1+\beta_1)u_{i-1}^n, \tag{6.14}$$

where $\alpha_1 = \frac{\Delta t}{\Delta x}$, $\beta_1 = \frac{\alpha_1}{\exp(\Delta x/\alpha)-1}$.

By the use of von Neumann stability criteria for stability analysis, we insert $u_i^n = f(\xi)^n e^{\iota\beta i}$ into Eq. (6.14). Then the square of the modulus of the amplification factor is given by

$$|f(\xi)|^2 = ((1-\alpha_1-2\beta_1)+(\alpha_1+2\beta_1)\cos(\omega))^2 + (\alpha_1\sin(\omega))^2.$$

For stability [42, 43], $0 < |f(\xi)| \leq 1$, which implies that $0 < |f(\xi)|^2 \leq 1$. Now to obtain the region of stability, consider the case when $\omega = \pi$.

Then

$$|f(\xi)|^2 = (1-2\alpha_1-4\beta_1)^2.$$

Therefore, $(1-2\alpha_1-4\beta_1) \leq 1$. Thus, for stability, we have the following inequality in simplified form

$$0 \leq \alpha_1 + 2\beta_1 \leq 1.$$

Here $\alpha_1 + 2\beta_1 \geq 0$ is the trivial inequality, as $\alpha_1, \beta_1 > 0$. Hence, we consider the inequality

$$\alpha_1 + 2\beta_1 \leq 1.$$

Since $\alpha_1 = \frac{\Delta t}{\Delta x}$ and $\beta_1 = \frac{\alpha_1}{\exp(\Delta x/\alpha)-1}$, we have

$$\frac{\Delta t}{\Delta x} + \frac{2\Delta t}{\Delta x(\exp(\Delta x/\alpha)-1)} \leq 1.$$

For stability, we need the condition, $\Delta t \leq \left(\frac{\exp(\Delta x/\alpha)-1}{\exp(\Delta x/\alpha)+1}\right)\Delta x$.

For small Δx, it becomes

$$\Delta t \leq \frac{\Delta x^2}{2\alpha}.$$

This is the relation that generally applies to the diffusion equation. For $\Delta t = 0.01$, $\Delta x = 0.04$, and $\Delta t = 0.005$, $\Delta x = 0.02$ the algorithm is least dissipative, and for $\Delta t = 0.02$, $\Delta t = 0.04$ the algorithm is least dispersive [44].

6.2.2.2 Applications to Modelled PDEs

Consider the diffusion PDE with a non-linear term

$$\frac{\partial u}{\partial t} = \frac{\partial^2 u}{\partial x^2} + u^2(1-u).$$

This is a combustion model equation in one dimension [9]. The physical solution of this equation should satisfy the relation $0 \leq u(x,0) \leq 1$, which implies that $0 \leq u(x,t) \leq 1$. The non-standard scheme for this model is

$$\begin{aligned}\frac{u_i^{n+1}-u_i^n}{\Delta t} &= (u_{i+1}^n)^2+(u_{i-1}^n)^2-[\frac{u_{i+1}^n+u_{i-1}^n}{2}]u_i^{n+1}\\ &\quad -[\frac{(u_{i+1}^n)^2+(u_{i-1}^n)^2}{2}]u_i^{n+1}+\frac{u_{i+1}-2u_i+u_{i-1}}{(\Delta x)^2}.\end{aligned}$$

This equation is linear in u_i^{n+1} and we get the following expression:

$$u_i^{n+1}=\frac{R(u_{i+1}^n+u_{i-1}^n)+\Delta t[(u_{i+1}^n)^2+(u_{i-1}^n)^2]+(1-2R)u_i^n}{1+(\Delta t/2)[u_{i+1}^n+(u_{i-1}^n)^2+u_{i-1}^n+(u_{i-1}^n)^2]},$$

where $R=\dfrac{\Delta t}{(\Delta x)^2}$. If $u_i^n \geq 0$; then for $R \leq \dfrac{1}{2}$, u_i^{n+1} will also be non-negative.

The equation gives the relation between the step sizes

$$\Delta t=\frac{(\Delta x)^2}{2}. \tag{6.15}$$

Thus, a non-standard finite difference scheme for the combustion model is

$$u_i^{n+1}=\frac{1/2(u_{i+1}^n+u_{i-1}^n)+\Delta t[(u_{i+1}^n)^2+(u_{i-1}^n)^2]}{1+(\Delta t/2)[u_{i+1}^n+(u_{i-1}^n)^2+u_{i-1}^n+(u_{i-1}^n)^2]},$$

provided the condition holds for Δt.

6.2.2.3 Applications to Modelled Fractional Differential Equations

The NSFD scheme for models describing fractional-order problems is quite a new subject and the available contributions [33] are very few. The solution of fractional-order equations cannot be attained in terms of a finite number of underlying functions. So, numerical methods have been used to assess approximate solutions by means of difference schemes. The integer-order differential operator is a local operator, whereas the fractional-order differential operator is non-local. This shows that the next state of a system also depends on all its previous states along with its current state, which is more rational. Generally, Grunwald-Letnikov and Riemann-Liouville derivatives are used in numerical analysis to approximate the variable order and fractional-order equations, and both are defined as follows:

The Riemann-Liouville fractional derivative:

$$D^\alpha f(x)=\frac{1}{\Gamma(n-\alpha)}\frac{d^n}{dx^n}\int_0^x\frac{f(\xi)}{(x-\xi)^{\alpha-n+1}}d\xi,\ \text{where } n-1\leq\alpha<n\in\mathbb{N}$$

The Grunwald-Letnikov fractional derivative:

$$D^\alpha f(x)=\lim_{h\to 0}h^{-\alpha}\sum_{m=0}^{[x/h]}(-1)^k\binom{\alpha}{k}f(x-mh),\ x\geq 0$$

where h is step size and $[x/h]$ means the integer part of x/h. The derivatives can be approximated using the Grunwald-Letnikov discretization method:
$\frac{\partial^\alpha x}{\partial t^\alpha} = \sum_{k=0}^{N} c_k^\alpha x(t_{n-k})$, $n = 1,2,3,\ldots$; where $t_n = nh$ and c_k^α are the Grunwald-Letnikov coefficient is defined as

$$c_k^\alpha = \left(1 - \frac{1+\alpha}{k}\right) c_{k-1}^\alpha \text{ and } c_0^\alpha = h^{-\alpha}.$$

Local Truncation Error: The local truncation error, τ_n, at step n is computed from the difference between the left- and the right-hand sides of the equation for the increment [45]

$$y_n \approx y_{n-1} + hA(t_{n-1}, y_{n-1}, h, f),$$

$$\tau_n = y(t_n) - y(t_{n-1}) - hA(t_{n-1}, y_{n-1}, h, f).$$

An algorithm is stable and consistent if and only if it is convergent. Also, if the local truncation error is of order h, then the algorithm is consistent. The local truncation error being of order h means that, for every $\varepsilon > 0$, there exists an H such that $|\tau_n| \leq \varepsilon h$ for all $h \leq H$.

6.3 CONCLUSIONS AND SCOPE

Numerical integration of ODEs using the conventional method could result in numerical instabilities, that is, solutions of the discrete derivatives could qualitatively become different from those of the original ODEs. The NSFD schemes are very efficient in overcoming this problem. In this chapter, we have demonstrated how the non-standard numerical methods are relatively easy to implement and have much greater computational efficiency as compared to standard numerical methods.

From the graphical results, it can be seen that the NSFD is more stable than the standard methods. The graphical result (Figure 6.4) shows that the classical h for the denominator function gives the best solution for the NSFD method. The solution with big time steps (Figure 6.4(c)) is more appropriate using the NSFD scheme, while it fails by the standard method and gives inappropriate and far from exact results (Figure 6.2(b), Figure 6.4(c)). Therefore, NSFD methods have a time step freedom. So, if the denominator functions are chosen inappropriate forms, the NSFD methods produce better results. It is concluded that by using the NSFD schemes, an appropriate form of discrete derivatives of the original differential equations could greatly reduce numerical instabilities while expediting computation time.

Further, application of the NSFD schemes to the challenging task of obtaining stable numerical solutions for highly non-linear and coupled differential equations is very popular. Progress in the creation/construction/understanding of the NSFD set of procedures requires the investigation of a set of related issues. Many of the concepts discussed in this chapter can be extended to significantly more general classes of linear and non-linear differential equations (integer and fractional) in various areas.

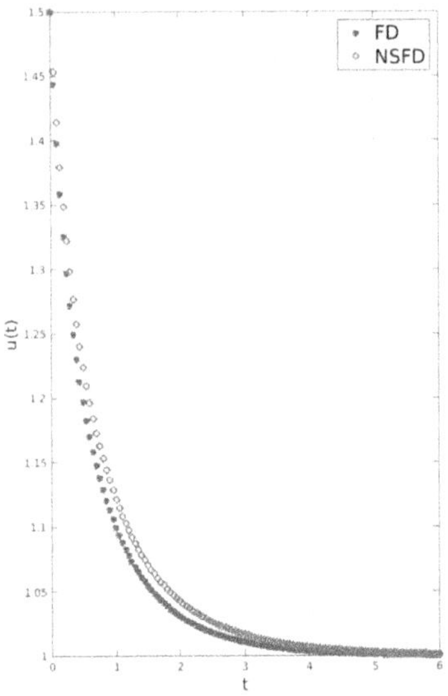

(a) Numerical approximation of the solution of Eq. (6.8) by FD and NSFD methods for $h = 0.05$.

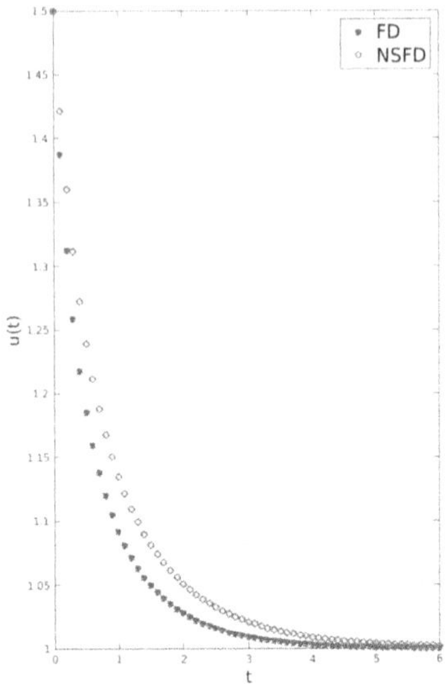

(b) Numerical approximation of the solution of Eq. (6.8) by FD and NSFD methods for $h = 0.1$.

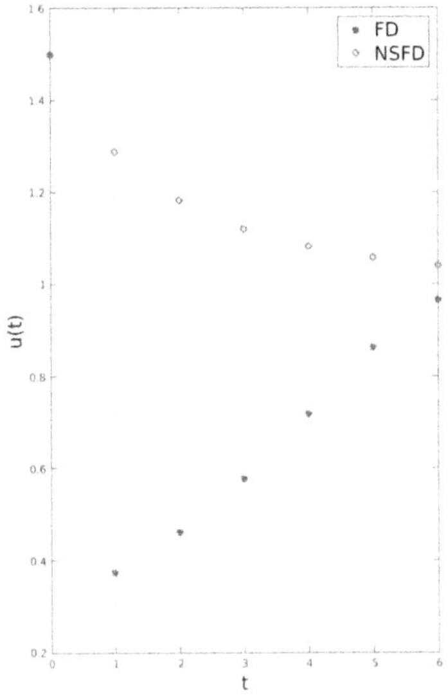

(c) Numerical approximation of the solution of Eq. (6.8) by FD and NSFD methods for $h = 1$.

Figure 6.4 Comparison of NSFD and FD methods for the solution of $\frac{du}{dt} = u^2(1-u)$ for $\varphi(h) = (1 - e^{-h})$ with different step sizes.

REFERENCES

1. Gumel, A. B., Maghadas, S. M., & Mickens, R. E. (2004). Effect of a preventive vaccine on the dynamics of the HIV transmission. *Communications in Nonlinear Science and Numerical Simulation*, 9, 649–659.
2. Berryman, A. A. (1992). The origins and evolution of predator-prey theory. *Ecology*, 73, 1530–1535.
3. Allen, L. J. S. (2007). *An Introduction to Mathematical Biology*. Prentice Hall, NJ.
4. Baek, H. (2018). Complex dynamics of a discrete-time predator-prey system with Ivlev functional response. *Mathematical Problems in Engineering*, 1–15.
5. Shabbir, M. S., Din, Q., Safeer, M., Khan, M. A. & Ahmad, K. (2019). A dynamically consistent nonstandard finite difference scheme for a predator-prey model. *Advances in Difference Equations*, 381–387.
6. Allen, M. B. III, Herrera, I., & Pinder, G. F. (1988). *Numerical Modeling in Science and Engineering*. Wiley-Interscience, New York.
7. Mickens, R. E. (1992). Finite-difference schemes having the correct linear stability properties for all finite step-sizes II. *Dynamic Systems and Applications* 1, 329–340.
8. Mickens, R. E. (1994). *Nonstandard Finite Difference Models of Differential Equations*. World Scientific, Singapore.
9. Mickens, R. E. (2000). *Applications of Nonstandard Finite Difference Schemes*. Addison-Clark Atlanta University, Atlanta, GA.
10. Mickens, R. E. (2002). Nonstandard finite difference schemes for differential equations. *Journal of Difference Equations and Applications*, 8(9), 823–847.
11. Mickens, R. E. (2005). *Advances in the Applications of Nonstandard Finite Difference Schemes*. World Scientific, Singapore, New Jersey.
12. Mickens, R. E. (1989). Exact solutions to a finite-difference model of a nonlinear reaction-advection equation: implications for numerical analysis. *Numerical Methods for Partial Differential Equations*, 5, 313–325.
13. Potter, D. (1997). *Computational Physics*, Wiley-Interscience, New York.
14. Yaghoubi, A. R., & Najafi, H. S. (2015). Comparison between standard and non-standard finite difference methods for solving first and second order ordinary differential equations. *International Journal of Applied Mathematical Research*, 4(2), 316–324.
15. Smith, G. D. (1985). *Numerical Solution of Partial Differential Equation: Finite Difference Methods*. Oxford University Press, New York.
16. Thomas, J. W. (1995). *Finite Difference Methods*. Department of Mathematics, Colorado State University Fort Collins, CO.
17. Mickens, R. E., & Gumel, A. B. (2002). Construction and analysis of a non-standard finite difference scheme for the Burgers Fisher equation. *Journal of Sound and Vibration (Letters to the Editor)*, 257(4), 791–797.

18. Sachdev, P. L. (2000). *Self-similarity and Beyond: Exact Solutions of Nonlinear Problems*. Chapman & Hall/CRC, New York.

19. Burgers, J. M. (1948). A mathematical model illustrating the theory of turbulence. *Advances in Applied Mechanics* 1, 171–199.

20. Murray, J. D. (1989). *Mathematical Biology*. Springer-Verlag, Berlin.

21. Mickens, R. E. (2003). A nonstandard finite difference scheme for a Fisher PDE having nonlinear diffusion. *Computers and Mathematics with Applications*, 45, 429–436.

22. Erdogan, U. & Ozis, T. (2011). A smart nonstandard finite difference scheme for second order nonlinear boundary value problems. *Journal of Computational Physics*, 230, 6464–6474.

23. Tirmizi, I. A., & Twizell, E. H. (2002). Higher-order finite difference methods for nonlinear second-order two-point boundary-value problems. *Applied Mathematics Letters*, 15, 897–902.

24. Elbarbary, E. M. E., & El-Kady, M. (2003). Chebyshev finite difference approximation for the boundary value problems. *Applied Mathematics and Computation*, 139(3), 513–523.

25. Ha, S. N. (2001). A nonlinear shooting method for two-point boundary value problems. *Computers and Mathematics with Applications*, 42, 1411–1420.

26. Liang, H., & Weng, P. (2007). Existence of positive solutions for boundary value problems of second order functional difference equations. *Nonlinear Analysis*, 326, 520–551.

27. Singh, H., Srivastava, H. M., & Kumar, D. (2017). A reliable numerical algorithm for the fractional vibration equation. *Chaos Solitons Fractals*, 103, 131–138.

28. Singh, H., Sahoo, M. R. & Singh, O. P. (2017). Numerical method based on Galerkin approximation for the fractional advection-dispersion equation, *International Journal of Applied and Computational Mathematics*, 3(3), 2171–2187.

29. Singh, H. (2018). Approximate solution of fractional vibration equation using Jacobi polynomials. *Applied Mathematics and Computation*, 317, 85–100.

30. Singh, H., Pandey, R. K. & Srivastava, H. M. (2019). Solving non-linear fractional variational problems using Jacobi polynomials. *Mathematics*, 7, 224.

31. Singh, H., Kumar, D., & Baleanu, D. (Eds.) (2019). *Methods of Mathematical Modelling: Fractional Differential Equations*. CRC Press (Taylor & Francis Group), Boca Raton, FL.

32. Singh, H., & Srivastava, H. M. (2020). Numerical simulation for fractional-order bloch equation arising in nuclear magnetic resonance by using the jacobi polynomials. *Applied Sciences*, 10, 2850.

33. Moaddya, K., Momanib, S., & Hashim, I. (2011). The non-standard finite difference scheme for linear fractional PDEs in fluid mechanics. *Computers and Mathematics with Applications*, 61, 1209–1216.

34. Dang, Q. A., & Hoang, M. T. (2019). Nonstandard finite difference schemes for a general predator-prey system. *Journal of Computational Science*, 36, 101015.

35. Mickens, R. E. (1990). *Difference Equations: Theory and Applications*. Chapman and Hall, New York.

36. Manning, P. M. & Margrave, G. F. (2006). Introduction to non-standard finite-difference modelling, *CREWES Research Report*, 18, 1–10.

37. Mickens, R. E. (1999). Non standard finite difference schemes for the reaction-diffusion equations. *Numerical Methods for Partial Differential Equations*, 15, 201–214.

38. Hildebrand, F. B. (1968). *Finite-Difference Equations and Simulations*. Prentice-Hall, Englewood Cliffs, NJ.

39. Ortega, J. M. & Poole, W. G. (1981). *An Introduction to Numerical Methods for Differential Equations*. Pitman, Marshfield, MA.

40. Greenspan, D. and Casulli, V. (1988). *Numerical Analysis for Applied Mathematics, Science and Engineering*. Addison-Wesley, Redwood City, CA.

41. Mickens, R. E. (1991). Analysis of a new finite-difference scheme for the linear advection-diffusion equation. *Journal of Sound and Vibration*, 146, 342–344.

42. Hindmarsh, A. C., Gresho, P. M., & Griffiths, D. F. (1984). The stability of explicit euler time integration for certain finite difference approximations of the multi-dimensional advection-diffusion equation. *International Journal for Numerical Methods in Fluids*, 4, 853–897.

43. Sousa, E. (2003). The contraversial stability analysis. *Applied Mathematics and Computation*, 145, 777–794.

44. Appadu, A. R. (2013). Numerical solution of the 1D advection-diffusion equation using standard and nonstandard finite difference schemes. *Journal of Applied Mathematics*, 1–14.

45. Gupta, G. K., Sacks-Davis, R., & Tescher (1985). A review of recent developments in solving ODEs. *ACM Computing Surveys*, 17, 5–47.

7 Solutions for Nonlinear Fractional Diffusion Equations with Reaction Terms

Giuliano G. La Guardia,[1] *Jocemar Q. Chagas,*[1] *Marcelo K. Lenzi,*[2] *and Ervin K. Lenzi*[3]
[1]Departamento de Matemática e Estatística, Universidade Estadual de Ponta Grossa – Ponta Grossa, Brazil
[2]Departamento de Engenharia Química, Universidade Federal do Paraná, Curitiba, Brazil
[3]Departamento de Física, Universidade Estadual de Ponta Grossa – Ponta Grossa, Brazil

CONTENTS

7.1 INTRODUCTION

Diffusion, one of the most fascinating scientific fields, essentially started with Robert Brown's observations of the random motion of particles suspended in a fluid (pollen grains in water). The importance of such a discovery is immeasurable and it has since been reported a multitude of natural contexts. A suitable explanation for this phenomenon can be found in pioneering works by Einstein [1], Smoluchowski [2], and Langevin [3]. The main feature of this random motion is the linear time dependence manifested by the mean square displacement, that is, $\langle(\Delta x)^2\rangle \sim t$, which is connected with the Markovian and ergodic properties of these systems. However, several experimental scenarios have reported effects which are not properly described in terms of standard diffusion, such as: dynamic processes in protein folding [4], transport in porous media [5–17], infiltration [18], single-particle tracking [19], electrical response [20–22], and diffusion on fractals [23], among others. These and other scenarios have motivated the extension of the approaches utilized by describing standard diffusion to include these situations and, consequently, their effects which are not yet suitably described. One of these extensions is the fractional Fokker-Planck equations [24–27], which essentially incorporate fractional differential operators [28] in the usual approach. These differential operators of non-integer order have also been utilized to investigate others scenarios such as wave phenomena by extending the vibration equations [29–31], nuclear magnetic resonance [32, 33], epidemiological models [34], and nonlinear problems related to Bratu's equation [35], among others. Other extensions are nonlinear Fokker-Planck equations [16,25], generalized Langevin equations [24], and continuous time random walks with long tailed distributions [24] for the probability density function. Beyond diffusive phenomena, interactions among particles may also occur, causing, for instance, the consumption or production of particles themselves (chemical reactions) or immobilizing the particles during the diffusion process. For a system composed of subsystems, such particles may also interact, in a different way, modifying the relaxation of parts of the system. Possible interactions with this characteristic are obtained when a portion of the particles act as a thermal bath to the other during the relaxation process [36].

In this chapter, we investigate and review the solutions, from the analytical and numerical point of view, of the following set of diffusion equations:

$$\frac{\partial}{\partial t}\rho_1(x,t) = \mathscr{D}_{\mu_1}(t)\frac{\partial^{\mu_1}}{\partial |x|^{\mu_1}}\rho_1^{\nu_1}(x,t)+\mathscr{R}_1(\rho_1,\rho_2,t) \tag{7.1}$$

$$\frac{\partial}{\partial t}\rho_2(x,t) = \mathscr{D}_{\mu_2}(t)\frac{\partial^{\mu_2}}{\partial |x|^{\mu_2}}\rho_2^{\nu_2}(x,t)+\mathscr{R}_2(\rho_1,\rho_2,t)\,, \tag{7.2}$$

which can be used to model the physical situations mentioned above. In Eqs. (7.1) and (7.2), $\mathscr{D}_{\mu_1}(t)$ and $\mathscr{D}_{\mu_2}(t)$ are the diffusion coefficients, and $\rho_1(x,t)$ and $\rho_2(x,t)$ represent the distributions related to different species (in this case, species 1 and 2). The fractional operators presented in the diffusive term of these equations are the Riesz-Welly fractional operators [28], where $1 < \mu_1, \mu_2 \leq 2$, $0 < \nu_1, \nu_2 \leq 2$, and $\mathscr{R}_1$ and $\mathscr{R}_2$ define the reaction processes. In particular, we analyze the cases related to the reaction processes $\mathscr{R}_1(\rho_1,\rho_2,t) = k_{11}\rho_1 + k_{12}\rho_2$ and $\mathscr{R}_2(\rho_1,\rho_2,t) = k_{21}\rho_1 + k_{22}\rho_2$. In this sense, from the analytical point of view, some of the solutions presented here revisit some cases analyzed in Refs. [37, 38]. We also extend some of them by incorporating an arbitrary time dependence on the parameter μ, i.e., $\mu \to \mu(t)$, as presented in Ref. [39], which enable us to obtain solutions with different long tailed behavior. In fact, the time dependence on μ may be used for modelling different behaviors. Therefore, the results obtained from this set of equations show a rich class of behaviors that can be utilized to model several interesting real-life problems.

7.2 REACTION DIFFUSION PROBLEM

7.2.1 LINEAR CASE

Let us start our analysis about the previous set of equations first by considering the linear case, that is, $\nu_1 = 1$ and $\nu_2 = 1$. After this, we investigate the nonlinear case by considering some cases which admit solutions in terms of scaled functions. For this case, the previous set of equations can be written as

$$\frac{\partial}{\partial t}\rho_1(x,t) = \mathscr{D}_{\mu_1}(t)\frac{\partial^{\mu_1}}{\partial|x|^{\mu_1}}\rho_1(x,t)+\mathscr{R}_1(\rho_1,\rho_2,t) \tag{7.3}$$

$$\frac{\partial}{\partial t}\rho_2(x,t) = \mathscr{D}_{\mu_2}(t)\frac{\partial^{\mu_2}}{\partial|x|^{\mu_2}}\rho_2(x,t)+\mathscr{R}_2(\rho_1,\rho_2,t), \tag{7.4}$$

with the reaction process defined by the constants k_{11}, k_{12}, k_{21}, and k_{22}. In fact, depending on the choice of these constants, we may have a scenario for which the substance is totally absorbed by the subtraction or may be used to produce another one by means of an irreversible or a reversible process. An irreversible process may be characterized, for example, by considering $k_{11} < 0$, $k_{12} = k_{21} = 0$, and $k_{22} < 0$, which implies that the substance characterized by $\rho_1(x,t)$ and $\rho_2(x,t)$ will be consumed by a reaction process. On the other hand, the case $k_{11} = -k_{21}$ with $k_{11} < 0$ and $k_{22} = -k_{21}$ and $k_{22} < 0$ may characterize a reversible case, where one species is utilized to produc another. In the following, we analyze a simple case of the irreversible process by considering $k_{12} \to 0$ and $k_{21} \to 0$, whose comportment is governed by the equation

$$\frac{\partial}{\partial t}\rho_i(x,t) = \mathscr{D}_{\mu_i}(t)\frac{\partial^{\mu_i}}{\partial|x|^{\mu_i}}\rho_i(x,t)-k_i\rho_i(x,t), \tag{7.5}$$

where $k_i = k_{ii} > 0$, for $i \in \{1,2\}$.

Remark 7.1: *Throughout this chapter, in order to simplify the reading, we adopt the notation* $1(2)$ *instead of considering* $i \in \{1,2\}$ *(sometimes with abuse of textual concordance).*

Equation (7.5), when restricted to the conditions $\lim_{x\to\pm\infty}\rho_{1(2)}(x,t) = 0$ and $\rho_{1(2)}(x,0) = \varphi_{1(2)}(x)$, has as solution

$$\rho_{1(2)}(x,t) = \int_{\infty}^{\infty} dx'\mathscr{G}_{1(2)}(x,x',t)\varphi_{1(2)}(x'), \tag{7.6}$$

where $\mathscr{G}_{1(2)}(x,x',t)$ is the Green function defined implicitly by

$$\frac{\partial}{\partial t}\mathscr{G}_{1(2)}(x,x',t)-\mathscr{D}_{\mu_{1(2)}}(t)\frac{\partial^{\mu_{1(2)}}}{\partial|x|^{\mu_{1(2)}}}\mathscr{G}_{1(2)}(x,x',t)+k_{1(2)}\mathscr{G}_{1(2)}(x,x',t) = \delta(x-x')\delta(t), \tag{7.7}$$

restricted to the conditions $\lim_{x\to\pm\infty}\mathscr{G}_{1(2)}(x,x',t) = 0$ and $\mathscr{G}_{1(2)}(x,x',t) = 0$ for $t < 0$. By applying both Laplace and Fourier transforms, the solution of Eq. (7.7) can be found and it is given by

$$\begin{aligned}\mathcal{G}_{1(2)}(x,x',t) &= \theta(t)e^{-k_{1(2)}t}\int_{-\infty}^{\infty}\frac{dk}{2\pi}e^{ik(x-x')}e^{-\bar{\mathcal{D}}_{\mu_{1(2)}}(t)|k|^{\mu_{1(2)}}}, \\ &= \frac{1}{\mu|x-x'|}H^{1,1}_{2,2}\left[\frac{|x-x'|}{\left[\bar{\mathcal{D}}_{\mu_{1(2)}}(t)\right]^{\frac{1}{\mu_{1(2)}}}}\left|\begin{matrix}\left(\frac{1}{\mu_{1(2)}},2\right)\left(1,\frac{1}{2}\right)\\ (1,1)\,(1,2)\end{matrix}\right.\right]\end{aligned} \tag{7.8}$$

where $\mathrm{H}^{m,n}_{p,q}\left[x\left|\begin{matrix}(a,A)\\(b,B)\end{matrix}\right.\right]$ is the Fox H function and $\bar{\mathcal{D}}_{\mu_{1(2)}}(t)=\int_0^t dt'\mathcal{D}_{\mu_{1(2)}}(t')$ and $\theta(t)$ is the step function. Equation (7.8) is a Lévy distribution, which is characterized asymptotically by a power-law as follows:

$$\mathcal{G}_{1(2)}(x,x',t)\sim\theta(t)e^{-k_{1(2)}t}\frac{\bar{\mathcal{D}}_{\mu_{1(2)}}(t)}{|x-x'|^{1+\mu_{1(2)}}}\,. \tag{7.9}$$

Figure 7.1 illustrates the Green function $\mathcal{G}_{1(2)}(x,x',t)$, for some values of $\mu_{1(2)}$, by considering the diffusion coefficient constant.

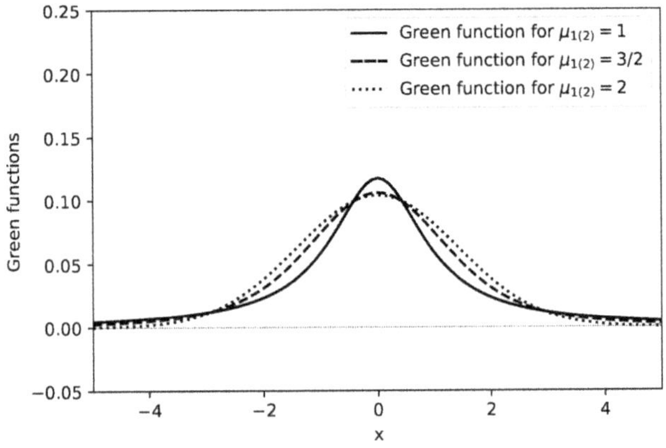

Figure 7.1 Behavior of the Green function, in Eq. (7.8), for some values of $\mu_{1(2)}$ ($\mathcal{D}_{\mu_{1(2)}}=1$, $k_{1(2)}=1$, and $t=1$).

It is worth mentioning that for this scenario, the fractional index μ may be extended to incorporate a time dependence, i.e., $\mu\to\mu(t)$. For this case, the solution of Eq. (7.5) may be also obtained in terms of the Green function approach. For simplicity, we consider that $\mu_1(t)=\mu_2(t)=\mu(t)$; the Green function is obtained from the equation

$$\frac{\partial}{\partial t}\mathcal{G}_{1(2)}(x,x',t)-\mathcal{D}_{\mu}(t)\frac{\partial^{\mu(t)}}{\partial|x|^{\mu(t)}}\mathcal{G}_{1(2)}(x,x',t)+k_{1(2)}\mathcal{G}_{1(2)}(x,x',t)=\delta(x-x')\delta(t)\,. \tag{7.10}$$

The solution for Eq. (7.10) is given by

$$\mathscr{G}_{1(2)}(x,x',t) = \theta(t)e^{-k_{1(2)}t}\int_{-\infty}^{\infty}\frac{dk}{2\pi}e^{ik(x-x')}e^{-\int_0^t dt'\mathscr{D}_{\mu}(t')|k|^{\mu(t')}} . \quad (7.11)$$

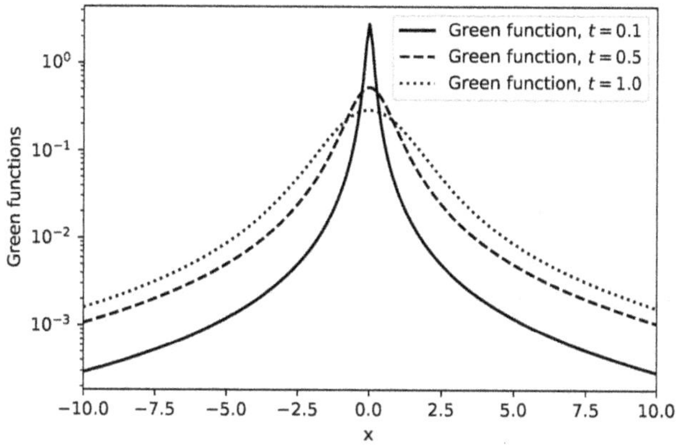

Figure 7.2 Behavior of the Green function, in Eq. (7.11), for different values of t with $\mu(t) = 2-e^{-t}$ ($\mathscr{D}_{\mu} = 1$, $k_{1(2)} = 0$).

We observe that for a suitable choice for $\mu(t)$, the usual Gaussian behavior is obtained. Additionally, changing the function $\mu(t)$, it is possible to obtain distributions with different behaviors when time evolves.

We next consider another situation, which characterizes a reversible reaction process. In such a case, we may have $k_{11} = -k_{21} = -k_1$, with $k_1 > 0$, $k_{22} = -k_{21} = -k_2$, and $k_2 > 0$, yielding

$$\frac{\partial}{\partial t}\rho_1(x,t) = \mathscr{D}_{\mu_1}(t)\frac{\partial^{\mu_1}}{\partial|x|^{\mu_1}}\rho_1(x,t) - k_1\rho_1(x,t) + k_2\rho_2(x,t) \quad (7.12)$$

$$\frac{\partial}{\partial t}\rho_2(x,t) = \mathscr{D}_{\mu_2}(t)\frac{\partial^{\mu_2}}{\partial|x|^{\mu_2}}\rho_2(x,t) + k_1\rho_1(x,t) - k_2\rho_2(x,t). \quad (7.13)$$

For simplicity, we first consider the case $\mu_1 = \mu_2 = \mu$ and $\mathscr{D}_{\mu_1}(t) = \mathscr{D}_{\mu_2}(t) = \mathscr{D}_{\mu}(t)$, which implies equal mobility for both species. Before utilizing integral transforms and the Green function approach in order to solve the system of Eqs. (7.12) and (7.13), we perform the following linear combinations of ρ_1 and ρ_2:

$$\rho_+(x,t) = \rho_1(x,t) + \rho_2(x,t) \quad \text{and} \quad (7.14)$$

$$\rho_-(x,t) = k_1\rho_1(x,t) - k_2\rho_2(x,t) . \quad (7.15)$$

Replacing Eqs. (7.14) and (7.15) in the previous two equations, we obtain

$$\frac{\partial}{\partial t}\rho_+(x,t) = \mathscr{D}_{\mu}(t)\frac{\partial^{\mu}}{\partial|x|^{\mu}}\rho_+(x,t) \quad (7.16)$$

and

$$\frac{\partial}{\partial t}\rho_-(x,t) = \mathscr{D}_\mu(t)\frac{\partial^\mu}{\partial|x|^\mu}\rho_-(x,t) - (k_1-k_2)\rho_-(x,t)\,. \tag{7.17}$$

Equations (7.16) and (7.17) are restricted to the conditions $\lim_{x\to\pm\infty}\rho_\pm(x,t)=0$, $\rho_+(x,0)=\varphi_1(x)+\varphi_2(x)$, and $\rho_-(x,0)=k_1\varphi_1(x)-k_2\varphi_2(x)$. The equations for $\rho_+(x,t)$ and $\rho_-(x,t)$ are not coupled and, consequently, may be solved independently by using the Green function approach, producing

$$\rho_+(x,t)=\int_\infty^\infty dx'\mathscr{G}_+(x,x',t)\left(\varphi_1(x')+\varphi_2(x')\right) \tag{7.18}$$

and

$$\rho_-(x,t)=\int_\infty^\infty dx'\mathscr{G}_-(x,x',t)\left(k_1\varphi_1(x')-k_2\varphi_2(x')\right), \tag{7.19}$$

with Green functions obtained from the equations:

$$\frac{\partial}{\partial t}\mathscr{G}_+(x,x',t)-\mathscr{D}_\mu(t)\frac{\partial^\mu}{\partial|x|^\mu}\mathscr{G}_+(x,x',t)=\delta(x-x')\delta(t)\,, \tag{7.20}$$

restricted to the conditions $\lim_{x\to\pm\infty}\mathscr{G}_+(x,x',t)=0$ and $\mathscr{G}_+(x,x',t)=0$ for $t<0$; and

$$\frac{\partial}{\partial t}\mathscr{G}_-(x,x',t)-\mathscr{D}_\mu(t)\frac{\partial^\mu}{\partial|x|^\mu}\mathscr{G}_-(x,x',t)-(k_1-k_2)\mathscr{G}_-(x,x',t)=\delta(x-x')\delta(t)\,, \tag{7.21}$$

restricted to the conditions $\lim_{x\to\pm\infty}\mathscr{G}_-(x,x',t)=0$ and $\mathscr{G}_-(x,x',t)=0$ for $t<0$. After performing some calculations, the solutions of Eqs. (7.20) and (7.21) are given by

$$\mathscr{G}_+(x,x',t)=\theta(t)\int_{-\infty}^{\infty}\frac{dk}{2\pi}e^{ik(x-x')}e^{-\bar{\mathscr{D}}_\mu(t)|k|^{\mu}1(2)}, \tag{7.22}$$

and

$$\mathscr{G}_-(x,x',t)=\theta(t)e^{-(k_1-k_2)t}\int_{-\infty}^{\infty}\frac{dk}{2\pi}e^{ik(x-x')}e^{-\bar{\mathscr{D}}_\mu(t)|k|^{\mu}}, \tag{7.23}$$

with $\bar{\mathscr{D}}_\mu(t)=\int_0^t dt'\mathscr{D}_\mu(t')$. This scenario can be also extended for incorporating a time dependence on the fractional parameter μ, as in the previous case. This extension implies that Eq. (7.20) and (7.21) are given by

$$\frac{\partial}{\partial t}\mathscr{G}_+(x,x',t)-\mathscr{D}_\mu(t)\frac{\partial^{\mu(t)}}{\partial|x|^{\mu(t)}}\mathscr{G}_+(x,x',t)=\delta(x-x')\delta(t)\,, \tag{7.24}$$

and

$$\frac{\partial}{\partial t}\mathscr{G}_-(x,x',t)-\mathscr{D}_\mu(t)\frac{\partial^{\mu(t)}}{\partial|x|^{\mu(t)}}\mathscr{G}_-(x,x',t)-(k_1-k_2)\mathscr{G}_-(x,x',t)=\delta(x-x')\delta(t)\,, \tag{7.25}$$

The solutions for these equations are

$$\mathscr{G}_+(x,x',t) = \theta(t)\int_{-\infty}^{\infty}\frac{dk}{2\pi}e^{ik(x-x')}e^{-\int_0^t dt'\mathscr{D}_\mu(t')|k|^{\mu(t')}}, \tag{7.26}$$

and

$$\mathscr{G}_-(x,x',t) = \theta(t)e^{-(k_1-k_2)t}\int_{-\infty}^{\infty}\frac{dk}{2\pi}e^{ik(x-x')}e^{-\int_0^t dt'\mathscr{D}_\mu(t')|k|^{\mu(t')}}. \tag{7.27}$$

Now, we consider the case $\mu_1 \neq \mu_2$, $\mathscr{D}_{\mu_1}(t) = \mathscr{D}_{\mu_1} = const$, and $\mathscr{D}_{\mu_2}(t) = \mathscr{D}_{\mu_2} = const$, in the previous reversible reaction process, where $k_{11} = -k_{21} = -k_1$, with $k_1 > 0$, and $k_{22} = -k_{21} = -k_2$ and $k_2 > 0$. For this case, the reversible process is governed by the following set of equations:

$$\frac{\partial}{\partial t}\rho_1(x,t) = \mathscr{D}_{\mu_1}\frac{\partial^{\mu_1}}{\partial|x|^{\mu_1}}\rho_1(x,t) - k_1\rho_1(x,t) + k_2\rho_2(x,t) \tag{7.28}$$

and

$$\frac{\partial}{\partial t}\rho_2(x,t) = \mathscr{D}_{\mu_2}\frac{\partial^{\mu_2}}{\partial|x|^{\mu_2}}\rho_2(x,t) + k_1\rho_1(x,t) - k_2\rho_2(x,t)\,. \tag{7.29}$$

In order to solve Eqs. (7.28) and (7.29), we apply Laplace and Fourier transforms, yielding

$$\left(s + k_1 + \mathscr{D}_{\mu_1}|k|^{\mu_1}\right)\rho_1(k,s) - k_2\rho_2(k,s) = \varphi_1(k)\,, \tag{7.30}$$

$$-k_1\rho_1(k,s) + \left(s + k_2 + \mathscr{D}_{\mu_2}|k|^{\mu_2}\right)\rho_2(k,s) = \varphi_2(k)\,, \tag{7.31}$$

which is a linear system of equations and can be solved by standard procedures. After some computations, we have

$$\begin{aligned}\rho_1(k,s) = {} & \frac{\varphi_1(k)}{s + k_1 + \mathscr{D}_{\mu_1}|k|^{\mu_1}} + \frac{k_2\varphi_2(k)}{\left(s + k_1 + \mathscr{D}_{\mu_1}|k|^{\mu_1}\right)\left(s + k_2 + \mathscr{D}_{\mu_2}|k|^{\mu_2}\right)} \\ & + \sum_{n=1}^{\infty}(-k_1k_2)^n\frac{\left(s + k_2 + \mathscr{D}_{\mu_2}|k|^{\mu_2}\right)\varphi_1(k) + k_2\varphi_2(k)}{\left[\left(s + k_1 + \mathscr{D}_{\mu_1}|k|^{\mu_1}\right)\left(s + k_2 + \mathscr{D}_{\mu_2}|k|^{\mu_2}\right)\right]^{n+1}}\end{aligned} \tag{7.32}$$

and

$$\begin{aligned}\rho_2(k,s) = {} & \frac{\varphi_2(k)}{s + k_2 + \mathscr{D}_{\mu_2}|k|^{\mu_2}} + \frac{k_1\varphi_1(k)}{\left(s + k_1 + \mathscr{D}_{\mu_1}|k|^{\mu_1}\right)\left(s + k_2 + \mathscr{D}_{\mu_2}|k|^{\mu_2}\right)} \\ & + \sum_{n=1}^{\infty}(-k_1k_2)^n\frac{\left(s + k_1 + \mathscr{D}_{\mu_1}|k|^{\mu_1}\right)\varphi_2(k) + k_1\varphi_1(k)}{\left[\left(s + k_1 + \mathscr{D}_{\mu_1}|k|^{\mu_1}\right)\left(s + k_2 + \mathscr{D}_{\mu_2}|k|^{\mu_2}\right)\right]^{n+1}}\,.\end{aligned} \tag{7.33}$$

The inverse Laplace and Fourier transforms of Eqs. (7.32) and (7.33) yields

$$\begin{aligned}
\rho_1(x,t) &= \int_{-\infty}^{\infty} dx' \mathscr{G}_1(x-x',t)\varphi_1(x') \\
&+ k_2 \int_0^t dt' \int_{-\infty}^{\infty} dx' \int_{-\infty}^{\infty} dx'' \mathscr{G}_1(x-x',t')\mathscr{G}_2(x'-x'',t-t')\varphi_2(x'') \\
&+ \sum_{n=1}^{\infty}(-k_1k_2)^n \int_0^t dt'(t-t')^n t'^{n-1} \\
&\times \int_{-\infty}^{\infty} dx' \int_{-\infty}^{\infty} dx'' \mathscr{G}_1(x-x',t-t')\mathscr{G}_2(x'-x'',t')\varphi_1(x'') \\
&+ k_2 \sum_{n=1}^{\infty}(-k_1k_2)^n \int_0^t dt' \left((t-t')t'\right)^n \\
&\times \int_{-\infty}^{\infty} dx' \int_{-\infty}^{\infty} dx'' \mathscr{G}_1(x-x',t-t')\mathscr{G}_2(x'-x'',t')\varphi_2(x'')
\end{aligned} \tag{7.34}$$

and

$$\begin{aligned}
\rho_2(x,t) &= \int_{-\infty}^{\infty} dx' \mathscr{G}_2(x-x',t)\varphi_1(x') \\
&+ k_2 \int_0^t dt' \int_{-\infty}^{\infty} dx' \int_{-\infty}^{\infty} dx'' \mathscr{G}_1(x-x',t')\mathscr{G}_2(x'-x'',t-t')\varphi_2(x'') \\
&+ \sum_{n=1}^{\infty}(-k_1k_2)^n \int_0^t dt'(t-t')^n t'^{n-1} \\
&\times \int_{-\infty}^{\infty} dx' \int_{-\infty}^{\infty} dx'' \mathscr{G}_1(x-x',t-t')\mathscr{G}_2(x'-x'',t')(t-t')\varphi_2(x'') \\
&+ k_1 \sum_{n=1}^{\infty}(-k_1k_2)^n \int_0^t dt' \left((t-t')t'\right)^n \\
&\times \int_{-\infty}^{\infty} dx' \int_{-\infty}^{\infty} dx'' \mathscr{G}_1(x-x',t-t')\mathscr{G}_2(x'-x'',t')\varphi_1(x'') .
\end{aligned} \tag{7.35}$$

Figure 7.3 illustrates the behavior of $\rho_1(x,t)$ and $\rho_2(x,t)$ for different values of μ_1 and μ_2. Figure 7.4 shows the behavior of $[1/\rho_1(x,t)]^{\mu_1}$ and $[1/\rho_2(x,t)]^{\mu_2}$.

7.2.2 NONLINEAR CASE

Let us consider from Eqs. (7.1) and (7.2) the nonlinear case with $\nu_1 = \nu \neq 1$, $k_{11} = -k_{21} = -k$ where $k > 0$, $k_{22} = 0$, and $k_{12} = 0$. We also consider that $\mu_1 = 2$, and $\mu_2 \to \mu(t)$ with $\nu_2 = 1$ $\mathscr{D}_{\mu_1}(t) = \mathscr{D}_2(t)$ and $\mathscr{D}_\mu(t) = \mathscr{D}_{\mu_2}(t)$, that is,

$$\frac{\partial}{\partial t}\rho_1(x,t) = \mathscr{D}_2(t)\frac{\partial^2}{\partial x^2}\rho_1^\nu(x,t) - k\rho_1(x,t) \tag{7.36}$$

$$\frac{\partial}{\partial t}\rho_2(x,t) = \mathscr{D}_\mu(t)\frac{\partial^{\mu(t)}}{\partial |x|^{\mu(t)}}\rho_2(x,t) + k\rho_1(x,t) . \tag{7.37}$$

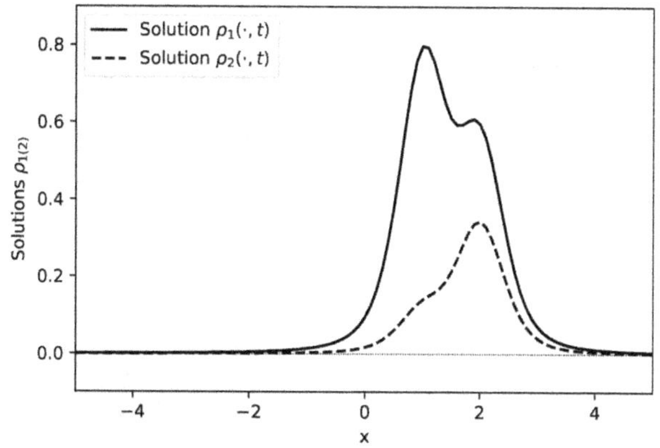

Figure 7.3 Behavior of the solutions $\rho_{1(2)}$ from Eqs. (7.34) and (7.35) ($\mathscr{D}_{\mu_1} = 1$, $\mathscr{D}_{\mu_2} = 1$, $k_1 = 2$, and $k_{22} = 1$).

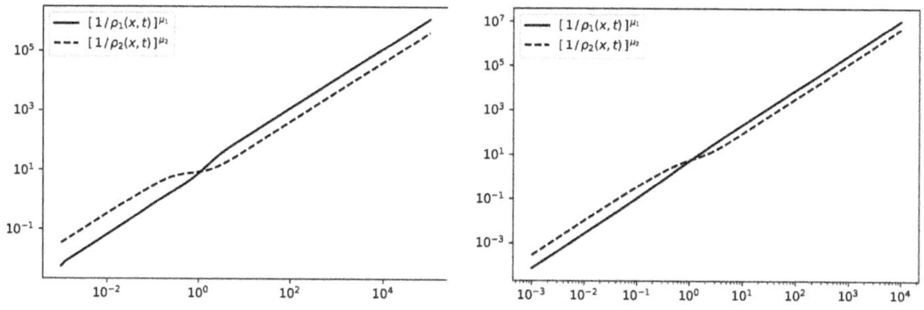

(a) $\mu_1 = \mu_2 = 3/2$ (b) $\mu_1 = \mu_2 = 5/4$

Figure 7.4 Behavior of $[1/\rho_1(x,t)]^{\mu_1}$ and $[1/\rho_2(x,t)]^{\mu_2}$ from Eqs. (7.34) and (7.35) ($\mathscr{D}_{\mu_1} = 1$, $\mathscr{D}_{\mu_2} = 1$, $k_1 = 2$, and $k_{22} = 1$).

This set of equations is interesting since the first one spreads the particles in terms of a nonlinear diffusion, which may have a distribution with compact or long-tailed behavior. Since Eqs. (7.36) and (7.37) are uncoupled, we first solve Eq. (7.36) by considering the change $\rho_1(x,t) = e^{-k_1 t}\bar{\rho}_1(x,t)$. Hence, Eq. (7.36) can be rewritten as

$$\frac{\partial}{\partial t}\bar{\rho}_1(x,t) = \bar{\mathscr{D}}_1(t)\frac{\partial^2}{\partial x^2}\bar{\rho}_1^{\nu}(x,t), \tag{7.38}$$

where $\bar{\mathscr{D}}_1(t) = \mathscr{D}_2(t)e^{(1-\nu)kt}$. The solution for this equation is found by considering that $\bar{\rho}_1(x,t)$ is a scaled function, that is,

$$\bar{\rho}_1(x,t) = \frac{1}{\Phi_1(t)}\mathscr{P}\left(\frac{x}{\Phi_1(t)}\right). \tag{7.39}$$

Substituting Eq. (7.39) in Eq. (7.38), we show that Eq. (7.38) can be separated into two equations, one for $\Phi_1(t)$ and another to $\mathscr{P}(z)$ (with $z = x/\Phi_1(t)$), as follows:

$$[\Phi_1(t)]^{\nu}\frac{d}{dt}\Phi_1(t) = \tilde{k}\bar{\mathscr{D}}_1(t), \tag{7.40}$$

$$-\tilde{k}\frac{d}{dz}[z\mathscr{P}(z)] = \frac{d^2}{dz^2}[\mathscr{P}(z)]^{\nu}, \tag{7.41}$$

where $\tilde{k}$ is a constant of separation. The solution for Eq. (7.40) is given by

$$\frac{\Phi_1(t)}{\Phi_1(0)} = \left\{1 + \frac{1+\nu}{[\Phi_1(0)]^{1+\nu}}\int_0^t dt'\bar{D}_1(t')\right\}^{\frac{1}{1+\nu}}. \tag{7.42}$$

It is interesting to note that $\Phi_1(0) \neq 0$ is related to an initial condition characterized by a distribution with an initial shape. The solution for Eq. (7.41) depends on ν and it is given by

$$\mathscr{P}(z) = \exp_q\left[-\frac{\tilde{k}}{2\nu}z^2\right], \tag{7.43}$$

where $\nu = 2 - q$ (see Figure 7.5) and $\exp_q[z]$ is the q-exponential presented in the Tsallis framework, defined as

$$\exp_q[z] = \begin{cases} (1+(1-q)z)^{\frac{1}{1-q}} & \text{if } z \geq -1/(1-q); \\ 0 & \text{if } z < -1/(1-q). \end{cases} \tag{7.44}$$

Note that there is a cut-off function in order to retain the probabilistic interpretation of the solution obtained for Eq. (7.41). It is worth to mentioning that Eq. (7.43) may also be obtained from the Tsallis entropy

$$S_q = \frac{1 - \int_{-\infty}^{\infty} dz\,[\mathscr{P}(z)]^q}{q-1} \tag{7.45}$$

by applying the principle of maximum entropy with suitable constraints. Therefore, the solutions given by Eq. (7.43) may exhibit a compact or a long-tailed behavior depending on the choice of the parameter ν_1. In particular, for $0 < \nu < 1$ (or $1 < q < 2$), the solutions may be asymptotically connected with the Lévy distributions [40, 41]. In this scenario, the solution for Eq. (7.36) is given by

$$\rho_1(x,t) = e^{-k_1 t}\frac{1}{\Phi_1(t)}\exp_q\left[-\frac{k}{2\nu}\left(\frac{|x|}{\Phi_1(t)}\right)^2\right], \tag{7.46}$$

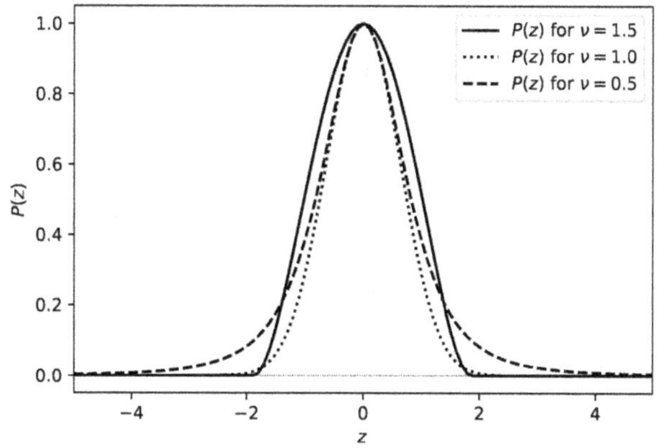

Figure 7.5 This figure illustrates the behavior of $\mathscr{P}(z)$ versus z for different values of ν.

and the mean square displacement related to Eq. (7.46) for $k_1 = 0$ is $\langle (x - \langle x \rangle)^2 \rangle \propto t^{2/(1+\nu)}$, typical of an anomalous diffusion. The solution for Eq. (7.37) can be found by applying integral transforms and the Green function approach. In this case, it is given by

$$\rho_2(x,t) = \int_{-\infty}^{\infty} dx' \mathscr{G}_+(x,x',t)\varphi_2(x') + k_1 \int_0^t dt' \int_{-\infty}^{\infty} dx' \mathscr{G}_+(x,x',t-t')\rho_1(x',t'), \quad (7.47)$$

with the Green function given by Eq. (7.26). Notice also that Eq. (7.47) may present different diffusive regimes, which will depend on the choice of $\mu(t)$ and the parameter $\nu_1 = \nu$. In particular, for the initial condition $\varphi_2(x) = 0$, the behavior of the system for small times is governed by Eq. (7.36); for long times, the system is governed by Eq. (7.37).

Although we have investigated specific cases here, we will discuss an arbitrary scenario in the next section by means of a numerical approach.

7.3 NUMERICAL METHOD

In order to obtain numerical solutions for Eqs. (7.1) and (7.2), we utilize a numerical approach based on finite differences, which is one of techniques available for searching for numerical solutions for linear and nonlinear partial differential equations (PDEs) as well as fractional partial differential equations (FPDEs). This approach employs suitable types of discretizations for each term of the equation (or system of equations) in a grid of points of its domain. For numerical analysis of PDEs, there exists a large and well-known theory available in the literature (see, e.g., [42,43]). In the case of FPDEs — where fractional derivatives are considered — some numerical methods are displayed in the literature (see, e.g., [28,44–47]).

In this section, we analyze two scenarios of FPDEs: the first considers linear equations and the second the nonlinear ones. Such scenarios can be solved by applying some methods based on finite differences. In the linear case of Eqs. (7.1) and (7.2), we apply the numerical method described in Theorem 7 of Ref. [46]; to solve the nonlinear case, we adapt the numerical method given in Theorem 3.1 of Ref. [47].

Let k and h be two positive real numbers called time-step and space-step, respectively. The exact solution $\rho_{1(2)}(x,t)$ when evaluated in the grid point (x_j,t^n) is denoted by $\rho_{1(2)}(x_j,t^n)$, or even $\rho_{1(2)}{}_j^n$. Here, we consider that $0 \leq j \leq M, 0 \leq n \leq N$, x_0 and x_M denote the boundary values of the spatial domain (artificial boundaries in our domain $\mathbb{R}$), and $t^N = t$ denotes the final time. We utilize the notation $v_{1(2)}(x_j,t^n)$, briefly, $v_{1(2)}{}_j^n$, for the approximation of the exact solution in the grid points. Therefore, our system of Eqs. (7.1) and (7.2) reads

$$\frac{\partial}{\partial t} v_1(x_j,t^n) = \mathscr{D}_{\mu_1}(t^n) \frac{\partial^{\mu_1}}{\partial |x|^{\mu_1}} v_1^{\nu_1}(x_j,t^n) + \mathscr{R}_1(v_1,v_2,t^n), \tag{7.48}$$

$$\frac{\partial}{\partial t} v_2(x_j,t^n) = \mathscr{D}_{\mu_2}(t^n) \frac{\partial^{\mu_2}}{\partial |x|^{\mu_2}} v_2^{\nu_2}(x_j,t^n) + \mathscr{R}_2(v_1,v_2,t^n). \tag{7.49}$$

The problem is to decide how to discretize each term of Eqs. (7.48) and (7.49) in such a way that the proposed numerical method is convergent. This will be presented in the following subsections: in subsection 7.3.1, we consider the linear case, that is, $\nu_1 = \nu_2 = 1$; and in subsection 7.3.2, nonlinear cases (where $\nu_1, \nu_2 > 0$) are investigated.

7.3.1 LINEAR CASE

In this subsection, as previously mentioned, we study the linear case ($\nu_1 = \nu_2 = 1$). We also consider the reaction terms $\mathscr{R}_1$ and $\mathscr{R}_2$ given by $\mathscr{R}_1(\rho_1,\rho_2,t) = k_{11}\rho_1 + k_{12}\rho_2$ and $\mathscr{R}_2(\rho_1,\rho_2,t) = k_{21}\rho_1 + k_{22}\rho_2$, where k_{11}, k_{12}, k_{21}, and k_{22} are constants. For simplicity, we consider that the diffusion coefficients $\mathscr{D}_{\mu_1}(t)$ and $\mathscr{D}_{\mu_2}(t)$ are equal to unity, yielding the following equations

$$\frac{\partial}{\partial t}\rho_1(x,t) = \frac{\partial^{\mu_1}}{\partial |x|^{\mu_1}}\rho_1(x,t) + k_{11}\rho_1(x,t) + k_{12}\rho_2(x,t), \tag{7.50}$$

$$\frac{\partial}{\partial t}\rho_2(x,t) = \frac{\partial^{\mu_2}}{\partial |x|^{\mu_2}}\rho_2(x,t) + k_{21}\rho_1(x,t) + k_{22}\rho_2(x,t). \tag{7.51}$$

The decision of how to discretize each term of Eqs. (7.50) and (7.51) is simple for the terms without differential operator: it suffices to take the exact values for these terms, that is, $\rho_{1(2)}(x_j,t^n) \approx v_{1(2)}{}_j^n$. For the term having the first-order time-derivative, it is sufficient to approximate the time-derivative for the standard backward difference:

$$\frac{\partial}{\partial t}\rho_{1(2)}(x,t) \approx \frac{\rho_{1(2)}(x_j t^{n+1}) - \rho_{1(2)}(x_j,t^n)}{k} = \frac{\rho_{1(2)}{}_j^{n+1} - \rho_{1(2)}{}_j^n}{k}. \tag{7.52}$$

For the term with the Riesz-Welly fractional operator, a first and natural idea is to utilize the standard Grünwald-Letnikov approximation given by ([28,45,46]):

$$\frac{\partial^{\mu_{1(2)}}}{\partial|x|^{\mu_{1(2)}}}\rho_{1(2)}(x,t) \approx \frac{1}{h^{\mu_{1(2)}}}\sum_{i=0}^{M}(-1)^i\frac{\Gamma(i-\mu_{1(2)})}{\Gamma(-\mu_{1(2)})\Gamma(i+1)}\rho_{1(2)}(x-ih,t), \quad (7.53)$$

when we utilize the notation

$$w_i^{\mu_{1(2)}} := (-1)^i\binom{\mu_{1(2)}}{i},$$

obtaining therefore

$$\frac{\partial^{\mu_{1(2)}}}{\partial|x|^{\mu_{1(2)}}}\rho_{1(2)}(x,t) \quad \approx \quad h^{-\mu_{1(2)}}\sum_{i=0}^{M}w_i^{\mu_{1(2)}}\rho_{1(2)}(x-ih,t)\,. \quad (7.54)$$

However, a scheme type Euler method constructed with these approximations (explicit or implicit) is unstable (see Propositions 2.1 and 2.3 in Ref. [46]) and, consequently, it does not work in our study of Eqs. (7.50) and (7.51). In order to overcome this problem of instability, it is necessary to consider the shifted Grünwald formula instead of considering the standard Grünwald-Letnikov approximation (see Theorems 2.4 and 2.7, Ref. [46]). Thus, we utilize the following equation

$$\frac{\partial^{\mu_{1(2)}}}{\partial|x|^{\mu_{1(2)}}}\rho_{1(2)}(x,t) \quad \approx \quad h^{-\mu_{1(2)}}\sum_{i=0}^{M+1}w_i^{\mu_{1(2)}}\rho_{1(2)}(x-(i-p)h,t), \quad (7.55)$$

where p is a non-negative integer. For $1<\mu_{1(2)}\leq 2$, the best choice for p is $p=1$ (see [46]).

To solve numerically Eqs. (7.50) and (7.51), based on all these previous choices, we propose semi-implicit and implicit schemes of accuracy order $O(k)+O(h)$. For the semi-implicit, we have

$$\frac{v_{1\,j}^{n+1}-v_{1\,j}^{n}}{k} = h^{-\mu_1}\sum_{i=0}^{j+1}w_i^{\mu_1}v_{1\,(j+1)-i}^{n+1}+k_{11}v_{1\,j}^{n}+k_{12}v_{2\,j}^{n}, \quad (7.56)$$

$$\frac{v_{2\,j}^{n+1}-v_{2\,j}^{n}}{k} = h^{-\mu_1}\sum_{i=0}^{j+1}w_i^{\mu_1}v_{2\,(j+1)-i}^{n+1}+k_{21}v_{1\,j}^{n}+k_{22}v_{2\,j}^{n}, \quad (7.57)$$

that can be rewritten as

$$-\lambda w_0^{\mu_1}v_{1\,j+1}^{n+1}+(1-\lambda w_1^{\mu_1})v_{1\,j}^{n+1}-\lambda\sum_{i=2}^{j+1}w_i^{\mu_1}v_{1\,(j+1)-i}^{n+1}$$
$$=(1+k\,k_{11})\,v_{1\,j}^{n}+k\,k_{12}v_{2\,j}^{n}, \quad (7.58)$$

$$-\lambda w_0^{\mu_2}v_{2\,j+1}^{n+1}+(1-\lambda w_1^{\mu_2})v_{2\,j}^{n+1}-\lambda\sum_{i=2}^{j+1}w_i^{\mu_2}v_{2\,(j+1)-i}^{n+1}$$
$$=k\,k_{21}v_{1\,j}^{n}+(1+k\,k_{22})\,v_{2\,j}^{n}, \quad (7.59)$$

where $\lambda = k/h^{\mu_{1(2)}}$ is a positive real number. For the implicit method, we get

$$-\lambda w_0^{\mu_1} v1_{j+1}^{n+1} + (1-\lambda w_1^{\mu_1}) v1_j^{n+1} b - \lambda \sum_{i=2}^{j+1} w_i^{\mu_1} v1_{(j+1)-i}^{n+1} = v1_j^n + k\,k_{11} v1_j^{n+1} + k\,k_{12} v2_j^{n+1}\,, \tag{7.60}$$

$$-\lambda w_0^{\mu_2} v2_{j+1}^{n+1} + (1-\lambda w_1^{\mu_2}) v2_j^{n+1} - \lambda \sum_{i=2}^{j+1} w_i^{\mu_2} v2_{(j+1)-i}^{n+1} = v2_j^n + k\,k_{21} v1_j^{n+1} + k\,k_{22} v2_j^{n+1}, \tag{7.61}$$

where the coupled terms are considered in the $(n+1)$-time.

In order to guarantee the convergence of the method and to avoid the error propagation in the iteration process, we propose the following form to iterate these schemes: in each time-step, we first solve the semi-implicit system of equations, Eqs. (7.58) and (7.59), to evaluate the values of $v_{1(2)}{}_j^{n+1}$ for using them in the coupled terms in the implicit schemes defined by Eqs. (7.60) and (7.61); then we return one time-step and solve the system (7.60) and (7.61) to obtain the desired values of the solutions $v_{1(2)}{}_j^{n+1}$, and so on.

Equations (7.60) and (7.61), together with the (artificial) boundary conditions $v_0^0 = v_M^0 = 0$, form linear systems of equations of the type $A_{1(2)}\vec{v}_{1(2)}^{\,n+1} = \vec{V}_{1(2)}^{\,n+1}$, where

$$\begin{aligned} \vec{V}_1^{\,n+1} &= v1_j^n + k\,k_{11} v1_j^{n+1} + k\,k_{12} v2_j^{n+1}, \quad 0 \le j \le M\,, & (7.62)\\ \vec{V}_2^{\,n+1} &= v2_j^n + k\,k_{21} v1_j^{n+1} + k\,k_{22} v2_j^{n+1}, \quad 0 \le j \le M\,, & (7.63) \end{aligned}$$

and $A_{1(2)} = [A_{1(2)_{i,j}}]$ is the matrix of coefficients of order $(M+1)$, given by the sum of a lower triangular with a super-diagonal matrices, described by

$$A_{1(2)_{i,j}} = \begin{cases} 0, & if \;\; j \ge i+2\,; \\ -\lambda w_0^{\mu_{1(2)}} & if \;\; j = i+1\,; \\ 1-\lambda w_1^{\mu_{1(2)}} & if \;\; j = i\,; \\ -\lambda w_i^{\mu_{1(2)}} & if \;\; j \le i-1\,, \end{cases} \tag{7.64}$$

for $0 \le i, j \le M$.

Remark 7.2: *The coefficients of the matrix A are similar to those that appear in the proof of Theorem 7.1, except by the coefficients $A_{0,0} = 1$ and $A_{0,j} = 0$ for $j = 1, 2, \cdots, M$ and $A_{M,M} = 1$ and $A_{M,j} = 0$ for $j = 0, 1, \cdots, M-1$ considered in Ref. [46]. Since our domain is $\mathbb{R}$, it suffices to take the boundary of domain sufficiently away of the interest points.*

The coefficients $w_i^{\mu_{1(2)}}$, for $i = 1, 2, 3, \cdots$, can be computed by using the following recurrence relation:

$$w_0^{\mu_{1(2)}} = 1; \;\; w_i^{\mu_{1(2)}} = \left(1 - \frac{\mu_{1(2)}+1}{i}\right) w_{i-1}^{\mu_{1(2)}}, \text{ for } i = 1, 2, 3, \cdots. \tag{7.65}$$

For $1 < \mu_{1(2)} \leq 2$, the unique negative coefficient in the sequence $\{w_i^{\mu_{1(2)}}\}$ is $w_1^{\mu_{1(2)}} = -\mu_{1(2)}$.

To generate the figures, given in the sequence, we used the implicit schemes (7.60) and (7.61) together with the initial profiles of Gaussian type $\rho_{10}(x)$ with mean $\langle x \rangle_1 = 2.5$ and standard deviation $\sigma_1 = 0.8$, and $\rho_{20}(x)$ has mean $\langle x \rangle_2 = 2.0$ with standard deviation $\sigma_2 = 0.3$, given by

$$\rho_{10}(x) = \frac{2}{\sqrt{2\pi\sigma_1{}^2}} e^{-\frac{(x-\langle x \rangle_1)^2}{2\sigma_1{}^2}} \text{ and } \rho_{20}(x) = \frac{1}{2.6\sqrt{2\pi\sigma_2^2}} e^{-\frac{(x-\langle x \rangle_2)^2}{2\sigma_2{}^2}}, \quad (7.66)$$

which were implemented by using the Python™ language in the Spyder software, distributed by Anaconda®, as is the case for the other figures in this chapter. Note that the initial conditions considered in this section are not normalized in order to evidence the effects of each parameter in the solutions — order of fractional derivatives and coupling coefficients. These initial profiles were chosen to evidence the iterations which occur in the evolution process of these distributions.

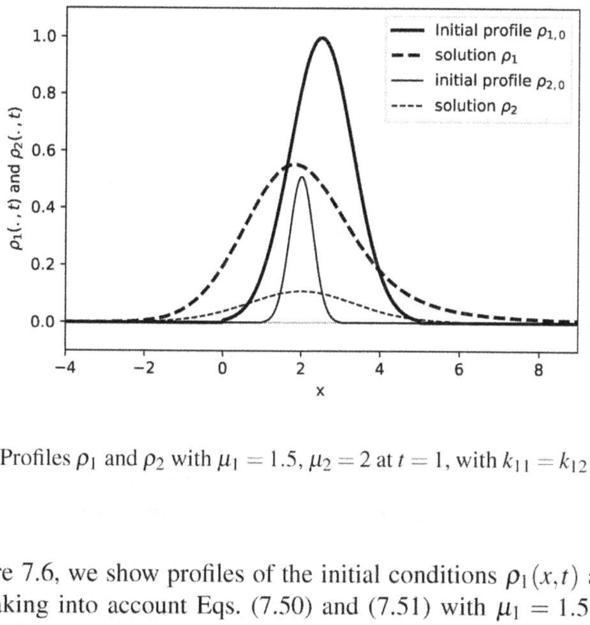

Figure 7.6 Profiles ρ_1 and ρ_2 with $\mu_1 = 1.5$, $\mu_2 = 2$ at $t = 1$, with $k_{11} = k_{12} = k_{21} = k_{22} = 0$.

In Figure 7.6, we show profiles of the initial conditions $\rho_1(x,t)$ and $\rho_2(x,t)$ for $t = 1$, by taking into account Eqs. (7.50) and (7.51) with $\mu_1 = 1.5$ and $\mu_2 = 2.0$, without coupling, that is, $k_{11} = k_{12} = k_{21} = k_{22} = 0$. The behaviors observed in the figure show that for $1 < \mu_{1(2)} \leq 2$, the long-time behaviors are essentially diffusion processes.

In Figure 7.7, we analyze the effects of the coupling terms k_{11}, k_{12}, k_{21}, and k_{22} in Eqs. (7.50) and (7.51). To evidence the effects of the coupling terms, we keep the initial profiles $\rho_{10}(x)$ and $\rho_{20}(x)$, and we consider $\mu_1 = 1.5$ and $\mu_2 = 1.9$ in $t = 1$, changing the coupling terms. It shows an irreversible reaction process for

$k_{1,1}, k_{22} < 0$, and $k_{12} = k_{21} = 0$, where $\rho_1(x,t)$ and $\rho_2(x,t)$ are consumed in this process. Since $k_{12} = k_{21} = 0$, these processes are uncoupled, and each equation (Eqs. (7.60) and (7.61)) evolves independently over time.

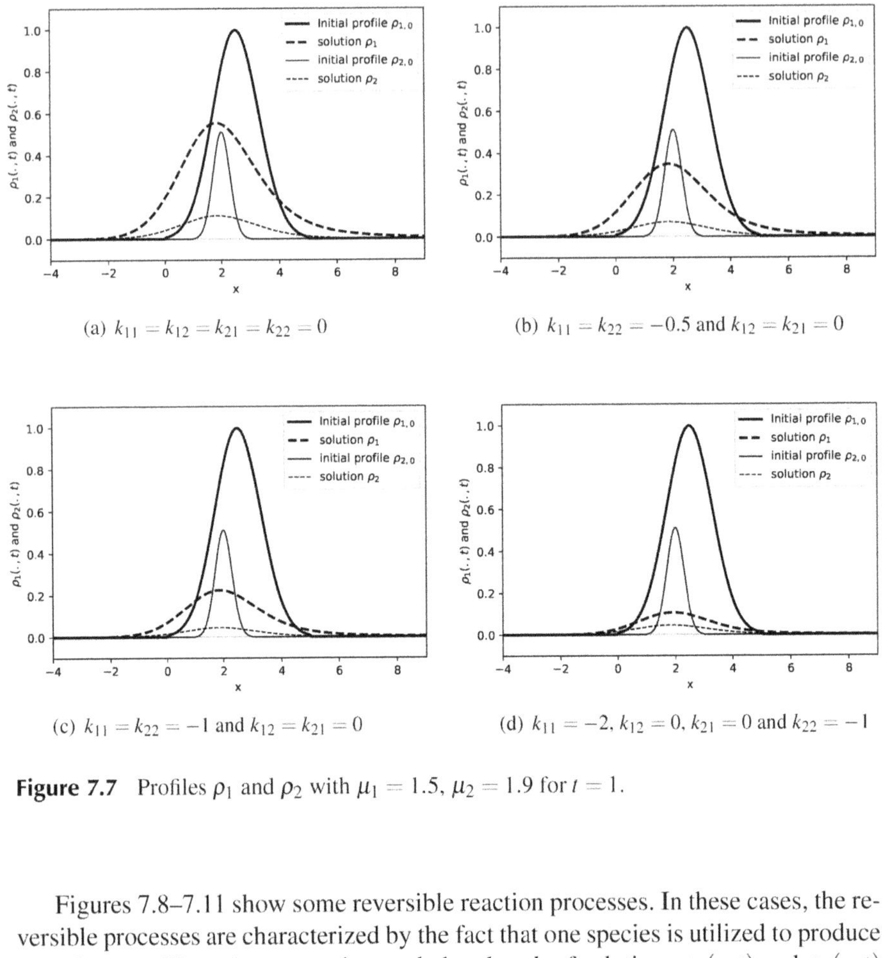

(a) $k_{11} = k_{12} = k_{21} = k_{22} = 0$

(b) $k_{11} = k_{22} = -0.5$ and $k_{12} = k_{21} = 0$

(c) $k_{11} = k_{22} = -1$ and $k_{12} = k_{21} = 0$

(d) $k_{11} = -2$, $k_{12} = 0$, $k_{21} = 0$ and $k_{22} = -1$

Figure 7.7 Profiles ρ_1 and ρ_2 with $\mu_1 = 1.5$, $\mu_2 = 1.9$ for $t = 1$.

Figures 7.8–7.11 show some reversible reaction processes. In these cases, the reversible processes are characterized by the fact that one species is utilized to produce the other one. Here, the system is coupled and each of solutions $\rho_1(x,t)$ and $\rho_2(x,t)$ influences the evolution of the other.

In Figs. 7.8 and 7.11, we can see the behaviours of $\rho_1(x,t)$ and $\rho_2(x,t)$, where $k_{11} = -k_{21}$, $k_{22} = -k_{21}$, and $k_{11}, k_{22} < 0$ (with different values). If the values of the coupling coefficients are symmetrical and not large (as in Figures 7.8 and 7.9), such process (where one species is utilized to produce the other) occurs until both species reach the equilibrium value. After this fact, both species evolve together and both have total dissipation for long times (see Figure 7.9). In Figure 7.10, we show the behavior of $1/[\rho_{1(2)}(x,t)]^{\mu_{1(2)}}$ in order to illustrate the behavior related to the spreading of the distributions $\rho_1(x,t)$ and $\rho_2(x,t)$. In particular, it is possible to observe in Fig. 7.10(a) that the initial spreading related to each distribution is different and for long periods the same behavior is obtained. Different behavior is exhibited in Fig. 7.10(b) for different values of $\mu_{1(2)}$.

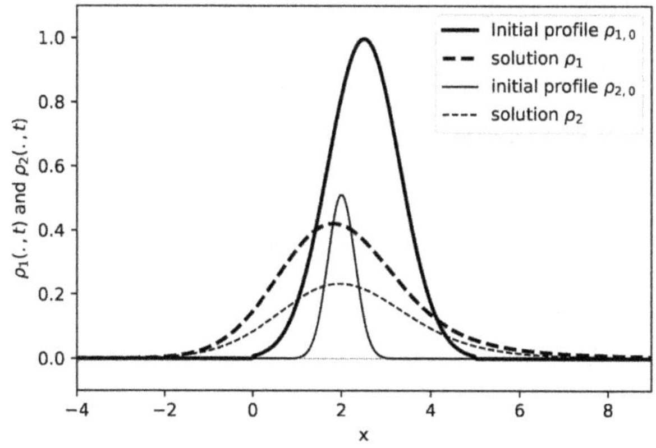

(a) $k_{11} = k_{22} = -0.5$ and $k_{12} = k_{21} = 0.5$.

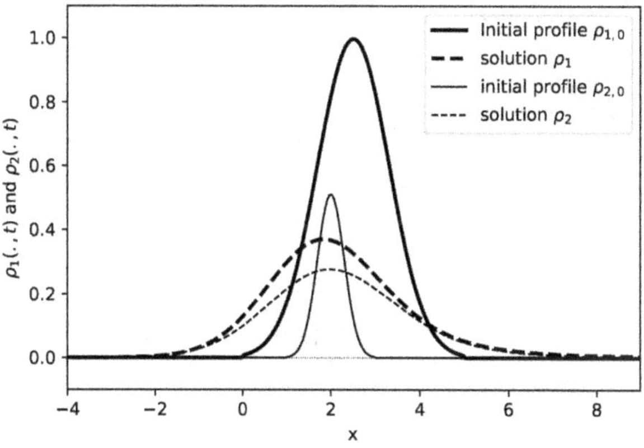

(b) $k_{11} = k_{22} = -1.0$ and $k_{12} = k_{21} = 1.0$.

Figure 7.8 Profiles ρ_1 and ρ_2 with $\mu_1 = 1.5$, $\mu_2 = 1.9$ at $t = 1$.

On the other hand, if the values of the coupling coefficients are different and not large (as in Figure 7.11), the process does not necessarily attain the equilibrium value between them, which implies that one species can increase more than the other, and therefore dominate the environment.

A third scenario that can occur is when one species (or both) increases: this situation can occur if the coefficients $k_{11}, k_{22} > 0$ or if the coupling coefficients $k_{12}, k_{21} > 0$

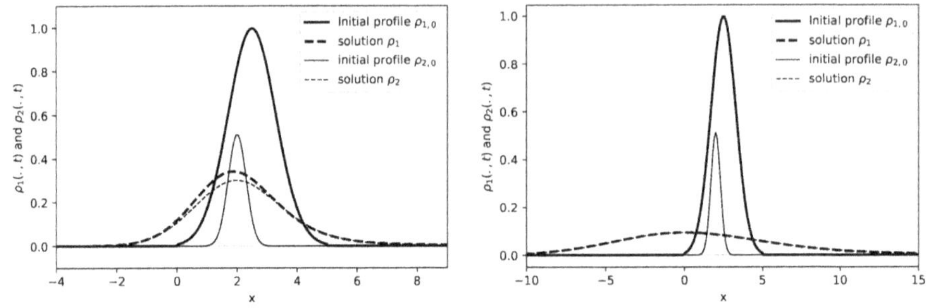

(a) $k_{11} = k_{22} = -2.0$ and $k_{12} = k_{21} = 2.0$, $t = 1$. (b) Profiles ρ_1 and ρ_2 with $\mu_1 = 1.5$, $\mu_2 = 1.9$ at $t = 10$.

Figure 7.9 Evolution profiles ρ_1 and ρ_2 at long-time, with $k_{11} = k_{22} = -2$ and $k_{12} = k_{21} = 2$.

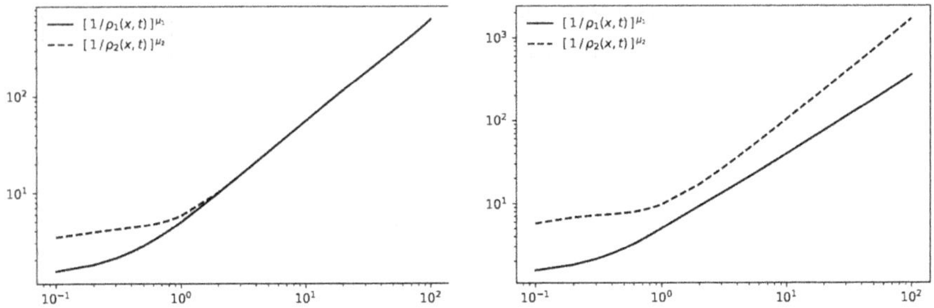

(a) $\mu_1 = 1.5$, $\mu_2 = 1.5$ (b) $\mu_1 = 1.5$, $\mu_2 = 1.9$

Figure 7.10 Behavior of $[1/\rho_1(x,t)]^{\mu_1}$ and $[1/\rho_2(x,t)]^{\mu_2}$, with $k_{11} = k_{22} = -2.0$, $k_{12} = k_{21} = 2.0$

are very large, because in these cases, the dominant behaviors of Eqs. (7.50) and (7.51) are linear and proportional to the absolute value of the solution — with proportionally constant greater than one. This behavior can be seen in Figure 7.12, where ρ_1 does not depend on ρ_2, and ρ_2 is proportional to ρ_1; this fact causes in ρ_2 a competition between increasing and decreasing (in $t = 1$ ρ_2 increases).

Another approach to solve numerically the system of linear equations (7.50) - (7.51) is by utilizing Galerkin approximation and weak asymptotic solution. More precisely, one can adapt the techniques utilized in [27] to deal with advection-dispersion equation.

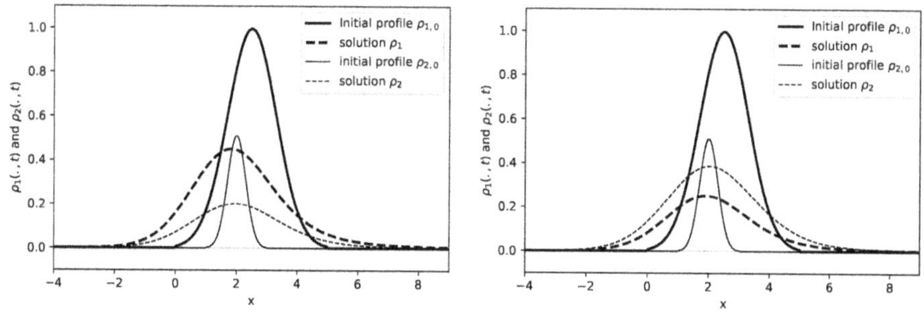

(a) $k_{11} = -1.0$, $k_{21} = 1.0$, $k_{12} = 2.0$, and $k_{22} = -2.0$. (b) $k_{11} = -2.0$, $k_{21} = 2.0$, $k_{12} = 1.0$, and $k_{22} = -1.0$.

Figure 7.11 Profiles ρ_1 and ρ_2 with $\mu_1 = 1.5$, $\mu_2 = 1.9$ at $t = 1$.

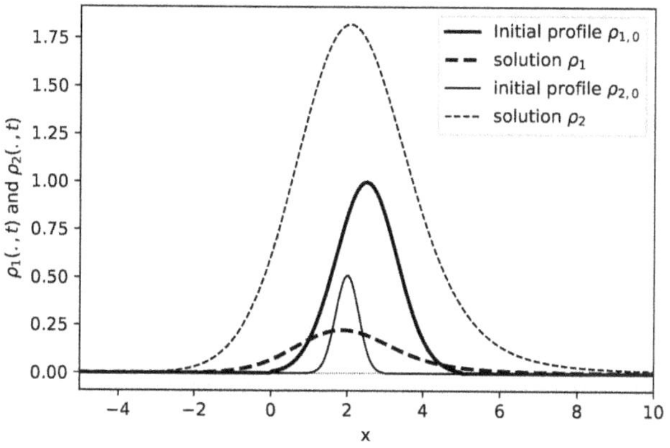

Figure 7.12 Profiles ρ_1 and ρ_2; $\mu_1 = 1.5$, $\mu_2 = 1.9$; $t = 1$; with $k_{11} = -1.0$, $k_{12} = 0.0$, $k_{21} = 10.0$, and $k_{22} = -1.0$.

7.3.2 NONLINEAR CASE

In this subsection, we propose a numerical method to solve the nonlinear system (7.1) and (7.2). Our implicit scheme is based on Theorem 7 in [46] and Theorem 3.1 in [47]. To present the approach introduced in [47], we consider first the simplest equation

$$\frac{\partial}{\partial t}\rho(x,t) = \mathscr{D}(t)\frac{\partial^{\mu}}{\partial |x|^{\mu}}\rho^{\nu}(x,t), \tag{7.67}$$

in the real interval $L \leq x \leq R$ and with $t > 0$, where $\mathscr{D}(t)$ denotes non-negative and bounded time-dependent diffusion coefficients. In Eq. (7.67), the functions under the derivatives action are different (by an exponent). To "correct" such difference, we consider $\nu = 1+\delta$, with $\delta > -1$, after multiplying Eq. (7.67) by $(1+\delta)\rho^{\delta}(x,t)$:

$$(1+\delta)\rho^{\delta}(x,t).\frac{\partial}{\partial t}\rho(x,t) = (1+\delta)\rho^{\delta}(x,t).\mathscr{D}(t)\frac{\partial^{\mu}}{\partial|x|^{\mu}}\rho^{\nu}(x,t). \quad (7.68)$$

Under this adjustment, we get

$$\frac{\partial}{\partial t}\rho^{1+\delta}(x,t) = (1+\delta)\rho^{\delta}(x,t)\,\mathscr{D}(t)\frac{\partial^{\mu}}{\partial|x|^{\mu}}\rho^{\nu}(x,t), \quad (7.69)$$

or

$$\frac{\partial}{\partial t}\rho^{\nu}(x,t) = \nu\rho^{\delta}(x,t)\,\mathscr{D}(t)\frac{\partial^{\mu}}{\partial|x|^{\mu}}\rho^{\nu}(x,t). \quad (7.70)$$

The functions under the derivatives action given in (7.70) are equal; hence, the approach for the linear case shown in Theorem 2.7 in [46] has a chance to be applied, since, in each time step, we can evaluate properly the weights $\nu\rho^{\delta}(x,t)$. In the grid points, the weights $\nu\rho^{\delta}(x_j,t^n)$ are denoted by $\delta_j^n := \nu\left(v^{\delta}\right)_j^n = \nu\, v^{\delta}(x_j,t^n)$, and the diffusion coefficient $\mathscr{D}(t^n)$ is denoted by $\mathscr{D}^n$. The main result is presented in the sequence.

Theorem 7.1: The implicit Euler method

$$-\lambda\mathscr{D}^n\delta_j^{n+1}w_0^{\mu}(v^{\nu})_{j+1}^{n+1} + (1-\lambda\mathscr{D}^n\delta_j^{n+1}w_1^{\mu})(v^{\nu})_j^{n+1} - \lambda\mathscr{D}^n\delta_j^{n+1}\sum_{i=2}^{j+1}w_i^{\mu}(v^{\nu})_{(j+1)-i}^{n+1} = (v^{\nu})_j^n, \quad (7.71)$$

where $\lambda = \frac{k\mathscr{D}^n}{h^{\mu}}$ and $\delta_j^n = \nu(v^{\delta})_j^n$, to solve Eq. (7.67), where $1 < \mu \le 2$, in the real interval $L \le x \le R$, with non-negative bounded initial condition $\rho(x,0) = \rho_0$ and boundary conditions $\rho(x=L,t) = 0 = \rho(x=R,t)$ for all $t \ge 0$, based on the shifted Grünwald approximation as in Eq. (7.55), where $h = (R-L)/M$, is consistent and unconditionally stable.

Proof. This proof adapts the argument utilized in the proof of Theorem 2.7 in [46]. For simplicity, and without loss of generality, we consider $\mathscr{D}^n = 1$, because, since $\mathscr{D}(t) > 0$ for all $t > 0$, the coefficients $\mathscr{D}^n$ do not cause any difficulty.

Initially, since the left boundary condition is equal to zero, it is possible to take an extension of $\rho(x,t)$ by zero for $x \le L$ and $t \ge 0$. Then, the Riesz-Welly fractional operator can be approximated by the sum approximation of shifted Grünwald type, given by

$$\frac{\partial^{\mu}}{\partial|x|^{\mu}}\rho(x,t) = \lim_{M\to\infty} h^{-\mu}\sum_{i=0}^{M} w_i^{\mu}\rho(x-ih,t), \quad (7.72)$$

where $w_i^{\mu} := (-1)^i\binom{\mu}{i}$ and $h = (R-L)/M$. Thus, adapting Theorem 2.4 in [46] to our context, we conclude that the accuracy order of this approximation is $O(h)$, and the implicit method shown in (7.71) is consistent with Eq. (7.70), having accuracy order $O(h)+O(k)$.

Applying in (7.70) discretizations properly (backward difference in time-derivative and shifted Grünwald in fractional space-derivative), we get

$$\frac{(v^{\nu})_j^{n+1}-(v^{\nu})_j^n}{k} = \delta_j^{n+1}h^{-\mu}\sum_{i=0}^{j+1} w_i^{\mu}(v^{\nu})_{(j+1)-i}^{n+1}\,. \tag{7.73}$$

Taking $\lambda = kh^{-\mu}$ ($\lambda > 0$), we obtain

$$(v^{\nu})_j^{n+1}-(v^{\nu})_j^n = \lambda\,\delta_j^{n+1}\sum_{i=0}^{j+1} w_i^{\mu}(v^{\nu})_{(j+1)-i}^{n+1}\,. \tag{7.74}$$

After a rearrangement of terms, we obtain

$$\begin{aligned}-\lambda\,\delta_j^{n+1}w_0^{\mu}(v^{\nu})_{j+1}^{n+1}+(1-\lambda\,\delta_j^{n+1}w_1^{\mu})(v^{\nu})_j^{n+1} &- \lambda\,\delta_j^{n+1}\sum_{i=2}^{j+1} w_i^{\mu}(v^{\nu})_{(j+1)-i}^{n+1}\\ &= (v^{\nu})_j^n\,.\end{aligned} \tag{7.75}$$

Equation (7.75) is a linear system of the type $A(\vec{v^{\mu}})^{n+1} = (\vec{v^{\mu}})^n$, where $A=[A_{i,j}]$ is the matrix of coefficients given by the sum of a lower triangular with a super-diagonal matrix, which is described by

$$A_{0,j} = \begin{cases} 1, & if \quad j=0\,;\\ 0, & if \quad 1\le j\le M\,,\end{cases} \tag{7.76}$$

$$A_{i,j} = \begin{cases} 0, & if \quad j\ge i+2\,, \quad 1\le j\le M-1\,;\\ -\lambda\,\delta_i^{n+1}w_0^{\mu} & if \quad j=i+1\,, \quad 1\le j\le M-1\,;\\ 1-\lambda\,\delta_i^{n+1}w_1^{\mu} & if \quad j=i\,, \quad 1\le j\le M-1\,;\\ -\lambda\,\delta_i^{n+1}w_i^{\mu} & if \quad j\le i-1\,, \quad 1\le j\le M-1\,,\end{cases} \tag{7.77}$$

$$A_{M,j} = \begin{cases} 0, & if \quad 0\le j\le M-1\,;\\ 1, & if \quad j=M\,.\end{cases} \tag{7.78}$$

To verify that A is invertible we take $z=-1$ in the well-known formula $(1+z)^{\mu} = \sum_{k=0}^{\infty}\binom{\mu}{k}z^k$ for the Binomial expansion (see e.g. [46] or [48,49]), concluding that $\sum_{k=0}^{\infty} w_k^{\mu} = 0$. Because $0\le\rho_0<\infty$, it follows that $\max\limits_{0\le i\le M}\{|\delta_i^0|\}<\infty$; since Eq. (7.70) describes a transport-diffusion phenomenon, we have $\max\limits_{0\le i\le M}\{|\delta_i^n|\}<\infty$ for all $n\in\{1,2,\cdots,N\}$. Therefore, $\sum\limits_{k=0}^{\infty}|\delta_i^{n+1}|\,w_k^{\mu} = 0$ for all $i=0,1,\cdots,M-1$. Moreover, as for $1<\mu\le 2$ the unique negative term in the sequence $\{w_k^{\mu}\}_{k\in\mathbb{N}}$ is $w_1^{\mu}=-\mu$, and $\delta\ge 0$, then it follows that $\delta_i^n\ge 0$ for all i. Hence,

$$-\delta_i^{n+1}w_1^{\mu} = \sum_{j=0,j\neq 1}^{\infty}\delta_i^{n+1}w_j^{\mu}\,, \tag{7.79}$$

which implies

$$-\delta_i^{n+1} w_1^\mu \geq \sum_{j=0, j\neq i}^{i+1} \delta_i^{n+1} w_{i-j+1}^\mu \,. \tag{7.80}$$

Since $-\delta_i^{n+1} w_1^\mu > 0$, we obtain

$$1 - \delta_i^{n+1} w_1^\mu > \sum_{j=0, j\neq i}^{i+1} \delta_i^{n+1} w_{i-j+1}^\mu \,. \tag{7.81}$$

The inequality (7.81) means that A is strictly diagonal-dominant by rows, therefore invertible.

On the other hand, let us consider $\widetilde{\lambda}$ be an eigenvalue of A. If we choose i such that $|y_i| = \max_{0\leq j\leq M}\{|y_j|\}$, it follows that $\sum_{j=0}^{M} A_{i,j} y_j = \widetilde{\lambda} y_i$; therefore,

$$\widetilde{\lambda} \;=\; A_{i,i} + \sum_{j=0, j\neq i}^{M} A_{i,j} \frac{y_j}{y_i} \,. \tag{7.82}$$

If $i = 0$ or $i = M$, we then get $\widetilde{\lambda} = 1$. In the other cases, if we replace $A_{i,j}$ in Eq. (7.82), we obtain

$$\widetilde{\lambda} \;=\; (1 - \lambda\,\delta_i^{n+1} w_1^\mu) - \lambda\,\delta_i^{n+1} w_0^\mu \frac{y_{i+1}}{y_i} - \lambda \sum_{j=0}^{i-1} \delta_i^{n+1} w_{i-j+1}^\mu \frac{y_j}{y_i} \,, \tag{7.83}$$

or, after a rearrangement of terms:

$$\widetilde{\lambda} \;=\; 1 - \lambda\left(\delta_i^{n+1} w_1^\mu + \sum_{j=0, j\neq i}^{i+1} \delta_i^{n+1} w_{i-j+1}^\mu \frac{y_j}{y_i}\right) . \tag{7.84}$$

Since i was chosen such that $\left|\frac{y_j}{y_i}\right| \leq 1$ for all $0 \leq j \leq M$, applying Eqs. (7.80) and (7.84), since $\lambda > 0$, we conclude that $\widetilde{\lambda} \geq 1$.

Thus, each eigenvalue $\widetilde{\widetilde{\lambda}}$ of A^{-1} satisfies $\widetilde{\widetilde{\lambda}} \leq 1$, and therefore the error ε_0 in $v^0 \approx \rho_0$ results in an error ε_1 in v^1 less than (or equal to) the error ε_0, and so on. This means that the error ε_n in the n-step is bounded by the initial error, which implies that the method (7.75) is unconditionally stable. □

In order to solve the nonlinear system (7.1) and (7.2), we propose an approach for evaluating the weights δ_j^n and we use them applied to a convergent method based on Theorem 7.1 and a two-step iteration process. We next adapt the convergent method given in Theorem 7.1 in order to analyse the nonlinear system (7.48) and (7.49), with $\nu_{1(2)} > 0$ (i.e., $\delta_{1(2)} > -1$). The first and natural idea is to propose the following implicit system of numerical schemes:

$$-\lambda \mathscr{D}_{\mu_1}^{n+1} {\delta_1}_j^{n+1} w_0^{\mu_1} (v_1^{\nu_1})_{j+1}^{n+1} + (1 - \lambda \mathscr{D}_{\mu_1}^{n+1} {\delta_1}_j^{n+1} w_1^{\mu_1})(v_1^{\nu_1})_j^{n+1}$$
$$- \lambda \mathscr{D}_{\mu_1}^{n+1} \sum_{i=2}^{j+1} {\delta_1}_j^{n+1} w_i^{\mu_1} (v_1^{\nu_1})_{(j+1)-i}^{n+1}$$
$$= (v_1^{\nu_1})_j^n + k\, k_{11}\, {\delta_1}_j^{n+1} (v_1)_j^{n+1} + k\, {\delta_1}_j^{n+1} k_{12} (v_2)_j^{n+1}, \tag{7.85}$$

$$-\lambda \mathscr{D}_{\mu_2}^{n+1} {\delta_2}_j^{n+1} w_0^{\mu_2} (v_2^{\nu_2})_{j+1}^{n+1} + (1 - \lambda \mathscr{D}_{\mu_2}^{n+1} {\delta_2}_j^{n+1} w_1^{\mu_2})(v_2^{\nu_2})_j^{n+1}$$
$$- \lambda \mathscr{D}_{\mu_2}^{n+1} \sum_{i=2}^{j+1} {\delta_2}_j^{n+1} w_i^{\mu_2} (v_2^{\nu_2})_{(j+1)-i}^{n+1}$$
$$= (v_2^{\nu_2})_j^n + k\, {\delta_2}_j^{n+1} k_{21} (v_1)_j^{n+1} + k\, k_{22}\, {\delta_2}_j^{n+1} (v_2)_j^{n+1} \tag{7.86}$$

where $\lambda = \frac{k}{h^{\mu_{1(2)}}}$, $\nu_{1(2)} = 1 + \delta_{1(2)}$, $\mathscr{D}_{\mu_{1(2)}}^{n+1} = \mathscr{D}_{\mu_{1(2)}}(t^{n+1})$, and ${\delta_{1(2)}}_j^{n+1} = \nu_{1(2)}\, v_{1(2)}{}^{\delta_{1(2)}}(x_j, t^{n+1})$. A problem arises: in the schemes (7.85) and (7.86), the weights ${\delta_{1(2)}}_j^{n+1}$ are considered in the $(n+1)$-step, inserting inside the coefficients of the matrix A the unknown values of the functions $\rho_{1(2)}$ in the $(n+1)$-step, which makes invalid our attempt to solve the problem. Moreover, the values $(v_{1(2)})_j^{n+1}$ are implicit, therefore making it harder to solve the problem. To circumvent these difficulties we propose evaluating the weights and the coupled terms in the previous time-step n: ${\delta_{1(2)}}_j^n$ and $(v_{1(2)})_j^n$ in the place of ${\delta_{1(2)}}_j^{n+1}$ and $(v_{1(2)})_j^{n+1}$, respectively. Thus, the semi-implicit schemes to be solved are

$$-\lambda \mathscr{D}_{\mu_1}^{n+1} {\delta_1}_j^n w_0^{\mu_1} (v_1^{\nu_1})_{j+1}^{n+1} + (1-\lambda \mathscr{D}_{\mu_1}^{n+1} {\delta_1}_j^n w_1^{\mu_1})(v_1^{\nu_1})_j^{n+1} - \lambda \mathscr{D}_{\mu_1}^{n+1} \sum_{i=2}^{j+1} {\delta_1}_j^n w_i^{\mu_1} (v_1^{\nu_1})_{(j+1)-i}^{n+1}$$
$$= (v_1^{\nu_1})_j^n + k\, k_{11}\, {\delta_1}_j^n (v_1)_j^n + k\, {\delta_1}_j^n k_{12} (v_2)_j^n \tag{7.87}$$

$$-\lambda \mathscr{D}_{\mu_2}^{n+1} {\delta_2}_j^n w_0^{\mu_2} (v_2^{\nu_2})_{j+1}^{n+1} + (1-\lambda \mathscr{D}_{\mu_2}^{n+1} {\delta_2}_j^n w_1^{\mu_2})(v_2^{\nu_2})_j^{n+1} - \lambda \mathscr{D}_{\mu_2}^{n+1} \sum_{i=2}^{j+1} {\delta_2}_j^n w_i^{\mu_2} (v_2^{\nu_2})_{(j+1)-i}^{n+1}$$
$$= (v_2^{\nu_2})_j^n + k\, {\delta_2}_j^n k_{21} (v_1)_j^n + k\, k_{22}\, {\delta_2}_j^n (v_2)_j^n. \tag{7.88}$$

These schemes also present a problem: when replacing the weights ${\delta_{1(2)}}_j$ in the time-step $n+1$ by the weights in the time-step n (and the values of $(v_{1(2)})_j^n$ at place of $(v_{1(2)})_j^{n+1}$ in the coupled terms), a small numerical error is introduced in each step, which will be propagated in each time-step.

To get around this new difficulty, we propose computing the solutions $v_{1(2)}^{\nu_{1(2)}}$ for each time-step twice: first, we replace the weights and the values of the coupled terms in the previous time-step n, after computing $(v_{1(2)}^{\nu_{1(2)}})_j^{n+1}$ by applying Eqs. (7.87) and (7.88), and evaluate ${\delta_{1(2)}}_j^{n+1}$. Then, with these new values for ${\delta_{1(2)}}_j^{n+1}$ and with $(v_{1(2)}^{\nu_{1(2)}})_j^{n+1}$ in the coupled terms evaluated in time-step $n+1$, we return to the previous time-step n, after computing $(v_{1(2)}^{\nu_{1(2)}})_j^{n+1}$ by applying Eqs. (7.85) and (7.86), and so on.

Summarizing, the algorithm is given by:

- solve the system (7.87) and (7.88);
- evaluate ${\delta_{1(2)}}_j^{n+1}$ (and $(v_{1(2)}^{\nu_{1(2)}})_j^{n+1}$ to be utilized in the coupled terms);
- return to the n-time step, after solving the system (7.85)-(7.86).

Figures 7.13–7.16 show the influence of non-linearity in the behaviors of the solutions $\rho_{1(2)}$. For simplicity, in these cases, we consider that $\mathscr{D}_{\mu_1}(t) = 1$ and $\mathscr{D}_{\mu_2}(t) = 1$. Figure 7.17 shows the influence of the different diffusion coefficients $\mathscr{D}_{\mu_{1(2)}}(t)$ in the behaviors of the solutions $\rho_{1(2)}$.

In order to generate these figures, we consider the initial profiles used in the linear case described in Eq. (7.66): we keep the initial profile as Gaussian type $\rho_{2_0}(x)$ with mean $\langle x\rangle_1 = 2.0$ and standard deviation $\sigma_1 = 0.3$, but we multiply by 3 the initial profile $\rho_{1_0}(x)$ with $\langle x\rangle_2 = 2.5$ and $\sigma_2 = 0.8$. This change is thought to facilitate the visualization of the iterations which occurs in the evolution processes. We observe that the initial conditions considered here are not normalized in order to evidence the effects of non-linearity in the solutions.

In Fig. 7.13 we present two graphics showing the effects of ν_1 and ν_2 in the evolution profiles ρ_1 and ρ_2, respectively: in Fig. 7.13(a) one has $\mu_1 = \mu_2 = 2.0$, and in Fig. 7.13(b), $\mu_1 = 1.5$ and $\mu_2 = 1.9$. Here, the system (7.48) and (7.49) is decoupled, that is, $k_{11} = k_{12} = k_{21} = k_{22} = 0$, and the solutions are evaluated in $t = 1$. These figures show two basic effects of the nonlinearity: if the exponents $\nu_{1(2)}$ are smaller than 1, the effects of diffusive terms and of the nonlinearities jointly decrease the magnitude of the solutions; if the exponents $\nu_{1(2)}$ are greater than 1, the nonlinearities act in order to increase the magnitude of the solutions, making the diffusion process slow. If the exponents $\nu_{1(2)}$ are large, we can have a solution increasing in opposition to the diffusion process, which generates a competition that is hard to analyze.

We next comment on the effects of ν_1 ν_2 in some cases of coupling equations (with terms k_{11}, k_{12}, k_{21}, and k_{22}, and at least one non-zero).

In Fig. 7.14 we show two graphics for a coupled system in Eqs. (7.48) and (7.49), for different values of ν_1 and ν_2, with the same values $\mu_1 = 1.5$ and $\mu_2 = 1.9$, evaluated in the same time $t = 1$. In Fig. 7.14(a) we consider that $k_{11} = -1.0$, $k_{12} = 2.0$, $k_{21} = 1.0$, and $k_{22} = -2.0$. Here, we can see a reaction process where, for each value of $\nu_{1(2)}$, the corresponding profiles $\rho_{1(2)}$ maintain the relative magnitude. In Fig. 7.14(b) we consider $k_{11} = -2.0$, $k_{12} = 1.0$, $k_{21} = 2.0$, and $k_{22} = -1.0$; here, the coupling terms produce a reversible reaction process that makes the solutions ρ_2 grow up and overcome the solutions ρ_1.

Figure 7.15 shows a reversible reaction process, for the same parameters: initial profiles $\rho_{0_{1(2)}}$, derivative orders $\mu_1 = 1.5$ and $\mu_2 = 1.9$, in Eqs. (7.48) and (7.49), evaluated in $t = 1$, for a coupled system with $k_{11} = -1$, $k_{12} = 0.0$, $k_{21} = 10.0$, and

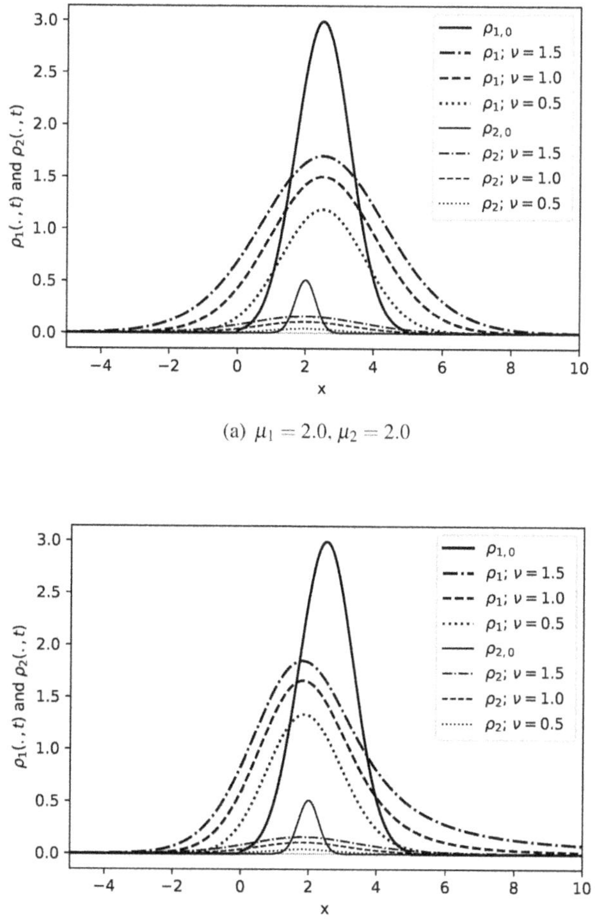

(a) $\mu_1 = 2.0, \mu_2 = 2.0$

(b) $\mu_1 = 1.5, \mu_2 = 1.9$

Figure 7.13 Profiles ρ_1 and ρ_2; with $\nu_{1(2)} = 0.5, 1.0, 1.5$ and $k_{11} = k_{12} = k_{21} = k_{22} = 0.0$; $t = 1$.

$k_{22} = -1.0$. This is an example of a case where the large value of k_{21} is implied in the growing up to species ρ_2, while species ρ_1 decreases. In this case, the reaction terms have a dominant effect on the behavior of the solution ρ_2 in Eqs. (7.48) and (7.49) for intermediate times, while for long times the behavior of the solutions is governed by the diffusive terms.

In Figure 7.16, we present an example of numerical solution for a case covered in subsection 7.2.2 by Eqs. (7.36) and (7.37), where the analytical solutions are given in Eqs. (7.46) and (7.47). The parameters chosen are: $\mu_1 = 2.0$, $\mu_2 = 1.5$, $\nu_1 = 1.5$,

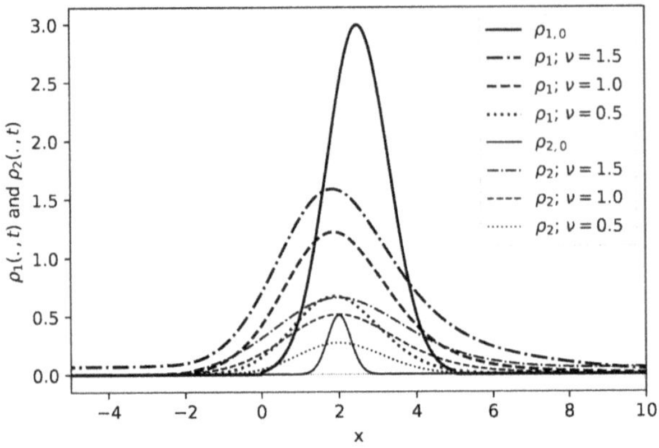

(a) $k_{11} = -1.0, k_{12} = 2.0, k_{21} = 1.0, k_{22} = -2.0.$

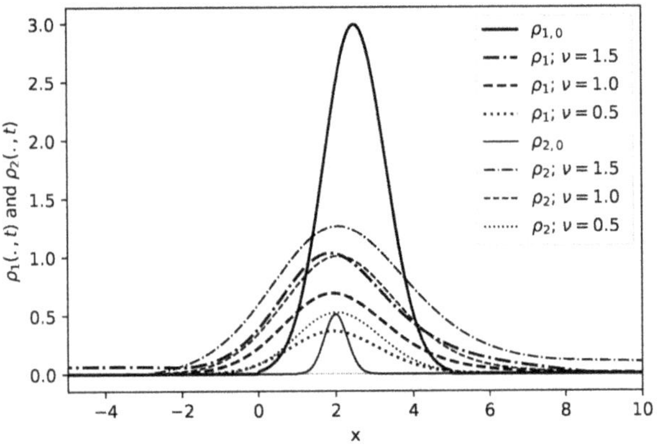

(b) $k_{11} = -2.0, k_{12} = 1.0, k_{21} = 2.0, k_{22} = -1.0.$

Figure 7.14 Profiles ρ_1 and ρ_2 with $\mu_1 = 1.5$, $\mu_2 = 1.9$ and $\nu_{1(2)} = 0.5, 1.0, 1.5$; $t = 1$.

$\nu_2 = 1.0$, and the solutions $\rho_{1(2)}$ are evaluated in $t = 1$, for a uncoupled system with $k_{11} - 1.0$, $k_{12} = 0.0$, $k_{21} = 1.0$, and $k_{22} = 0.0$.

Finally, in Figure 7.17 we show the effects of different diffusion coefficients $\mathscr{D}_{1(2)}(t)$ in the behaviors of the solutions $\rho_{1(2)}$, in $t = 5$. The coefficients considered are $\mathscr{D}_{1(2)}(t) = 1$, $\mathscr{D}_{1(2)}(t) = t^3$, and $\mathscr{D}_{1(2)}(t) = t^5$. In other words, a time dependence on the diffusion coefficient modifies the spreading of the system, which for the choices performed in Figure 7.17 leads us a super-diffusive behavior.

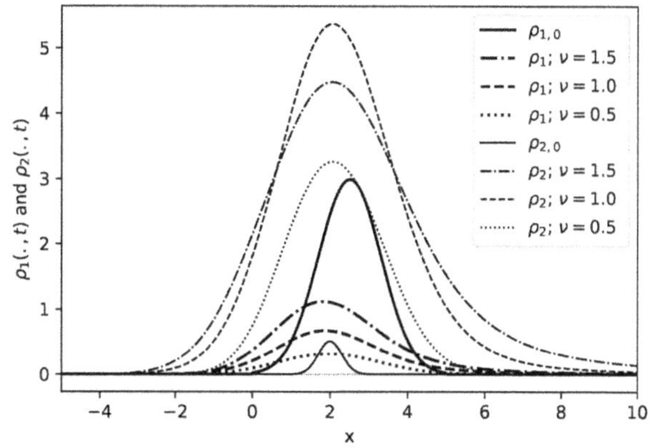

Figure 7.15 Profiles ρ_1 and ρ_2; $\mu_1 = 1.5$, $\mu_2 = 1.9$; with $\nu_{1(2)} = 0.5, 1.0$ and 1.5 and $k_{11} = 1.0$, $k_{12} = 0.0$, $k_{21} = 10.0$, $k_{22} = -1.0$; $t = 1$.

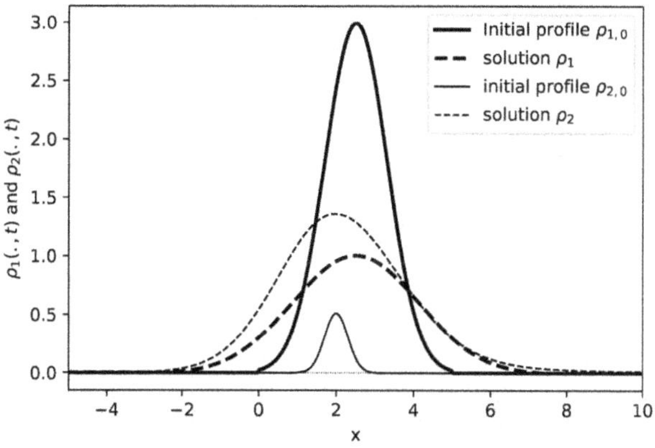

Figure 7.16 Profiles ρ_1 and ρ_2; $\mu_1 = 2.0$, $\mu_2 = 1.5$; with $\nu_1 = 1.5$, $\nu_2 = 1.0$ and $k_{11} = 1.0$, $k_{12} = 0.0$, $k_{21} = 1.0$, $k_{22} = 0.0$; $t = 1$.

7.4 FINAL REMARKS

In this chapter, we have investigated nonlinear fractional diffusion equations in the presence of reaction terms from analytical and also from numerical points of view. The reaction terms incorporated in the nonlinear fractional diffusion are linear and may cover irreversible as well as reversible reaction processes.

We have started our analytical investigation by considering the linear case with irreversible reaction terms. For this case, we have computed the Green function and we shown that the asymptotic behavior is governed by a power-law function. After

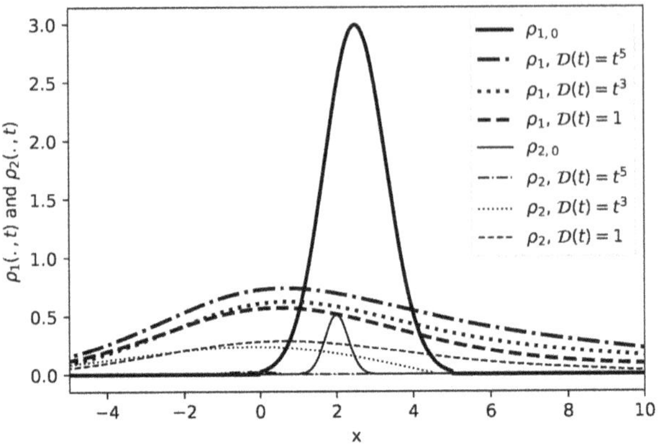

Figure 7.17 Profiles ρ_1 and ρ_2; $\mu_1 = 1.5$, $\mu_2 = 1.9$; with $\nu_1 = 1.0$, $\nu_2 = 1.5$ and $k_{11} = 1.0$, $k_{12} = 2.0$, $k_{21} = 1.0$, $k_{22} = -2.0$; $t = 5$.

this, we have considered reversible reaction terms in this context. For such case, we have also obtained the solutions in terms of Green functions. We have also shown the spreading of the solutions and the influence of the reaction terms which have a dominant effect for intermediate times. In fact, for small and long times, the diffusive terms govern these solutions. In our attempt to obtain analytical solutions for specific nonlinear cases, we have obtained solutions which are connected to the q-exponential presented in the Tsallis formalism. This feature is interesting in the sense that it allows one to give thermostatistic context for the situations investigated in this scenario. In the numerical point of view, these equations have been analyzed by utilizing schemes of implicit finite difference. For the linear case, we have applied the approach described in Theorem 7 of Ref. [46]. To solve the nonlinear case, we have adapted the numerical method given in Theorem 3.1 of Ref. [47]. The numerical procedure developed here is consistent and unconditionally stable, and therefore, convergent with accuracy order $O(h) + O(k)$. It is interesting to observe that Theorem 7.1, as well as the numerical method proposed in (7.85) and (7.86) and (7.87) and (7.88), covers the cases $0.5 \leq \nu \leq 1.5$, which are related to physical situations. For future developments, one can consider - from the analytical point of view - the presence of drift terms in Eq. (7.1) and Eq. (7.2). In terms of numerical approach, one can incorporate in Eq. (7.1) and Eq. (7.2) drift terms and/or time dependence in the fractional order of derivative. Both future approaches, when performed, should provide natural generalizations of the results shown here. We hope that the results presented here can be useful in analyzing the equations which appear in the context of anomalous diffusion.

ACKNOWLEDGMENTS

We would like to thank the Brazilian Agencies CNPq and CAPES for their financial support.

REFERENCES

1. Einstein, A. Über die von der molekularkinetischen Theorie der Wärme geforderte Bewegung von in ruhenden Flüssigkeiten suspendierten Teilchen, *Annalen der Physik* **1905**, *322*, 549–560.

2. von Smoluchowski, M. Zur kinetischen Theorie der Brownschen Molekularbewegung und der Suspensionen, *Annalen der Physik* **1906**, *326* (14), 756–780.

3. Langevin, P. Sur la théorie du mouvement brownien, *Comptes-Rendus de l'Académie des Sciences Paris* **1908**, *146*, 530.

4. McGuffee, S.R.; Elcock, A.H. Diffusion, crowding & protein stability in a dynamic molecular model of the bacterial cytoplasm, *PLoS Computational Biology* **2010**, *6*, e1000694.

5. Silva, A.T.; Lenzi, E.K.; Evangelista, L.R.; Lenzi, M.K.; da Silva, L.R. Fractional nonlinear diffusion equation, solutions and anomalous diffusion, *Physica A* **2007**, *375*, 65–71.

6. Alibaud, N.; Cifani, S.; Jakobsen, E.R. Continuous dependence estimates for nonlinear fractional convection-diffusion equations, *SIAM Journal on Mathematical Analysis* **2012**, *44*, 603–632.

7. Pascal, H. A nonlinear model of heat conduction, *Journal of Physics A: Mathematical and General* **1992**, *25*, 939.

8. Daly, E.; Porporato, A. Similarity solutions of nonlinear diffusion problems related to mathematical hydraulics and the Fokker-Planck equation, *Physical Review E* **2004**, *70*, 056303.

9. Troncoso, P.; Fierro, O.; Curilef, S.; Plastino, A. A family of evolution equations with nonlinear diffusion, Verhulst growth, and global regulation: Exact time-dependent solutions, *Physica A* **2007**, *375*, 457–466.

10. Plastino, A.R.; Plastino, A. Non-extensive statistical mechanics and generalized Fokker-Planck equation, *Physica A* **1995**, *222*, 347–354.

11. Lenzi, E.K.; Mendes, G.A.; Mendes, R.S.; da Silva, L.R.; Lucena, L.S. Exact solutions to nonlinear nonautonomous space-fractional diffusion equations with absorption, *Physical Review E* **2003**, *67*, 051109.

12. Assis Jr, P.C.; da Silva, L.R.; Lenzi, E.K.; Malacarne, L.C.; Mendes, R.S. Nonlinear diffusion equation, Tsallis formalism and exact solutions, *Journal of Mathematical Physics* **2005**, *46*, 123303.

13. Plastino, A.R.; Casas, M.; Plastino, A. A nonextensive maximum entropy approach to a family of nonlinear reaction–diffusion equations, *Physica A* **2000**, *280*, 289–303.

14. Wada, T.; Scarfone, A.M. On the non-linear Fokker-Planck equation associated with κ-entropy, In *AIP Conference Proceedings*, (AIP, 2007), Vol. 965, pp 177–180.

15. Wada, T.; Scarfone, A.M. Asymptotic solutions of a nonlinear diffusive equation in the framework of κ-generalized statistical mechanics, *The European Physical Journal B* **2009**, *70*, 65–71.

16. Frank, T.D. *Nonlinear Fokker-Planck equations: fundamentals and applications*; Springer Science & Business Media, 2005.

17. Hristov, J. An approximate analytical (integral-balance) solution to a nonlinear heat diffusion equation, *Thermal Science* **2015**, *19* (2), 723–733.

18. Filipovitch, N.; Hill, K.M.; Longjas, A.; Voller, V.R. The modelling of heat, mass and solute transport in solidification systems, *Water Resources Research* **2016**, *52*, 5167–5178.

19. Gmachowski, L. Fractal model of anomalous diffusion, *European Biophysics Journal* **2015**, *44*, 613–621.

20. Santoro, P.A.; de Paula, J.L.; Lenzi, E.K.; Evangelista, L.R. Anomalous diffusion governed by a fractional diffusion equation and the electrical response of an electrolytic cell, *Journal of Chemical Physics* **2011**, *135*, 114704.

21. Lenzi, E.K.; Fernandes, P.R.G.; Petrucci, T.; Mukai, H.; Ribeiro, H.V. Anomalous-diffusion approach applied to the electrical response of water, *Physical Review E* **2011**, *84*, 041128.

22. Lenzi, E.K.; de Paula, J.L.; Silva, F.R.G.B.; Evangelista, L.R. A Connection Between Anomalous Poisson–Nernst–Planck Model and Equivalent Circuits with Constant Phase Elements, *The Journal of Physical Chemistry C* **2013**, *117* (45), 23685–23690.

23. Sokolov, I.M. Anomalous Diffusion on Fractal Networks. In *Mathematics of Complexity and Dynamical Systems*; Springer, 2012; pp 13–25.

24. Metzler, R.; Klafter, J. The random walk's guide to anomalous diffusion: a fractional dynamics approach, *Physics Reports* **2000**, *339* (1), 1–77.

25. Evangelista, L.R.; Lenzi, E.K. *Fractional Diffusion Equations and Anomalous Diffusion*; Cambridge University Press, 2018.

26. Singh, H.; Kumar, D.; Baleanu, D. *Methods of Mathematical Modelling*; Boca Raton: CRC Press, 2020.

27. Singh, H.; Sahoo, M.R.; Singh, O.P. Numerical Method Based on Galerkin Approximation for the Fractional Advection-Dispersion Equation, *International Journal of Applied and Computational Mathematics* **2017**, *3* (3), 2171–2187.

28. Podlubny, I. *Fractional differential equations: an introduction to fractional derivatives, fractional differential equations, to methods of their solution and some of their applications*; Academic press, 1998; Vol. 198.

29. Singh, H.; Srivastava, H.; Kumar, D. A reliable numerical algorithm for the fractional vibration equation, *Chaos, Solitons & Fractals* **2017**, *103*, 131–138.

30. Singh, H. Approximate solution of fractional vibration equation using Jacobi polynomials, *Applied Mathematics and Computation* **2018**, *317*, 85–100.

31. Singh, H.; Pandey, R.K.; Srivastava, H.M. Solving Non-Linear Fractional Variational Problems Using Jacobi Polynomials, *Mathematics* **2019**, *7* (3), 224.

32. Singh, H.; Srivastava, H.M. Numerical Simulation for Fractional-Order Bloch Equation Arising in Nuclear Magnetic Resonance by Using the Jacobi Polynomials, *Applied Sciences* **2020**, *10* (8), 2850.

33. Singh, H.; Singh, C. A reliable method based on second kind Chebyshev polynomial for the fractional model of Bloch equation, *Alexandria Engineering Journal* **2018**, *57* (3), 1425–1432.

34. Singh, H. Analysis for fractional dynamics of Ebola virus model, *Chaos, Solitons & Fractals* **2020**, *138*, 109992.

35. Singh, H.; Akhavan Ghassabzadeh, F.; Tohidi, E.; Cattani, C. Legendre spectral method for the fractional Bratu problem, *Mathematical Methods in the Applied Sciences* **2020**, *43* (9), 5941–5952.

36. Marin, D.; Ribeiro, M.A.; Ribeiro, H.V.; Lenzi, E.K. A nonlinear Fokker–Planck equation approach for interacting systems: Anomalous diffusion and Tsallis statistics, *Physics Letters A* **2018**, *382*, 1903–1907.

37. Lenzi, E.; Menechini Neto, R.; Tateishi, A.; Lenzi, M.; Ribeiro, H. Fractional diffusion equations coupled by reaction terms, *Physica A* **2016**, *458*, 9–16.

38. Lenzi, E.; Ribeiro, M.; Fuziki, M.; Lenzi, M.; Ribeiro, H. Nonlinear diffusion equation with reaction terms: Analytical and numerical results, *Applied Mathematics and Computation* **2018**, *330*, 254–265.

39. Bhrawy, A.; Zaky, M. An improved collocation method for multi-dimensional space–time variable-order fractional Schrödinger equations, *Applied Numerical Mathematics* **2017**, *111*, 197–218.

40. Tsallis, C.; Levy, S.V.F.; Souza, A.M.C.; Maynard, R. Statistical-Mechanical Foundation of the Ubiquity of Lévy Distributions in Nature, *Physical Review Letters* **1995**, *75*, 3589–3593.

41. Tsallis, C. Lévy distributions, *Physics World* **1997**, *10* (7), 42–46.

42. Strikwerda, J.C. *Finite Difference Schemes and Partial Differential Equations*; Society for Industrial and Applied Mathematics, 2004.

43. Sod, G.A. *Numerical methods in fluid dynamics: initial and initial boundary-value problems*; Cambridge University Press, 1985.

44. Li, C.; Zeng, F. *Numerical Methods for Fractional Calculus*; CRC Press, 2015.

45. Karniadakis, G.E. *Handbook of Fractional Calculus with Application*; De Gruyter, 2019; Vol. 3: Numerical Methods.

46. Meerschaert, M.M.; Tadjeran, C. Finite difference approximations for fractional advection-dispersion flow equations, *Journal of Computational and Applied Mathematics* **2004**, *172*, 65–77.

47. Chagas, J.Q.; La Guardia, G.G.; Lenzi, E.K. A numerical method for a class of nonlinear fractional advection-diffusion equations, *arXiv preprint arXiv:2003.08015v1* **2020**.

48. Kochubei, A.; Luchko, Y. *Handbook of Fractional Calculus with Application*; De Gruyter, 2019; Vol. 1: Basic Theory.

49. Euler, L. Nova demonstratio quod evolutio potestatvm Binomii Newtoniana etiam pro exponentibus fractis valeat, *Nova Acta Academiae Scientiarvm Imperialis Petropolitanae* **1787**, *Tomvs V*, 52–58 (Mathematica).

8 Convergence of Some High-Order Iterative Methods with Applications to Differential Equations

I.K. Argyros,[1] *M.I. Argyros,*[2] *Á.A. Magreñán,*[3] *J.A. Sicilia,*[4] *and Í. Sarría*[5]

[1]Cameron University, Department of Mathematics Sciences, Lawton, OK, USA
[2]Cameron University, Department of Computing and Technology, Lawton, OK, USA
[3]Universidad de la Rioja, Departamento de Matemáticas y Computación, Logroño, La Rioja, Spain
[4]UNIR, Universidad Internacional de la Rioja, Escuela Superior de Ingeniería y Tecnología, Logroño, La Rioja, Spain
[5]UNIR, Universidad Internacional de la Rioja, Escuela Superior de Ingeniería y Tecnología, Logroño, La Rioja, Spain

CONTENTS

8.1 INTRODUCTION

In this chapter, we are concerned with the problem of approximating a solution x^* of the equation

$$F(x) = 0, \tag{8.1}$$

where F is a Fréchet-differentiable operator defined on a convex subset Ω of a Banach space B_1 with values in a Banach space B_2 [1,3,5,14,21].

Numerous problems in engineering and applied sciences [4,5] reduce to finding solutions of equations in the form of (8.1). As an example we use difference equations or differential equations to analyze dynamic systems. Then, the solutions represent the state of the system. To find solutions to these equations, we usually use iterative methods because it is feasible to find solutions in a closed form only in certain cases.

The practice of numerical functional analysis for finding such solutions is essentially connected to Newton-like methods [1–9, 11–19, 21, 23–25]. Newton's method converges quadratically to x^*, if the initial guess is close enough to the solution. Iterative methods of convergence order higher than two such as Chebyshev-Halley-type methods [6, 14, 15, 21] require the evaluation of the second Fréchet-derivative, which is very expensive in general. However, there are integral equations where the second Fréchet-derivative is diagonal by blocks and inexpensive [7–9, 11–14]. Moreover, for quadratic equations, the second Fréchet-derivative is constant. Furthermore, in some applications involving stiff systems, high-order methods are useful. That is why it is important to study the convergence of high-order methods.

We introduce the iterative method defined for all $n = 0, 1, 2, \ldots$ by

$$
\begin{aligned}
y_n &= x_n - F'(x_n)^{-1}F(x_n), \\
z_n &= y_n + \frac{1}{2}F'(x_n)^{-1}F(x_n) - (3F'(y_n) - F'(x_n))^{-1}F(x_n) \\
x_{n+1} &= z_n - (bF'(x_n) + cF'(y_n))^{-1}(F'(x_n) + aF'(y_n))F'(x_n)^{-1}F(z_n),
\end{aligned}
\tag{8.2}
$$

to generate a sequence $\{x_n\}$ approximating x^*, where x_0 is an initial point, S is $\mathbb{R}$ or $\mathbb{C}$, and $a, b, c \in S$ are given parameters.

Method (8.2) has been studied in [22] in the special case, when $B_1 = B_2 = \mathbb{R}$, and $a, b, c \in \mathbb{R}$. In particular, it was shown that the convergence order of method (8.2) is six, if

$$a \neq -1, \quad b = -\frac{1}{2}(3a+1),$$

$$c = \frac{1}{2}(5a+3)$$

and of order fifth, if

$$a = -1.$$

The convergence order was based on Taylor expansions and hypotheses reaching up to the seventh derivative of F although only the first derivative appears in these methods (8.2). Moreover, no computable radius of convergence or error bounds on the distances $\|x_n - x^*\|$ or uniqueness of the solution results were given.

The condition that F must be seven times Fréchet-differentiable limits the applicability of method (8.2). As a motivational example, define function f on $\Omega = [-\frac{1}{2}, \frac{3}{2}]$ by

$$f(x) = \begin{cases} x^3 lnx^2 + x^5 - x^4, & x \neq 0 \\ 0, & x = 0 \end{cases}. \tag{8.3}$$

Notice that $x^* = 1$. We also have that

$$f'''(x) = 6lnx^2 + 60x^2 - 24x + 22.$$

So, $f'''(x)$ is unbounded on Ω. Hence, the results in [13] cannot apply in this case.

In this chapter we address these problems. In particular, we present a local convergence analysis using only the first derivative which actually appears in method (8.2). Moreover, we provide the radius of convergence and error bounds on $\|x_n - x^*\|$ using Lipschitz constants. Notice that finding the radius of convergence gives us the choices for initial points, whereas the error bounds help us predetermine the number of iterates to achieve a certain error tolerance. This way, we extend the applicability of the method. Our technique can be used to extend the applicability of other methods along the same lines [15, 16].

The rest of the chapter is structured as follows. Section 8.2 contains the local convergence analysis. Applications and numerical examples appear in Sections 8.3 and 8.4. The conclusion is given in Section 8.5.

8.2 LOCAL CONVERGENCE ANALYSIS

Let us consider real functions and parameters to help with our local convergence analysis of method (8.2). Let $T = [0, \infty)$.

Assume there exists continuous and nondecreasing function $\omega_0 : T \to T$ such that equation

$$\omega_0(t) - 1 = 0 \tag{8.4}$$

has a least positive zero called r_0.

Let $T_0 = [0, r_0)$.

Assume there exists continuous nondecreasing function $\omega : T_0 \to T$. Define functions g_1 and h_1 on T_0 by

$$g_1(t) = \frac{\int_0^1 \omega((1-\theta)t)}{1 - \omega_0(t)}$$

and

$$h_1(t) = g_1(t) - 1.$$

Assume equation

$$h_1(t) = 0 \tag{8.5}$$

has a least positive zero called $R_1 \in (0, r_0)$.

Assume equation

$$p(t) = 0 \tag{8.6}$$

has a least positive zero called $r_p \in (0, r_0)$, where

$$p(t) = \frac{1}{3}(3\omega_0(g_1(t)t) + \omega_0(t)).$$

Define functions g_2 and h_2 on $[0, r_p)$ by

$$g_2(t) = g_1(t) + \varphi(t) \int_0^1 \omega_1(\theta t) d\theta,$$

and

$$h_2(t) = g_2(t) - 1,$$

where

$$\varphi(t) = \frac{3}{4} \frac{\omega_0(g_1(t)t) + \omega_0(t)}{(1 - \omega_0(t))(1 - p(t))}$$

and function $\omega_1 : [0, r_p) \to T$ is continuous and non-decreasing.

Assume equation

$$h_2(t) = 0 \tag{8.7}$$

has a least solution $R_2 \in (0, r_p)$.

Assume equation

$$q(t) - 1 = 0 \tag{8.8}$$

has a least solution $r_q \in (0, r_0)$, where

$$q(t) = \frac{1}{|b+c|} (|b|\omega_0(t) + |c|\omega_0(g_1(t)t)) \text{ for } b + c \neq 0.$$

Define functions g_3 and h_3 on the interval $[0, r_p)$ by

$$g_3(t) = \left[g_1(g_2(t)t) + \frac{(\omega_1(t) + |a|\omega_1(g_1(t))) \int_0^1 \omega_1(\theta g_2(t)t) d\theta}{|b+c|(1 - q(t))(1 - \omega_0(t))} \right] g_2(t)$$

and

$$h_3(t) = g_3(t) - 1.$$

Assume equation

$$h_3(t) = 0 \tag{8.9}$$

has a least solution $R_3 \in (0, r_p)$.

The choices for these scalar functions are completely justified in the proof of Theorem 8.1.

Let R be as

$$R = \min\{R_m\}, \quad m = 1, 2, 3. \tag{8.10}$$

We shall show R to be a radius of convergence for method (8.2). By this definition of R and for all $t \in [0, R)$

$$0 \leq w_0(t) < 1 \tag{8.11}$$

$$0 \leq p(t) < 1 \tag{8.12}$$

$$0 \leq q(t) < 1 \tag{8.13}$$

and

$$0 \leq g_m(t) < 1. \tag{8.14}$$

From now on we assume r_0, R_1, r_p, R_2, r_q, and R_3 exist as defined previously.

Moreover, the following set of conditions (C) is required for the study of method (8.2).

(C_1): $F : D \to B_2$ is differentiable, and x^* is a simple solution of equation (8.1).

(C_2): There exists function $w_0 : T \to T$ continuous and nondecreasing such that for all $x \in \Omega$

$$\|F'(x^*)^{-1}(F'(x) - F'(x^*))\| \leq \omega_0(\|x - x^*\|).$$

Let $\Omega_0 = \Omega \cap U(x^*, r_0)$.

(C_3): There exists function $w : T_0 \to T$ continuous and nondecreasing such that for all $x, y \in \Omega_0$

$$\|F'(x^*)^{-1}(F'(y) - F'(x))\| \leq \omega(\|y - x\|).$$

Let $\Omega_1 = \Omega \cap U(x^*, r_p)$.

(C_4): There exists function $w_1 : [0, r_p) \to T$ continuous and nondecreasing such that for all $x \in \Omega_1$

$$\|F'(x^*)^{-1}F'(x)\| \leq \omega_1(\|x - x^*\|).$$

(C_5): $\overline{U}(x^*, R) \subset \Omega$, where R is given in (8.10).

(C_6): There exists $\lambda \geq R$ such that

$$\int_0^1 \omega_0(\theta\lambda)d\theta \leq 1.$$

Let $\Omega_2 = \Omega \cap U(x^*\lambda)$.

Next, we present the local convergence analysis of method (8.2) using the aforementioned information. The trial functions appearing in conditions (C) are completely justified if one looks at the proof of Theorem 8.1. Moreover, such conditions are standard in the study of iterative methods [15, 16].

Theorem 8.1: *Assume conditions (C) hold, $b + c \neq 0$, and $x_0 \in U(x^*; R) \setminus \{x^*\}$. Then, sequence $\{x_n\}$ developed by method (8.2) exists, $\{x_n\} \subset U(x^*, R) \setminus \{x^*\}$ and $\lim_{n\to\infty} x_n = x^*$. Moreover, the following assertions hold:*

$$\begin{aligned}||y_n - x^*|| &\leq g_1(||x_n - x^*||)||x_n - x^*|| \\ &\leq ||x_n - x^*|| < R,\end{aligned} \tag{8.15}$$

$$\begin{aligned}||z_n - x^*|| &\leq g_2(||x_n - x^*||)||x_n - x^*|| \\ &\leq ||x_n - x^*||\end{aligned} \tag{8.16}$$

and

$$\begin{aligned}||x_{n+1} - x^*|| &\leq g_3(||x_n - x^*||)||x_n - x^*|| \\ &\leq ||x_n - x^*||,\end{aligned} \tag{8.17}$$

and x^* solves uniquely equation (8.1) *in the set Ω_2 introduced in condition* (C_6).

Proof. Let $v \in U(x^*, R) \setminus \{x^*\}$. In view of (8.10), (8.11), (C_1) and (C_2), we obtain

$$\begin{aligned}||F'(x^*)^{-1}(F'(v) - F'(x^*)|| &\leq \omega_0(||v - x^*||) \\ &\leq \omega_0(r_0) < 1,\end{aligned}$$

□

which along with a result due to Banach for invertible operators [14] implies

$$||F'(v)^{1-}F'(x^*)|| \leq \frac{1}{1 - \omega_0(||v - x^*||)}. \tag{8.18}$$

It also follows that y_0 exists (see first substep of method (8.2) for $n = 0$). Using (8.10), (8.14) (for $m = 1$), method (8.2), (C_1), (C_3) and (8.32) for $v = x_0$, we have

$$\begin{aligned}||y_0 - x^*|| &= ||x_0 - x^* - F'(x_0)^{-1}F(x_0)|| \\ &\leq ||F'(x_0)^{-1}F'(x^*)|| \times \\ &\quad || \textstyle\int_0^1 F'(x^*)^{-1}(F'(x^* + \theta(x_0 - x^*)) - F'(x_0))d\theta(x_0 - x^*)|| \\ &\leq \frac{\int_0^1 \omega(1-\theta)||x_0 - x^*||)d\theta||x_0 - x^*||}{1 - \omega_0(||x_0 - x^*||)} \\ &= g_1(||x_0 - x^*||)||x_0 - x^*|| \\ &\leq ||x_0 - x^*|| < R,\end{aligned} \tag{8.19}$$

verifying (8.15) (for $n=0$) and $y_0 \in U(x^*,R)$. Set $A_0 = 3F'(y_0) - F'(x_0)$. We shall show A_0 is invertible, so z_0 will exist. By (8.10), (8.12), (C_1) and (8.19), we get

$$\begin{aligned}
&\|(2F'(x^*))^{-1}(A_0 - 2F'(x_0))\| \\
&\leq \frac{1}{2}\left[3\|F'(x^*)^{-1}(F'(y_0)-F'(x^*))\| + \|F'(x^*)^{-1}(F'(x_0)-F'(x^*))\|\right] \\
&\leq \frac{1}{2}(\omega_0(\|y_0-x^*\|) + \omega(\|x_0-x^*\|)) \\
&\leq p(\|x_0-x^*\|) < 1,
\end{aligned}$$

so

$$\|A_0^{-1}F'(x^*)\| \leq \frac{1}{2(1-p(\|x_0-x^*\|))}. \tag{8.20}$$

Then, we can write by (8.10), (8.14) (for $m=2$), $(C_1)-(C_4)$, (8.32) (for $v = x_0$), (8.19), (8.20), and the second substep of method (8.2) for $n=0$:

$$\begin{aligned}
\|z_0 - z^*\| &= \|y_0 - x^* + \left[\frac{1}{2}F'(x_0)^{-1} - A_0^{-1}\right] F(x_0)\| \\
&\leq \left[g_1(\|x_0-x^*\|) + \varphi(\|x_0-x^*\|)\textstyle\int_0^1 \omega_1(\theta\|x_0-x^*\|)d\theta\right]\|x_0-x^*\| \\
&= g_2(\|x_0-x^*\|)\|x_0-x^*\| \\
&\leq \|x_0-x^*\|,
\end{aligned} \tag{8.21}$$

verifying (8.16) for $n=0$ and $z_0 \in U(x^*,R)$, where we also used

$$\begin{aligned}
&\|\frac{1}{2}F'(x_0)^{-1} - A_0^{-1})\| \\
&= \|\frac{3}{2}F'(x_0)^{-1}[(F'(y_0)-F'(x^*)) + (F'(x^*)-F'(x_0))]A_0^{-1}\| \\
&\leq \frac{3}{4}\frac{\omega_0(\|y_0-x^*\|)+\omega_0(\|x_0-x^*\|)}{(1-\omega_0(\|x_0-x^*\|))(1-p(\|x_0-x^*\|))} \\
&\leq \varphi(\|x_0-x^*\|)
\end{aligned}$$

and

$$F(x_0) = F(x_0) - F(x^*) = \int_0^1 F'(x^*) + \theta(x_0-x^*))d\theta(x_0-x^*),$$

so

$$\|F'(x^*)^{-1}F'(x)\| \leq \int_0^1 \omega_1(\theta\|x_0-x^*\|)d\theta\|x_0-x^*\|. \tag{8.22}$$

Similarly, using the third substep of method (8.2) for $n=0$, (8.10), (8.14) and (8.18)–(8.22), we obtain

$$\begin{aligned}\|x_1-x^*\| &\leq \left[g_1(\|z_0-z^*\|)+\frac{(\omega_1(\|x_0-x^*\|)+|a|\omega_1(\|y_0-x^*\|))\int_0^1\omega_1(\theta)\|z_0-x^*\|d\theta}{|b+c|(1-q(\|x_0-x^*\|))(1-\omega_0(\|x_0-x^*\|)}\right]\\ &\quad\times\|z_0-x^*\|\\ &\leq g_3(\|x_0-x^*\|)\|x_0-x^*\|\\ &\leq \|x_0-x^*\|,\end{aligned} \tag{8.23}$$

verifying (8.17) for $n=0$ and $x_1\in U(x^*,R)$, where we also used

$$\begin{aligned}\|(b+c)F'(x^*)^{-1}\| &= \frac{1}{|b+c|}\|b(F'(x_0)-F'(x^*))+c(F'(y_0)-F'(x^*))\|\\ &\leq \frac{1}{|b+c|}(|b|\omega_0(\|x_0-x^*\|)+|c|\omega_0(\|y_0-x^*\|))\\ &\leq q(\|x_0-x^*\|),\end{aligned} \tag{8.24}$$

so

$$\|\|(bF'(x_0)+cF'(y_0))F'(x^*))\|\leq\frac{1}{|b+c|(1-q(\|x_0-x^*\|))}, \tag{8.25}$$

$$\|F'(x^*)^{-1}(F'(x_0)+aF'(y_0)\|\leq\omega_1(\|x_0-x^*\|)+|a|\omega_1(\|y_0-x^*\|), \tag{8.26}$$

and

$$\|(F'(x_0)^{-1}F'(x^*))(F'(x^*)^{-1}F(z_0))\|\leq\frac{\int_0^1\omega_1(\theta\|z_0-z^*\|)d\theta\|z_0-x^*\|}{1-\omega_0(\|x_0-x^*\|)}. \tag{8.27}$$

So far the induction for estimates (8.15)–(8.17) is shown for $n=0$. Assume (8.15)–(8.17) hold for all $j=0,1,2,\ldots,n-1$. Then, by repeating these calculations, we show (8.15)–(8.17) fold for $j=n$. Next, using the estimate

$$\|x_{j+1}-x^*\|\leq\mu\|x_j-x^*\|<R, \tag{8.28}$$

where

$$\mu=g_3(\|x_0-x^*\|)\in[0,1),$$

we conclude

$$\lim_{j\to+\infty}x_j=x^*$$

and $x_{j+1} \in U(x^*, R)$. Then, let $d \in \Omega_2$ with $F(d) = 0$. Set $M = \int_0^1 F'(x^* + \theta(d - x^*))d\theta$. By (C_2) and (C_6), we get

$$\begin{aligned} \|F'(x^*)^{-1}(M - F'(x^*))\| &\leq \int_0^1 \omega_0(\theta(\|d - x^*\|)d\theta \\ &< \int_0^1 \omega_0(\theta\lambda)d\theta \\ &\leq 1, \end{aligned}$$

so $x^* = d$ follows from the invertibility of M and

$$0 = F(d) = -F(x^*) = m(d - x^*). \quad \square$$

To deal with the convergence of method (8.2) when $b + c = 0$, $b \neq 0$, assume equation

$$l(t) - 1 = 0$$

has a least solution $r_l \in (0, r_p)$, where

$$l(t) = \omega((1 + g_1(t))t). \tag{8.29}$$

Define functions $\overline{g}_3, \overline{h}_3$ by

$$\overline{g}_3(t) = \left[g_1(t)(g_2(t)t) + \frac{1}{|b|(1 - l(t))} \frac{(\omega_1(t) + |a|\omega_1(g_1(t)t) \int_0^1 \omega_1(\theta g_2(t)t)d\theta}{1 - \omega_0(t)}\right] g_2(t)$$

and

$$\overline{h}_3(t) = \overline{g}_3(t) - 1.$$

Assume equation

$$\overline{h}_3(t) = 0$$

has a least solution $\overline{R}_3 \in (0, r_p)$.

Then, by replacing

$$\frac{1}{|b + c|(1 - q(t))}$$

with

$$\frac{1}{|b|} \frac{1}{1 - l(t)},$$

we for the computation on the upper bound obtain results for this case too.

Let $\overline{R}$ be a radius of convergence for method (8.2) in this case given by

$$\overline{R} = \min\{R_1, R_2, \overline{R}_3\}. \tag{8.30}$$

Notice that we have

$$\begin{aligned}||F'(x^*)^{-1}(F'(x_n)-F'(y_n)) &\le \omega(||x_n-y_n||)\\ &\le \omega(||x_n-x^*||+||y_n-x^*||)\\ &\le \omega(||x_n-x^*||+g_1(||x_n-x^*||)||x_n-x^*||)\\ &= \omega(1+g_1(||x_n-x^*||)||x_n-x^*||)\\ &= l(||x_n-x^*||).\end{aligned} \tag{8.31}$$

This justifies the choice of l.

Hence, we arrive at

Theorem 8.2: *Assume conditions (C) hold, for $b+c=0$, and R, g_3 replaced by $\overline{R}$, $\overline{g}_3$ respectively. Then, the conclusions of Theorem 8.1. for method (8.2) hold.*

Remark 8.1: **(a)** *Assume ω_0, ω, are constant functions given as*

$$\omega_0(t)=L_0t$$

and

$$\omega(t)=Lt.$$

The estimate

$$r_A=\frac{2}{2L_0+L}$$

was obtained by Argyros in [3] as the convergence radius for Newton's method under conditions (C_1)–(C_3). We have $L_0\le L_1$ and $L\le L_1$ where L_1 is the Lipschitz constant on Ω. Notice that the convergence radius for Newton's method given independently by Rheinboldt [20] and Traub [21] is given by

$$\rho=\frac{2}{3L_1}<r_A. \tag{8.32}$$

As an example, let us consider the function $f(x)=e^x-1$. Then $x^*=0$. Set $\Omega=U(0,1)$. Then, we have that $L_0=e-1<L=e^{\frac{1}{e-1}}<L_1=e$, so $\rho=0.24252961<r_0=0.384947231$. Moreover, the new error bounds [3] are:

$$||x_{n+1}-x^*||\le\frac{L}{1-L_0||x_n-x^*||}||x_n-x^*||^2,$$

whereas the old ones [15, 17] are:

$$||x_{n+1}-x^*||\le\frac{L_1}{1-L_1||x_n-x^*||}||x_n-x^*||^2.$$

Clearly, the new error bounds are more precise if $L_0 < L_1$ *or* $L < L_1$. *Moreover, the radius of convergence of method* (8.2) *given by* R *or* $\overline{R}$ *is not larger than* r_A.

(b) *The local results can be used for projection methods such as Arnoldi's method, the generalized minimum residual method (GMREM), the generalized conjugate method (GCM) for combined Newton/finite projection methods, and in connection to the mesh independence principle in order to develop the cheapest and most efficient mesh refinement strategy [3, 4, 6].*

(c) *The results can be also be used to solve equations where the operator* F' *satisfies the autonomous differential equation [3, 4, 6]:*

$$F'(x) = P(F(x)),$$

where P *is a known continuous operator. Since* $F'(x^*) = P(F(x^*)) = P(0)$, *we can apply the results without actually knowing the solution* x^*. *Let as an example* $F(x) = e^x - 1$. *Then, we can choose* $P(x) = x + 1$ *and* $x^* = 0$.

(d) *It is worth noticing that method* (8.2) *is not changing if we use the* $(\mathscr{C})$ *instead of the stronger conditions in [21]. Moreover, for the error bounds in practice we can use the approximated computational order of convergence (ACOC) [9]*

$$\xi = \sup \frac{ln \frac{\|x_{n+2} - x_{n+1}\|}{\|x_{n+1} - x_n\|}}{ln \frac{\|x_{n+1} - x_n\|}{\|x_n - x_{n-1}\|}}, \qquad \text{for each } n = 1, 2, \ldots$$

or the computational order of convergence (COC)

$$\xi^* = \sup \frac{ln \frac{\|x_{n+2} - x^*\|}{\|x_{n+1} - x^*\|}}{ln \frac{\|x_{n+1} - x^*\|}{\|x_n - x^*\|}}, \qquad \text{for each } n = 0, 1, 2, \ldots,$$

instead of the error bounds obtained in Theorem 8.1.

(e) *In view of the estimate*

$$\begin{aligned}
\|F'(x^*)^{-1}F'(x)\| &= \|F'(x^*)^{-1}(F'(x) - F'(x^*)) + I)\| \\
&\leq 1 + \|F'(x^*)^1(F'(x) - F'(x^*))\| \\
&\leq 1 + \omega_0(\|x - x^*\|)
\end{aligned}$$

condition (C_4) *can be dropped, and* ω_1 *can be replaced by*

$$\omega_1(t) = 1 + \omega_0(t)$$

or

$$\omega_1(t) = 1 + \omega_0(r_0).$$

8.3 APPLICATION

In order to show the applicability of this method, we consider a Planck's radiation law problem.

Planck's law of radiation is a mathematical relationship formulated by German physicist Max Planck in 1900 to explain the spectral energy distribution emitted by a blackbody (a non-real body that absorbs all radiant energy). Planck assumes that the radiation source is an atom in an oscillating state, and the vibrational energy of each oscillator may have any one of a series of discrete values, but there is no value between them. Planck further assumes that when the oscillator changes from the energy state E_1 to the lower energy state E_2, the discrete energy of $E_1 — E_2$ or the amount of radiation is equal to the product of the radiation frequency, expressed as the Greek letter v and a constant P called the Planck constant, determined from the blackbody radiation data.

The problem we are going to solve can be found in [10]:

Application 1.

$$\varphi(\lambda) = \frac{8\pi cP\lambda^{-5}}{e^{\frac{cP}{\lambda BT}} - 1}, \tag{8.33}$$

which calculates the energy density within an isothermal blackbody, where

- λ is the wavelength of the radiation
- T is the absolute temperature of the blackbody
- B is the Boltzmann constant
- P is the Planck constant
- c is the speed of light.

Suppose we would like to determine wavelength λ which corresponds to maximum energy density $\varphi(\lambda)$. We get from (8.33)

$$\varphi'(\lambda) = \left(\frac{8\pi cP\lambda^{-6}}{e^{\frac{cP}{\lambda BT}} - 1}\right)\left(\frac{\left(\frac{cP}{\lambda BT}\right)e^{\frac{cP}{\lambda BT}} - 1}{e^{\frac{cP}{\lambda kT}} - 1} - 5\right). \tag{8.34}$$

The maxima for φ occurs, when

$$\frac{\left(\frac{cP}{\lambda BT}\right)e^{\frac{cP}{\lambda BT}} - 1}{e^{\frac{cP}{\lambda BT}} - 1} = 5. \tag{8.35}$$

Here, if $x = \frac{cP}{\lambda BT}$, the above equation becomes

$$1 - \frac{x}{5} = e^{-x}. \tag{8.36}$$

Let us define

$$f(x) = e^{-x} - 1 + \frac{x}{5}. \tag{8.37}$$

As a consequence, finding the roots (8.37) gives us the maximum wavelength of radiation (λ) by means of the following formula:

$$\lambda \approx \frac{cP}{x^* BT}. \tag{8.38}$$

It is easy to see that function $f(x)$ is continuous, $f(4) = -0.181684\ldots$ and $f(6) = 0.202479\ldots$. Then, it follows from the Intermediate Value Theorem that $f(x)$ has zeros in the interval $(4,6)$.

We consider $\Omega = [4,6]$ and $x^* = 4.965114\ldots$. Then, we can choose

$$\omega_0(t) = t,$$

$$\omega(t) = t,$$

and

$$\omega_1(t) = 1 + t.$$

Next, we are going to see different values of a, b, and c for which our results can be applied:

- Case 1:

  $$\{a,b,c\} = \{1,1,1\}.$$

  We obtain that:

  $$r_0 = 1,$$

  $$R_1 = \frac{2}{3},$$

  $$r_p = 0.6904157598234297\cdots,$$

  $$R_2 = 0.3449666245403301\cdots$$

  $$r_q = 0.7639320225002102\cdots$$

  and

  $$R_3 = 0.270686403223383$$

  Consequently,

  $$R = R_3 = 0.270686403223383$$

  and by choosing

  $$\lambda = \frac{1}{3},$$

  all conditions of Theorem 8.1 are satisfied.

- Case 2

$$a, b, c = 2, 1, 1.$$

We obtain that

$$r_0 = 1,$$
$$R_1 = \frac{2}{3},$$
$$r_p = 0.6904157598234297\cdots,$$
$$R_2 = 0.3449666245403301\cdots$$
$$r_q = 0.7639320225002102\cdots$$

and

$$R_3 = 0.229243$$

Consequently,

$$R = R_3 = 0.229243\cdots$$

and by choosing

$$\lambda = \frac{1}{3},$$

all conditions of Theorem 8.1 are satisfied.

8.4 NUMERICAL EXAMPLE

Example Let us consider a system of differential equations governing the motion of an object and given by

$$H_1'(t) = e^t$$
$$H_2'(t) = (e-1)t$$
$$H_3'(t) = 1$$

with initial conditions

$$H_1(0) = H_2(0) = H_3(0) = 0.$$

Let

$$H = (H_1, H_2, H_3) = (e^t, \frac{(e-1)t^2}{2} + t, t),$$

Choose $B_1 = B_2 = \mathbb{R}^3$, $\Omega = \overline{U}(0,1)$. Then, we have

$$x^* = \begin{pmatrix} 0 \\ 0 \\ 0 \end{pmatrix}.$$

Define function H on Ω for

$$w(t) = \begin{pmatrix} x(t) \\ y(t) \\ z(t) \end{pmatrix}$$

by

$$H(w) = \begin{pmatrix} e^t - 1 \\ \frac{e-1}{2}t^2 + t \\ t \end{pmatrix}. \tag{8.39}$$

Then, the Fréchet-derivative is given by

$$H'(v) = \begin{bmatrix} e^t & 0 & 0 \\ 0 & (e-1)t+1 & 0 \\ 0 & 0 & 1 \end{bmatrix}.$$

Notice that

$$H'(x^*) = \begin{pmatrix} 1 & 0 & 0 \\ 0 & 1 & 0 \\ 0 & 0 & 1 \end{pmatrix}.$$

Choose

$$a = 1,$$
$$b = 1,$$

and

$$c = 1.$$

We obtain that:

$$r_0 = 1,$$
$$R_1 = 0.382692$$
$$r_p = 0.398189\cdots,$$
$$R_2 = 0.691833\cdots$$
$$r_q = 0.441487\cdots\cdots$$

and

$$R_3 = 0.0728892\cdots$$

Hence, we can guarantee the convergence of the method (8.2) to the solution x^* of equation $H(w) = 0$ by Theorem 8.1.

8.5 CONCLUSION

A local convergence analysis is presented for a family of sixth-order methods based only on the first derivative which only appears in them. Our technique can be used to extend the applicability of other methods along the same lines. It turns out that there are many sixth-order methods in the literature. This is the case since no one of them is shown to perform better than all the others. That is why there is a need for finding one sixth-order method that outperforms all the others. The same is expected of other convergence order methods. Our present work is a step in this direction because other sixth-order methods can be studied with the same set of conditions.

REFERENCES

1. Amat, S., Busquier, S., Gutiérrez, J.M., Geometric constructions of iterative functions to solve nonlinear equations, J. Comput. Appl. Math., 157 (2003), 197–205.
2. Argyros, I.K., Magreñán, Á.A., Iterative Methods and Their Dynamics With Applications, CRC Press, New York, 2017.
3. Argyros, I.K., A unifying local–semilocal convergence analysis and applications for two-point Newton-like methods in Banach space, J. Math. Anal. Appl., 298 (2004), 374–397.
4. Argyros, I.K., Magreñán, Á.A., Iterative Algorithms I, Nova Science Publishers Inc., New York, 2016.
5. Argyros, I.K., Hilout, S., Weaker conditions for the convergence of Newton's method. J. Complexity, 28 (2012) 364–387.
6. Argyros, I.K., Hilout, S., Numerical Methods in Nonlinear Analysis, World Scientific Publ. Comp., New Jersey, 2013.
7. Candela, V., Marquina, A., Recurrence relations for rational cubic methods II: The Chebyshev method, Computing, 45 (1990), 355–367.
8. Chun, C., Stanica, P., Neta, B., Third order family of methods in Banach spaces, Comput. Math. Appl., 61 (2011), 1665-1675.
9. Cordero, A., Torregrosa, J.R., Variants of Newton's method using fifth-order quadrature formulas. Appl. Math. Comput., 190(1), (2007), 686–698.
10. Divya, J., Families of Newton-like methods with fourth-order convergence, Int. J. Comput. Math. 90(5) (2013), 1072–1082.
11. Hernández, M.A., Salanova, M.A., Modification of the Kantorovich assumptions for semilocal convergence of the Chebyshev method, J. Comput. Appl. Math., 126 (2000), 131–143.
12. Hernández M.A., Chebyshev's approximation algorithms and applications, Computers Math. Applic. 41 (2001), 433–455.

13. Hueso, J.L., Martínez, E. Teruel, C. Convergence, efficiency and dynamics of new fourth and sixth order families of iterative methods for nonlinear systems. J. Comput. Appl. Math., 275 (2015), 412–420.

14. Kantorovich, L.V., Akilov, G.P., Functional Analysis, Pergamon Press, Oxford, 1982.

15. Magreñán, Á.A., Argyros, I.K., Ball convergence theorems and the convergence planes of an iterative method for nonlinear equations, SeMA Journal, Boletń de la Sociedad Española de Matemática Aplicada, 71 (2015), 39–55.

16. Magreñán, Á. A. and Argyros, I.K., A Contemporary Study of Iterative Methods: Convergence, Dynamics and Applications. Academic Press, London, 2018.

17. Magreñán, Á.A., Different anomalies in a Jarratt family of iterative root-finding methods, Appl. Math. Comput., 233 (2014), 29–38.

18. Magreñán, Á.A., A new tool to study real dynamics: The convergence plane, Appl. Math. Comput., 248 (2014), 215–224.

19. Ortega, J.M., Rheinboldt, W.C., Iterative Solution of Nonlinear Equations in Several Variables, Academic Press, New York, 1970.

20. Rheinboldt, W.C., An adaptive continuation process for solving systems of nonlinear equations, Polish Academy of Science, Banach Ctr. Publ. 3 (1978) 129–142.

21. Traub, J.F., Iterative Methods for the Solution of Equations, Prentice-Hall Series in Automatic Computation, Englewood Cliffs, NJ, 1964.

22. Sharma, R., Some fifth and sixth order iterative methods for solving nonlinear equations, Int. J. Eng. Res. Appl., 4 (2014) 268–273.

23. Singh, H., Srivastava, H.M., and Kumar, D., A reliable numerical algorithm for the fractional vibration equation, Chaos, Solitons Fractals, 103 (2017), 131–138.

24. Singh, H., Kumar, D., Baleanu, D., Methods of Mathematical Modelling: Fractional Differential Equations, CRC Press, New York, 2019.

25. Singh, H., Approximate solution of fractional vibration equation using Jacobi polynomials, Appl. Math. Comput., 317 (2018), 85–100.

9 Fractional Derivative Operator on Quarantine and Isolation Principle for COVID-19

Albert Shikongo,[1] *Samuel M. Nuugulu,*[1] *David Elago,*[1] *Andreas T. Salom,*[2] *Kolade M. Owolabi*[3,4]

[1]Department of Mathematics, University of Namibia, Namibia

[2]CERENA-Polo FEUP, Faculty of Engineering, University of Porto, Portugal

[3]Institute for Groundwater Studies, University of the Free State, Bloemfontein, South Africa

[4]Department of Mathematical Sciences, Federal University of Technology Akure, Ondo State, Nigeria

CONTENTS

9.1 INTRODUCTION

Many businesses across all activities have been at a halt due to the unprecedented outbreak of a pneumonia of unknown aetiology in Wuhan City, Hubei province in China in December 2019. Consequently, more than 200 hundreds thousands people have been confirmed dead worldwide due to this deadly disease. However, in the absence of a cure and vaccine, it has been established that quarantine and isolation are fundamental factors contributing to a successful recovery of a COVID-19 patient, providing the infection is detected as early as by the day the infected individual contracted the infection. Globally, it has been established that many people have been recovering from the deadly infection of COVID-19 based on the principle of quarantine and isolation, and some business activities have resumed, by means of practicing social distancing. Quarantine is a systemic method of keeping the exposed people who do not have symptoms away from others so they do not unknowingly contract the disease. In view of COVID-19 spread, quarantine is referred to as staying indoors for 14 days from the date of contact with a confirmed positive case or a returner from a high-risk area, as well as avoiding contact with other persons, infected or not. In contrast, isolation serves the same purpose as quarantine, but it is reserved for those who have already tested positive for the disease. Consequently, social distancing is required: to avoid large gatherings, and to keep at least 2 m of social distance between each others whenever possible. Thus, in this chapter we highlight most of the recent developments on quarantine and isolation to counter COVID-19 spread.

Here we consider how, motivated by the rapid spread of COVID-19 that started in mainland China, in December 2019, [1] proposed a global meta-population disease transmission model to project the impact of travel limitations on the national and international spread of the epidemic. Their model is calibrated on the basis of internationally reported cases, which presents the start of the travel ban from Wuhan on 23 January 2020. By that time, most Chinese cities had already received many infected travellers, and the travel quarantine of Wuhan delayed the overall epidemic progression by only 3–5 days in mainland China, though it had a more marked effect on the international scale where the cases of importations were reduced by nearly 80% until mid-February. The authors in [1] also state that their results present that 90% travel restrictions to and from mainland China only modestly affected the epidemic trajectory unless combined with a 50% or higher reduction of community transmission.

Zhao and Chen [2] mentioned that COVID-19 was rapidly spreading in China and over 30 other countries over the last 4 months. They also identified that COVID-19 has multiple characteristics distinct from other infectious diseases, including high infectivity during incubation, and the time delay between real dynamics and daily numbers of confirmed cases.

Hassan et al. [3] defined COVID-19 as an enveloped RNA virus that is diversely found in humans and wildlife. Furthermore, they mentioned that a total of six species have been identified to cause disease in humans, known to infect the neurological, respiratory, enteric, and hepatic systems. They compared COVID-19 spread to the form of Middle East respiratory syndrome corona virus (MERS-CoV) which has occurred over the past few decades, and their review is believed to introduce a general overview of COVID-19 and present the clinical features, evaluation, and treatment of COVID-19 patients.

COVID-19 is an infectious disease caused by a new type of virus called SARS-CoV-2, and by the beginning of 2020 had spread throughout the world [4]. Thus, in [4] it is mentioned that a high rate of spread of COVID-19 causes the number of infected cases to increase significantly. Hence, their study's model is designed to predict the number of COVID-19 patients and the duration of the COVID-19 pandemic in Indonesia. Consequently, they modified SEIR model (Susceptible, Exposed, Infected, Recovered) with several assumptions such as a constant homogeneous population, patients who have recovered cannot be re-infected, and that COVID-19 spread only occurs from human to human. In addition, their dynamics assumed that there are individuals who carry out quarantine and isolation. Their results further presented that the peak of the COVID-19 pandemic in Indonesia had occurred by the middle of May 2020, and that the number of infected patients was about 15,000. Furthermore, they mentioned that such a number can be drastically reduced if the quarantine and isolation process is carried out optimally.

Therefore, in the context of COVID-19 spread, it has become an interest for researchers to investigate the spread of COVID-19 and its implications for society at large. Therefore, firstly, this chapter presents the design of a mathematical model to demonstrate the impact of quarantine and isolation on the spread of COVID-19. Thus, based on the understanding of recovering from COVID-19 using the principle of quarantine and isolation, we let $\mathscr{C}$ denote the COVID-19 spread at any time t, and so the dynamics of quarantine and isolation for COVID-19 spread ($\mathscr{C}$) is

$$\frac{d\mathscr{C}}{dt} = \xi_c \xi_{an} x^{\delta} \mathscr{C}, \tag{9.1}$$

where x, ξ_c, ξ_{an} denote the number of days a host tested positive for COVID-19 within an interval $[a,b]$ for $0 \leq a < b$, $a,b \in \mathbb{N}$, COVID-19 proliferation effect and anti-proliferation effect for the virus due to the immune system of a host, $\delta \in \mathbb{R} \setminus \{(0,1)\}$ denotes the number of admissible days for quarantine and isolation when the process is ignored, and $\delta \in (0,1)$ denotes the inversion of the number of admissible days for quarantine when the process is applied. Therefore, parameter value for quarantine and isolation splits the ordinary differential equation (ODE) in Eq. (9.1) into two components. That is, the component for quarantine and isolation ignores $\mathscr{C}_1$ and the component for quarantine and isolation applies $\mathscr{C}_2$. Thus, henceforth Eq. (9.1) becomes a system of ODEs

$$\dot{\mathscr{C}_{\mathbf{t}}} = \xi_c \xi_{an} \mathbf{x}^{\delta} \mathscr{C}, \tag{9.2}$$

where $\dot{\mathscr{C}}$ denotes the partial derivative with respect to time $\mathbf{t}$. Equation (9.2) comprises a system of ODEs [5], hence, an initial condition for the system of ODEs in Eq. (9.2) is required. Thus, the initial condition for the system of ODEs in Eq. (9.2) is $\mathscr{C}(\mathbf{a}) \geq \mathbf{0}$, where, $\mathbf{a} = [a,a]^T$. Many real-life phenomena have been modelled by ODEs, see for instance in [6] and therefore, here we highlight most of the recent development in modelling the COVID-19 spread by means of ODEs.

The authors of [7] developed a bats-hosts-reservoir-people transmission network dynamic for simulating the potential transmission from the infection source to the human infection. Since the bats-hosts reservoir network was hard to explore clearly and public concerns were focusing on the transmission from the Huanan seafood wholesale market (reservoir) to people, they simplified the model as a reservoir-people (RP) transmission network. Hence, they were able to adopt the matrix approach to calculate the basic reproduction number (R_0) from the RP model to assess the transmissibility of SARS-CoV-2.

At the beginning of COVID-19 infection, there is a period of time known as the exposed or latency period, before an infected person transmits the infection to another person [8]. In [8] they developed two differential equation models, the first of which incorporates infected persons in the exposed class, before transmission is possible. The second model incorporates a time delay in infected persons, before transmission is possible.

In [9] the temporal dynamics of the COVID-19 outbreak in China, Italy, and France in the time window January 22 to March 15, 2020 is analyzed by means of a system of ordinary differential equations using a simple susceptible-infected-recovered-deaths model. Their analysis for this model presents that the kinetic parameter that describes the rate of recovery seems to be the same irrespective of the country, while the infection and death rates appear to be more variable. Thus, their model was able to present that the peak in Italy as around March 21st, 2020, with a peak number of infected individuals of about 26,000 (not including recovered and dead) and a number of deaths at the end of the epidemic of about 18,000. Since the confirmed cases are believed to be between 10% and 20% of the real number of individuals who eventually get infected, the apparent mortality rate of COVID-19 falls between 4% and 8% in Italy, while it appears substantially lower, between 1% and 3%, in China. Eventually, the authors of [9] mentioned that the effects of drastic containment measures on the outbreak in Italy indicate that a reduction of the infection rate does indeed cause a quench of the epidemic peak. However, they were able to note that the infection rate needs to be cut down drastically and quickly to observe an appreciable decrease of the epidemic peak and mortality rate, and this only appears possible through a concerted and disciplined, albeit painful, effort of the population.

A model considering susceptible, exposed, infected, asymptomatic, quarantine, isolation, and recovered classes as in the case of COVID-19 disease was considered in [10], in which the facility of quarantine and isolation has been provided to both exposed and infected classes. The authors mentioned that asymptomatic individuals have either recovered without undergoing treatment or moved into the infected class after some duration. Thus, they formulated the reproduction number for the proposed dynamics, in which elasticity and sensitivity analysis present that their formulation is more sensitive towards the transmission rate from the exposed to infected classes rather than from the susceptible to exposed classes.

Since the process of quarantine and isolation is a mandatory process worldwide, we extend the system of ODEs in Eq. (9.2) to a system of time-dependent parabolic partial differential equations (PPDEs)

$$\dot{\mathscr{C}}_{\mathbf{t}} - \mathscr{D}\ddot{\mathscr{C}}_{\mathbf{xx}} = \xi_c \xi_{an} \mathbf{x}^{\delta} \mathscr{C}, \tag{9.3}$$

in order to include the constant spatial distribution $\mathscr{D} = [\mathscr{D}_1, \mathscr{D}_2]^T$, (where T denotes transpose) of the COVID-19 spread at any given time $\mathbf{t}$ in a given host. Since quarantine and isolation are mandatory for any infected individual and/or community, the boundary conditions are Neumann boundary conditions and the initial conditions remain the same as for the system of ODEs in Eq. (9.2). Hence, it suffices to mention the recent developments carried out by means of time-dependent PPDEs, with regard to COVID-19.

Thus, a system of PPDEs in [11] is developed describing the spread of the COVID-19 virus, considering the mean daily movement of susceptible, exposed, and asymptomatic individuals. Their model is calibrated using data on the confirmed infection and death rates from France as well as their initial spatial distribution. First, the system of partial differential equations is studied, then the basic reproduction number, R_0, is derived.

The significance of the solution to our formulation is critical, and therefore we extend the system of PPDEs in Eq. (9.3) to a system of fractional parabolic partial differential equation (FPPDEs) in the Caputo sense as

$$ {}^{C}D^{\alpha}_{a,t}\mathscr{C} - \frac{\mathscr{D}}{2}\left[{}^{C}D^{\alpha}_{a,x}\mathscr{C} + {}^{C}D^{\alpha}_{b,x}\mathscr{C}\right] = \xi_c \xi_{an}\mathbf{x}^{\delta}\mathscr{C}, \tag{9.4} $$

where $\alpha, \in (0,1)$, ${}^{C}D^{\alpha}_{a,t}, {}^{C}D^{\alpha}_{a,x}$ denote the Caputo fractional derivative with respect to time $\mathbf{t}$, left-side and right-side Caputo fractional derivative with respect to space $\mathbf{x}$ [12, 13] respectively. We refer to Eq. (9.4) as a system of FPPDEs. Fractional calculus has played a significant role in dynamical systems such as the one under our consideration in Eq. (9.1) to Eq. (9.4). Fractional derivatives occur in real-life phenomena; thus, here we highlight some of the recent contributions, with respect to COVID-19.

In [14] the dynamics of a novel COVID-19 (2019-nCoV) are presented in terms of interaction among the bats and unknown hosts, the people and the infections reservoir (seafood market). Since the purchasing of items from the seafood market by people has the ability to infect either asymptotically or symptomatically, the dynamics are reduced to those that assume that the seafood market has enough source of infection to be effective at infecting people. Thus, Khan and Atangana [14] were able to formulate a fractional dynamic and solve their dynamics based on the available infection cases for the period January 21, 2020, till January 28, 2020.

In [15] the study of a fractional Ebola virus model is considered and an efficient computational method based on iterative scheme is proposed to solve the fractional Ebola model numerically. Singh [15] presented that the solution varies for fractional and integer order Ebola virus models.

The dreadful impact of COVID-19 is enormous both socially and economically worldwide. Therefore, the second aim of this chapter is to design a novel numerical method to solve the system of FPPDEs in Eq. (9.4). The novel numerical method enables us to unify the solution of an ordinary integer with a fractional order differential equation and vice versa [16]. In [17] and [18] the novel numerical method has been applied to drug resistance phenomena and on intrinsic resistance phenomena arising in multi-mutation of cancer dynamics, respectively. The novel numerical method successfully captured the qualitative features of both phenomena presented in each of the experiments.

Al-Refai [19] defined a numerical method based on the weighted Atangana-Baleanu fractional operators of Caputo sense. They were able to obtain the solution related to linear fractional differential equation in a closed form, in terms of expressing the weighted Atangana-Baleanu fractional derivative in a convergent series of

Riemann-Liouville fractional integrals, and to establish commutative results of the weighted Atangana-Baleanu fractional operators.

A numerical algorithm for the solution of the fractional vibration equation (FVE) based on the applications of the operational matrices of the Legendre scaling functions is considered in [20]. They mentioned that the numerical algorithm reduces the FVE into Sylvester form, which in turn significantly simplifies the FVE. Further information can be traced in [21] and [22].

The rest of the chapter is organized as follows. In Section 9.2 we analyze the formulation in (9.4), and in Section 9.3, we design the novel numerical method. We present the results and discussion in Section 9.4, and conclude the chapter with Section 9.5.

9.2 MATHEMATICAL ANALYSIS OF THE DYNAMICS

In this section, we establish the existence of uniqueness of solution, and continuous dependence of the solution for one component to the system in Eq. (9.4).

Since left-side Caputo fractional derivative of order α with respect to time t is

$$ {}^{C}D^{\alpha}_{a,t}\mathscr{C}_1(t) = \begin{cases} \frac{1}{\Gamma(1-\alpha)}\int_a^t (t-\tau)^{-\alpha}\mathscr{C}_1'(\tau)d\tau, \text{ if } 0<\alpha<1, \\ \\ \mathscr{C}_1', \text{ if } \alpha = 1, \end{cases} \tag{9.5}$$

and for $n-1<\alpha\leq n$ where $n\in\mathbb{N}\leq 2$ (in our case) on $[a,b]$ we have the left-side Caputo fractional operator as

$$ {}^{C}D^{\alpha}_{a,x}\mathscr{C}_1(x,t) = \begin{cases} \frac{1}{\Gamma(n-\alpha)}\int_a^x (x-\varphi)^{n-\alpha-1}\frac{\partial^n\mathscr{C}_1(\varphi,t)}{\partial\varphi^n})d\varphi, \text{ if } n-1<\alpha<n, \\ \\ \frac{\partial^n\mathscr{C}_1(\varphi,t)}{\partial\varphi^n})d\varphi,, \text{ if } \alpha = n, \end{cases} \tag{9.6}$$

and the right-side Caputo fractional operator as

$$ {}^{C}D^{\alpha}_{x,b}\mathscr{C}_1(x,t) = \begin{cases} \frac{(-1)^n}{\Gamma(n-\alpha)}\int_x^b (\varphi-x)^{n-\alpha-1}\frac{\partial^n\mathscr{C}_1(\varphi,t)}{\partial\varphi^n})d\varphi, \text{ if } n-1<\alpha<n, \\ \\ (-1)^n\frac{\partial^n\mathscr{C}_1(\varphi,t)}{\partial\varphi^n})d\varphi, \text{ if } \alpha = n. \end{cases} \tag{9.7}$$

For the system of FPPDEs in Eq. (9.4), we assumed that

$$ R(\Omega) = \left\{ \mathscr{C}_1(x,t) \left| \frac{d^2\mathscr{C}_1}{dx^2} \in C_1^2(\bar{\Omega}) \text{ and } \frac{d\mathscr{C}_1}{dx} \in C(\bar{\Omega}) \right. \right\}, \tag{9.8}$$

where $\bar{\Omega} = [0,T]\times[a,b]$ for some $T\in\mathbb{N}<\infty$. In view of [23], we state the following results.

Lemma 9.1: If a function $f\in C^1[0,T]$ attains its maximum over the interval $[0,T]$ at a point $\tau = t_0 \in (0,T]$, then

$$ 0 \leq {}^{C}D^{\alpha}_{a,t_0} f(t_0), \forall\alpha\in(0,1]. \tag{9.9}$$

Lemma 9.2: If a function $f \in C^1[a,b]$ attains its maximum over the interval $[a,b]$ at a point $x = x_0 \in (a,b]$, then

$$0 \leq {}^C D^{\gamma}_{a,x_0} f(x_0), \forall \gamma \in (0,1]. \tag{9.10}$$

Consequently,

$$\left[{}^C D^{\gamma}_{a,x} \mathscr{C} + {}^C D^{\gamma}_{b,x} \mathscr{C}\right] f(x_0) \leq 0, \ \forall \gamma \in (0,1). \tag{9.11}$$

Proof: The proving of Lemma 9.2 suffices. Thus, we note that when $\gamma = 1$ Lemma 9.2 holds. Thus, when $\gamma \in (0,1)$, we let

$$g(x) = f(x_0) - f(x), \ \forall x_0 \in [a,b]. \tag{9.12}$$

Thus, Eq. (9.12) implies that

$$g(x) \geq 0, \text{ and } {}^C D^{\gamma}_{x,b} g(x) = -{}^C D^{\gamma}_{x,b} f(x_0), \ \forall x_0 \in [a,b].$$

By Lemma 9.2, we see that

$$g(x) \geq 0, \text{ and } {}^C D^{\gamma}_{x,b} g(x) = -{}^C D^{\gamma}_{x,b} f(x), \ \forall x \in [a,b].$$

On the other hand, we have

$$\begin{aligned}
{}^C D^{\gamma}_{x_0,b} g(x_0) &= -\frac{1}{\Gamma(1-\gamma)} \int_{x_0}^{b} (\varphi - x_0)^{-\gamma} g'(\varphi) d\varphi, \\
&= -\frac{1}{\Gamma(1-\gamma)} \int_{x_0}^{b-\varepsilon} (\varphi - x_0)^{-\gamma} g'(\varphi) d\varphi \\
&\quad -\frac{1}{\Gamma(1-\gamma)} \int_{b-\varepsilon}^{b} (\varphi - x_0)^{-\gamma} g'(\varphi) d\varphi, \text{ for each, } 0 < \varepsilon < b - \varepsilon.
\end{aligned}$$

Since $f \in C^1[a,b]$, then $g' \in L([a,b])$. Thus, for any $\delta > 0$, there exists $\varepsilon > 0$ such that

$$\left| -\frac{1}{\Gamma(1-\gamma)} \int_{b-\varepsilon}^{b} (\varphi - x_0)^{-\gamma} g'(\varphi) d\varphi \right| \leq \delta.$$

But,

$$\begin{aligned}
-\frac{1}{\Gamma(1-\gamma)} \int_{x_0}^{b-\varepsilon} (\varphi - x_0)^{-\gamma} g'(\varphi) d\varphi &= -\frac{1}{\Gamma(1-\gamma)} (b - \varepsilon - x_0)^{-\gamma} g(b-\varepsilon) \\
&\quad -\frac{\gamma}{\Gamma(1-\gamma)} \int_{x_0}^{b-\varepsilon} (\varphi - x_0)^{-\gamma-1} g'(\varphi) d\varphi \leq 0. \square
\end{aligned}$$

In view of the work of Mohammed [24], we have the following results.

Lemma 9.3: *Let $f \in C^2([a,b])$, such that it attains its maximum over the interval $[a,b]$ at a point $x_0 \in [a,b]$ and $f'(x_0) \geq 0$. Then*

$${}^C D^{\beta}_{0,x_0} f(x_0) \leq 0, \forall \beta \in (0,2). \tag{9.13}$$

Lemma 9.4: *Let $f \in C^2([a,b])$, such that it attains its maximum over the interval $[a,b]$ at a point $x_0 \in [a,b]$ and $f'(x_0) \leq 0$. Then*

$$\left[{}^C D^{\beta}_{a,x} f(x_0) + {}^C D^{\beta}_{b,x} f(x_0)\right] \leq 0, \forall \beta \in (0,2]. \tag{9.14}$$

Proof: When $\beta = 2$ the proof follows directly from the second derivative tests. However, for $\beta \in (0,2)$, we have

$${}^C D^{\beta}_{0,x_0} f(x_0) \leq 0, \forall \beta \in (0,2), \tag{9.15}$$

by Lemma 9.3 and using the results in [24], we have

$${}^C D^{\beta}_{a,b} f(x_0) \leq 0, \forall \beta \in (0,2). \tag{9.16}$$

□

Theorem 9.1: Let $\mathscr{C}_1 \in R(\Omega)$ denote the solution for one component of the system of FPPDEs in Eq. (9.4) in Ω, in which $\frac{d\mathscr{C}_1}{dx}|_{x=a} \geq 0$ and $\frac{d\mathscr{C}_1}{dx}|_{x=b} \leq 0$. Then either $\mathscr{C}_1(x,t) \leq 0, \forall (x,t) \in \bar{\Omega}$, or $\mathscr{C}_1$ attains its positive maximum on the bottom or back-side parts of the sides $S = \{[a,b] \times \{a\} \cup \{a\} \times [a,T] \cup \{b\} \times [a,T]\} S$ of the boundary of the domain Ω.

Proof: Let a point (x_0,t_0) exist for $x_0 \in (a,b)$ and $t_0 \in (0,T]$ such that

$$\mathscr{C}_1(x_0,t_0) \geq \left\{0, \max_{a \leq x \leq b} \mathscr{C}_1(0), 0, 0\right\} = M \geq 0.$$

Let $\varepsilon = \mathscr{C}_1(x_0,t_0) - M > 0$, such that

$$z(x,t) = \mathscr{C}_1(x,t) + \frac{\varepsilon(T-t)}{2T}, \forall (x,t) \in \bar{\Omega},$$

which is equivalent to

$$z(x,t) \leq \mathscr{C}_1(x,t) + \frac{\varepsilon}{2}, \forall (x,t) \in \bar{\Omega}.$$

Thus,

$$z(x_0,t_0) \geq \mathscr{C}_1(x_0,t_0) = \varepsilon + M \geq \frac{\varepsilon}{2} + z(x,t), \forall (x,t) \in S$$

implies that z cannot attain its maximum on S. Hence, we let (x_1,t_1) denote the maximum point of z over $\bar{\Omega}$, such that $x_1 \in (a,b), t_1 \in (a,T]$ and

$$z(x_1,t_1) \geq z(x_0,t_0) \geq \varepsilon + M > \varepsilon.$$

In view of Lemma 9.1, Lemma 9.2, and Lemma 9.4 we obtain

$${}^C D^{\alpha}_{a,t} w|_{(x_1,t_1)} \geq 0,\ \forall \alpha \in (0,1),\ \left[{}^C D^{\gamma}_{a,x} w|_{(x_1,t_1)} + {}^C D^{\gamma}_{b,x} w|_{(x_1,t_1)}\right] \leq 0, \forall \gamma \in (0,1),$$
$$\left[{}^C D^{\beta}_{a,x} w|_{(x_1,t_1)} + {}^C D^{\beta}_{b,x} w|_{(x_1,t_1)}\right] \leq 0, \forall \beta \in (0,2]. \tag{9.17}$$

Applying the properties of the Caputo fractional derivative operator to z, we find

$$P(D_t)\mathscr{C}_1 = P(D_t)z + \frac{\varepsilon}{2T}\left(\frac{t^{1-\alpha}}{\Gamma(2-\alpha)} + \sum_{i=1}^{m}\varpi_i \frac{t^{1-\alpha}}{\Gamma(2-\alpha)}\right),$$

where $\varpi_i \geq 0$ and $i = 1,2,\ldots,m \in \mathbb{N}$. Therefore, in view of each component of the system in Eq. (9.4) we have

$$\begin{aligned}
&\left(P(D_t)\mathscr{C}_1 - \mathscr{D}_1\left[{}^{C}D^{\gamma}_{a,x}\mathscr{C}_1 + {}^{C}D^{\gamma}_{b,x}\mathscr{C}_1\right] - \xi_c\xi_{an}x^{\delta}\mathscr{C}_1\right)|_{(x_1,t_1)} = P(D_t)z \\
&+\frac{\varepsilon}{2T}\left(\frac{t_1^{1-\alpha}}{\Gamma(2-\alpha)} + \sum_{i=1}^{m}\varpi_i \frac{t_1^{1-\alpha}}{\Gamma(2-\alpha)}\right) \\
&-\xi_c\xi_{an}x^{\delta}\left(z(x_1,t_1) - \frac{\varepsilon(T-t_1)}{2T}\right) \\
&\geq \frac{\varepsilon}{2T}\left(\frac{t_1^{1-\alpha}}{\Gamma(2-\alpha)} + \sum_{i=1}^{m}\varpi_i \frac{t_1^{1-\alpha}}{\Gamma(2-\alpha)}\right) \\
&-\xi_c\xi_{an}x^{\delta}\varepsilon\left(1 - \frac{T-t_1}{2T}\right) > 0,
\end{aligned}$$

which is a contradiction. □

By replacing $\mathscr{C}_1 = -\mathscr{C}_1$ into Theorem 9.1, the minimum principle follows.

Theorem 9.2: Let $\mathscr{C}_1 \in R(\Omega)$ denote the solution for one component of the system of FPPDEs in Eq. (9.4) in Ω, in which $\frac{d\mathscr{C}_1}{dx}|_{x=a} \geq 0$ and $\frac{d\mathscr{C}_1}{dx}|_{x=b} \leq 0$. Then either $\mathscr{C}_1(x,t) \leq 0, \forall (x,t) \in \bar{\Omega}$, or $\mathscr{C}_1$ attains its positive minimum on the bottom or back-side parts of the sides $S = \{[a,b]\times\{a\}\cup\{a\}\times[a,T]\cup\{b\}\times[a,T]\}S$ of the boundary of the domain Ω.

9.2.1 UNIQUENESS AND CONTINUOUS DEPENDENCE OF THE SOLUTION

The maximum principle and minimum principle are applied to the system in Eq. (9.4) by means of the following theorem.

Theorem 9.3: Let $\mathscr{C}_1$ denote a classical solution to the system in Eq. (9.4), $F \in C(\bar{\Omega})$, $\frac{d\mathscr{C}_1}{dx}|_{x=a} = 0$ and $\frac{d\mathscr{C}_1}{dx}|_{x=b} = 0$. Then,

$$\|\mathscr{C}_1\|_{C(\bar{\Omega})} \leq \max\{M_0, M_1, M_2\} + \frac{2T^{\alpha}M}{\Gamma(1+\alpha)}, \tag{9.18}$$

where $M_1 := \|\mathscr{C}_1(0)\|_{C(\bar{\Omega})}$, $M_1 = \|\mathscr{C}_1'(a,t)\|$, $M_2 = \|\mathscr{C}_1'(b,t)\|$, and since $M = \|F\|$.

Proof: This follows easily since the boundary conditions are Neumann boundary conditions and the source term for the system in Eq. (9.4) is identically zero.

Theorem 9.4: Let $\frac{d\mathscr{C}_1}{dx}|_{x=a} = 0$ and $\frac{d\mathscr{C}_1}{dx}|_{x=b} = 0$. Then the system in Eq. (9.4) possesses at most one classical solution. This solution, if it exists, continuously depends on the data associated with the system in Eq. (9.4) in the sense that if

$$\|F - \tilde{F}\| \leq \varepsilon,\ \|\mathscr{C}_1(0) - \tilde{\mathscr{C}_1}(0)\|_{C(\bar{\Omega})} \leq \varepsilon_0,\ \|\mathscr{C}_1'(a,t) - \tilde{\mathscr{C}_1'}(a,t)\| \leq \varepsilon_1,\ \|\mathscr{C}_1'(b,t) - \tilde{\mathscr{C}_1'}(b,t)\| \leq \varepsilon_2,$$

then

$$\|\mathscr{C}_1 - \tilde{\mathscr{C}_1}\|_{C(\bar{\Omega})} \leq \max\{\varepsilon, \varepsilon_0, \varepsilon_1, \varepsilon_2\} + \frac{2T^\alpha \varepsilon}{\Gamma(1+\alpha)}$$

holds.

9.2.2 EQUILIBRIUM FOR THE DYNAMICS

At steady state the dynamics in Eq. (9.4) for one component becomes

$$\mathscr{D}\frac{\partial^2 \mathscr{C}_1(x)}{\partial x^2} + \xi_c \xi_{an} x^\delta \mathscr{C}_1 = 0. \tag{9.19}$$

Theorem 9.5: The steady-state solution to the equation in Eq. (9.19) is

$$\mathscr{C}_1(x,t) = c_j \sin(j\pi x), \quad \text{for } j \in \mathbb{N}. \tag{9.20}$$

Proof: Thus, corresponding characteristic equation [5] for the second-order linear ODE in Eq. (9.19) is

$$\mathscr{D}\mathfrak{C}^2 + \xi_c \xi_{an} x^\delta = 0. \tag{9.21}$$

Solving the characteristic Eq. (9.21) we obtain

$$\mathfrak{C}^2 = -\frac{\xi_c \xi_{an} x^\delta}{\mathscr{D}}. \tag{9.22}$$

Therefore, from Eq. (9.22) we have

$$\mathfrak{C} = \pm\sqrt{-\frac{\xi_c \xi_{an} x^\delta}{\mathscr{D}}} \in \mathbb{C}.$$

Thus, the general solution for second-order linear ODE in Eq. (9.19) is

$$\mathscr{C}_1(x) = c_1 \cos\left(\frac{\xi_c \xi_b x^\delta}{\mathscr{D}} x\right) + c_2 \sin\left(\frac{\xi_c \xi_{an} x^\delta}{\mathscr{D}} x\right), \tag{9.23}$$

where c_1, c_2 are constants to be determined. Applying the Neumann boundary condition to Eq. (9.23), we have

$$\mathscr{C}_1'(x) = c_2 \frac{(\delta+1)\xi_c \xi_{an} x^\delta}{\mathscr{D}} \cos\left(\frac{A_c \xi_{an} x^\delta}{\mathscr{D}} x\right) - c_1 \frac{(\delta+1)\xi_c \xi_{an} x^\delta}{\mathscr{D}} \sin\left(\frac{\xi_c \xi_{an} x^\delta}{\mathscr{D}} x\right). \tag{9.24}$$

When $x = 0$, the equations in (9.24) reduce to

$$c_2 \frac{(\delta+1)\xi_c \xi_{an} x^\delta}{\mathscr{D}} = 0, \Rightarrow c_2 = 0.$$

Consequently, when $x = b$, we have

$$-c_1 \frac{(\delta+1)\xi_c \xi_{an} b^\delta}{\mathscr{D}} \sin\left(\frac{\xi_c \xi_{an}}{\mathscr{D}} b^{\delta+1}\right) = 0, \Rightarrow \left(\frac{\xi_c \xi_{an}}{\mathscr{D}} b^{\delta+1}\right) = \pm j\pi, \forall j \in \mathbb{N}. \tag{9.25}$$

Thus, in view of Eq. (9.25) Theorem 9.6 follows. □

Theorem 9.6: The eigenvalue of Eq. (9.19) is negative.

In order to prove Theorem 9.6, we first have to prove the following lemma.

Lemma 9.5: *Let $f, g \in C_1^2[a,b]$ and $\lambda \in \mathbb{R}^+$. Then,*

$$\int_a^b [f''(x) + \lambda f(x)]g(x)dx = \int_a^b f(x)[g''(x) + \lambda g(x)]dx. \tag{9.26}$$

Proof: Since

$$\frac{d[f'(x)g(x)]}{dx} = f''(x)g(x) + f'(x)g'(x), \tag{9.27}$$

then, integrating the equation in (9.27), we obtain

$$\begin{aligned}
\int_a^b \frac{d[f'(x)g(x)]}{dx} dx &= \int_a^b [f''(x)g(x) + f'(x)g'(x)]dx, \\
\Rightarrow [f'(x)g(x)]|_a^b &= \int_a^b [f''(x)g(x) + f'(x)g'(x)]dx, \\
\Rightarrow 0 &= \int_a^b [f''(x)g(x) + f'(x)g'(x)]dx, \\
\Rightarrow \int_a^b f''(x)g(x))dx &= -\int_a^b f'(x)g'(x)dx.
\end{aligned} \tag{9.28}$$

Similarly,

$$\frac{d[f(x)g'(x)]}{dx} = f'(x)g'(x) + f(x)g''(x). \tag{9.29}$$

Integrating the equation in (9.29), we obtain

$$\int_a^b f(x)g''(x))dx = -\int_a^b f'(x)g'(x)dx. \tag{9.30}$$

Combining the equation in (9.28) with Eq. (9.30), we obtain

$$\int_a^b [f''(x)g(x)]dx = \int_a^b [f(x)g''(x)]dx,$$

$$\Rightarrow \int_a^b [f''(x) + \lambda f(x)]g(x)dx = \int_a^b f(x)[g''(x) + \lambda g(x)]dx, \tag{9.31}$$

which conclude the proof. □

Now we are in a position to prove Theorem 9.6.

Proof of Theorem 9.6: It suffices to prove for one component of $\mathscr{C}$, so for the second component of $\mathscr{C}$ it follows similarly. Thus, we let $f = g = \mathscr{C}_1$ and $\lambda = \frac{\xi_c \xi_{an} x^\delta}{\mathscr{D}}$ in Lemma 9.5. Then,

$$\int_a^s [\mathscr{C}_1'' + \frac{\xi_c \xi_{an} x^\delta}{\mathscr{D}} \mathscr{C}_1]\mathscr{C}_1 dx = -\int_a^s [(\mathscr{C}_1')^2 + \frac{\xi_c \xi_{an} x^\delta}{\mathscr{D}} (\mathscr{C}_1)^2]dx = 0. \tag{9.32}$$

If $\lambda > 0$, this implies that $(\mathscr{C}_1')^2 = 0$ and $(\mathscr{C}_1)^2 = 0, \forall\, x$. Then $\mathscr{C}_1$ cannot be an eigenvalue. If $\lambda = 0$ then $\mathscr{C}_1' = 0$ by the Neumann boundary condition. Hence λ is the only possibility. □

9.3 DERIVATION OF THE NUMERICAL METHOD

In this section we derive the method which enables us to capture the dynamics of the formulation in Eq. (9.4). The method is one of the well-known methods [25]. Thus, we are able to discretize the fractional and/or integer derivative operators present in the dynamics of Eq. (9.4), and for each component of $\mathscr{C}$ in Eq. (9.4) we discretize the interval $[a, t_f], t_f \in \mathbb{N}$ through the points

$$a = t_0 < t_1 < t_2 < \cdots < t_f,$$

where the step size $\Delta t = t_{j+1} - t_j = t_f / S_t$, for $j = 0, 1, \ldots, S_t$, where $S_t \in \mathbb{N}$. Applying the backward fractional difference approximation operator for the Caputo fractional derivative ${}_a^C D_t^\alpha (\mathscr{C}_1)(t)$ to a component of the system in Eq. (9.4), then

$${}_a^C D_t^\alpha (\mathscr{C}_1)(t) \approx \frac{\nabla_-^\alpha (\mathscr{C}_1)(t_k)}{h^\alpha} = h^{-\alpha} \sum_{j=0}^{k} (-1)^j \begin{pmatrix} \alpha \\ j \end{pmatrix} (\mathscr{C}_1)_{k-j}, k = 0, 1, \cdots, S_t, \tag{9.33}$$

which can be re-expressed as

$$\begin{bmatrix} h^{-\alpha}\nabla_-^\alpha(\mathscr{C}_1)(t_0) \\ h^{-\alpha}\nabla_-^\alpha(\mathscr{C}_1)(t_1) \\ h^{-\alpha}\nabla_-^\alpha(\mathscr{C}_1)(t_2) \\ \vdots \\ h^{-\alpha}\nabla_-^\alpha(\mathscr{C}_1)(t_{S_t-1}) \\ h^{-\alpha}\nabla_-^\alpha(\mathscr{C}_1)(t_{S_t}) \end{bmatrix} = \frac{1}{h^\alpha} \begin{bmatrix} \omega_0^{(\alpha)} & 0 & 0 & 0 & \cdots & 0 \\ \omega_1^{(\alpha)} & \omega_0^{(\alpha)} & 0 & 0 & \cdots & 0 \\ \ddots & \ddots & \ddots & \ddots & \cdots & \cdots \\ \omega_{S_t-1}^{(\alpha)} & \ddots & \omega_2^{(\alpha)} & \omega_1^{(\alpha)} & \omega_0^{(\alpha)} & 0 \\ \omega_{S_t}^{(\alpha)} & \omega_{S_t-1}^{(\alpha)} & \ddots & \omega_2^{(\alpha)} & \omega_1^{(\alpha)} & \omega_0^{(\alpha)} \end{bmatrix} \begin{bmatrix} (\mathscr{C}_1)_0 \\ (\mathscr{C}_1)_1 \\ (\mathscr{C}_1)_2 \\ \vdots \\ (\mathscr{C}_1)_{S_t-1} \\ (\mathscr{C}_1)_{S_t} \end{bmatrix}, \tag{9.34}$$

where

$$\omega_j^{(\alpha)} = (-1)^j \left(\begin{array}{c} \alpha \\ j \end{array} \right), j = 0, 1, \cdots, S_t.$$

Similarly, the forward fractional difference approximation for the Caputo fractional derivative operator ${}_a^C D_t^\alpha(\cdot)(t)$ for a component of the system in Eq. (9.4) is

$$\begin{bmatrix} h^{-\alpha}\nabla_+^\alpha(\mathscr{C}_1)(t_0) \\ h^{-\alpha}\nabla_+^\alpha(\mathscr{C}_1)(t_1) \\ h^{-\alpha}\nabla_+^\alpha(\mathscr{C}_1)(t_2) \\ \vdots \\ h^{-\alpha}\nabla_+^\alpha(\mathscr{C}_1)(t_{S_t-1}) \\ h^{-\alpha}\nabla_+^\alpha(\mathscr{C}_1)(t_{S_t}) \end{bmatrix} = \frac{1}{h^\alpha} \begin{bmatrix} \omega_0^{(\alpha)} & \omega_1^{(\alpha)} & \omega_2^{(\alpha)} & \ddots & \omega_{S_t-1}^{(\alpha)} & \omega_{S_t}^{(\alpha)} \\ 0 & \omega_0^{(\alpha)} & \omega_1^{(\alpha)} & \ddots & \ddots & \omega_{S_t-1}^{(\alpha)} \\ 0 & 0 & \omega_0^{(\alpha)} & \ddots & \omega_2^{(\alpha)} & \ddots \\ 0 & 0 & 0 & \ddots & \omega_1^{(\alpha)} & \omega_2^{(\alpha)} \\ \cdots & \cdots & \cdots & \cdots & \omega_0^{(\alpha)} & \omega_1^{(\alpha)} \\ 0 & 0 & 0 & \cdots & 0 & \omega_0^{(\alpha)} \end{bmatrix} \begin{bmatrix} (\mathscr{C}_1)_0 \\ (\mathscr{C}_1)_1 \\ (\mathscr{C}_1)_2 \\ \vdots \\ (\mathscr{C}_1)_{S_t-2} \\ (\mathscr{C}_1)_{S_t-1} \\ (\mathscr{C}_1)_{S_t} \end{bmatrix}. \tag{9.35}$$

For the spatial distribution, we discretize the interval $[a, b]$ as

$$a = x_0 < x_1 < x_2 < \cdots < x_{S_x} = x_f = b,$$

in which the step size is given by $\Delta x = x_{i+1} - x_i = x_f / S_x$, $i = 0, 1, \ldots, S_x$. According to [25–27], for one component of $\mathscr{C}$, we have that adding equation (9.34) to the equation in (9.35) yields

$$\begin{bmatrix} \frac{1}{h^\alpha}(\nabla_+^\alpha + \nabla_-^\alpha)(\mathscr{C}_1)(t_0) \\ \frac{1}{h^\alpha}(\nabla_+^\alpha + \nabla_-^\alpha)(\mathscr{C}_1)(t_1) \\ \frac{1}{h^\alpha}(\nabla_+^\alpha + \nabla_-^\alpha)(\mathscr{C}_1)(t_2) \\ \vdots \\ \frac{1}{h^\alpha}(\nabla_+^\alpha + \nabla_-^\alpha)(\mathscr{C}_1)(t_{S_t-1}) \\ \frac{1}{h^\alpha}(\nabla_+^\alpha + \nabla_-^\alpha)(\mathscr{C}_1)(t_{S_t}) \end{bmatrix} = \frac{1}{h^\beta} \begin{bmatrix} \omega_0^{(\beta)} & \omega_1^{(\beta)} & \omega_2^{(\beta)} & \omega_3^{(\beta)} & \cdots & \omega_m^{(\beta)} \\ \omega_1^{(\beta)} & \omega_0^{(\beta)} & \omega_1^{(\beta)} & \omega_2^{(\beta)} & \cdots & \omega_{m-1}^{(\beta)} \\ \omega_2^{(\beta)} & \omega_1^{(\beta)} & \omega_0^{(\beta)} & \omega_1^{(\beta)} & \cdots & \omega_{m-2}^{(\beta)} \\ \ddots & \ddots & \ddots & \ddots & \cdots & \cdots \\ \omega_{m-1}^{(\beta)} & \ddots & \omega_2^{(\beta)} & \omega_1^{(\beta)} & \omega_0^{(\beta)} & \omega_1^{(\beta)} \\ \omega_m^{(\beta)} & \omega_{m-1}^{(\beta)} & \ddots & \omega_2^{(\beta)} & \omega_1^{(\beta)} & \omega_0^{(\beta)} \end{bmatrix} \begin{bmatrix} (\mathscr{C}_1)_0 \\ (\mathscr{C}_1)_1 \\ (\mathscr{C}_1)_2 \\ \vdots \\ (\mathscr{C}_1)_{S_t-2} \\ (\mathscr{C}_1)_{S_t-1} \\ (\mathscr{C}_1)_{S_t} \end{bmatrix}.$$

In [25], it is mentioned that the simultaneous approximations of αth-order time derivative of $\mathscr{C}_1(x,t)$ in all nodes are obtained when all the function values $(\mathscr{C}_1)_{ij}$ at the discretization nodes form a column vector of the form

$$(\mathscr{C}_1)_{t_f x_f} = \begin{bmatrix} (\mathscr{C}_1)_{t_f,x_f}(\mathscr{C}_1)_{x_{f-1},t_f} \cdots (\mathscr{C}_1)_{1,x_f}(\mathscr{C}_1)_{0,x_f} \\ (\mathscr{C}_1)_{t_f,x_{f-1}}(\mathscr{C}_1)_{t_{f-1},x_{f-1}} \cdots (\mathscr{C}_1)_{1,x_{f-1}}(\mathscr{C}_1)_{0,x_{f-1}} \\ \cdots\cdots \\ (\mathscr{C}_1)_{t_f,1}(\mathscr{C}_1)_{t_{f-1},1} \cdots (\mathscr{C}_1)_{1,1}(\mathscr{C}_1)_{0,1} \\ (\mathscr{C}_1)_{t_f,0}(\mathscr{C}_1)_{t_{f-1},0} \cdots (\mathscr{C}_1)_{1,0}(\mathscr{C}_1)_{0,0} \end{bmatrix}.$$

Consequently, the matrix that transforms the vector $(\mathscr{C}_1)_{t_f x_f}$ to the vector $(\mathscr{C}_1)_t^{(\alpha)}$ of the partial fractional derivative of order α with respect to time variable can be obtained as a Kronecker product of the matrix [25]

$$(\mathbf{B}_1)^{(\alpha)}_{t_f\times t_f}=\frac{1}{h^{\alpha}}\begin{bmatrix}\omega_0^{(\alpha)} & 0 & 0 & 0 & \cdots & 0\\ \omega_1^{(\alpha)} & \omega_0^{(\alpha)} & 0 & 0 & \cdots & 0\\ \ddots & \ddots & \ddots & \ddots & \cdots & \cdots\\ \omega_{S_t-1}^{(\alpha)} & \ddots & \omega_2^{(\alpha)} & \omega_1^{(\alpha)} & \omega_0^{(\alpha)} & 0\\ \omega_{S_t}^{(\alpha)} & \omega_{S_t-1}^{(\alpha)} & \ddots & \omega_2^{(\alpha)} & \omega_1^{(\alpha)} & \omega_0^{(\alpha)}\end{bmatrix},$$

which corresponds to the ordinary fractional derivative of order α and the unit matrix $\mathbf{I}_{x_f\times x_f}$. Thus, through the Kronecker product [25], the time discretization of $(\mathscr{C}_1)_{t_f x_f}$ can be expressed as

$$(\mathbf{B}_1)^{(\alpha)}_{t_f\times t_f}\otimes\mathbf{I}_{x_f\times x_f}. \tag{9.36}$$

Similarly, the matrix that transforms the vector $\mathscr{C}$ to the vector $\mathscr{C}_x^{(\beta)}$ of the partial fractional derivative of order β with respect to a spatial variable is obtained through a Kronecker product of the unit matrix $\mathbf{I}_{x_f\times x_f}$ and the matrix

$$(\mathbf{B}_1)^{(\beta)}_{x_f\times x_f}=\frac{1}{h^{\beta}}\begin{bmatrix}\omega_0^{(\beta)} & \omega_1^{(\beta)} & \omega_2^{(\beta)} & \omega_3^{(\beta)} & \cdots & \omega_m^{(\beta)}\\ \omega_1^{(\beta)} & \omega_0^{(\beta)} & \omega_1^{(\beta)} & \omega_2^{(\beta)} & \cdots & \omega_{m-1}^{(\beta)}\\ \omega_2^{(\beta)} & \omega_1^{(\beta)} & \omega_0^{(\beta)} & \omega_1^{(\beta)} & \cdots & \omega_{m-2}^{(\beta)}\\ \ddots & \ddots & \ddots & \ddots & \cdots & \cdots\\ \omega_{m-1}^{(\beta)} & \ddots & \omega_2^{(\beta)} & \omega_1^{(\beta)} & \omega_0^{(\beta)} & \omega_1^{(\beta)}\\ \omega_m^{(\beta)} & \omega_{m-1}^{(\beta)} & \ddots & \omega_2^{(\beta)} & \omega_1^{(\beta)} & \omega_0^{(\beta)}\end{bmatrix}.$$

We see that Eq. (9.36) implies that a symmetric Riesz of the ordinary derivative of order β is given by

$$\mathbf{I}_{t_f\times t_f}\otimes(\mathbf{B}_1)^{(\beta)}_{x_f\times x_f}. \tag{9.37}$$

Thus, in view of systems in Eqs. (9.36) and (9.37), we see that the discretization for the dynamics in Eq. (9.4) is achieved by replacing the derivatives with their discrete analogues in Eqs. (9.36) and (9.37). Hence, our novel numerical method for the system of FPPDEs in Eq. (9.4) is a system of well-posed linear discrete equations [25]:

$$\left.\begin{aligned}\left[(\mathbf{B}_1)^{(\alpha)}_{t_f\times t_f}\otimes\mathbf{I}_{x_f\times x_f}-\mathscr{D}_1\mathbf{I}_{t_f\times t_f}\otimes(\mathbf{B}_1)^{(\beta)}_{x_f\times x_f}\right](\mathscr{C}_1)_{t_f,x_f}&=\xi_c\xi_b x^{\delta}_{x_f}(\mathscr{C}_1)_{t_f,x_f},\\ \left[(\mathbf{B}_2)^{(\alpha)}_{t_f\times t_f}\otimes\mathbf{I}_{x_f\times x_f}-\mathscr{D}_2\mathbf{I}_{t_f\times t_f}\otimes(\mathbf{B}_2)^{(\beta)}_{x_f\times x_f}\right](\mathscr{C}_2)_{t_f,x_f}&=\xi_c\xi_b x^{\delta}_{x_f}(\mathscr{C}_2)_{t_f,x_f},\end{aligned}\right\}, \tag{9.38}$$

where we are imposing the auxiliary initial conditions [25]

$$\left.\begin{aligned}(\mathscr{E}_1)_{t_f,x_f}(x,t)&=(\mathscr{C}_1)_{t_f,x_f}(x,t)-(\mathscr{C}_1)_{t_f,x_f}(x,0),\\ (\mathscr{E}_2)_{t_f,x_f}(x,t)&=(\mathscr{C}_2)_{t_f,x_f}(x,t)-(\mathscr{C}_2)_{t_f,x_f}(x,0),\end{aligned}\right\} \tag{9.39}$$

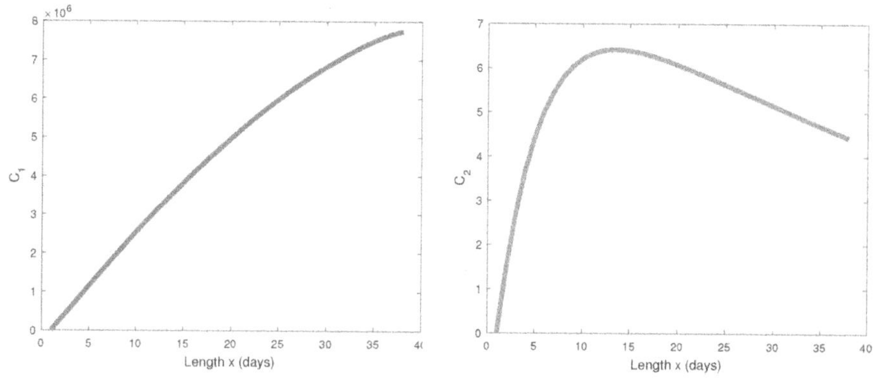

(a) (b)

Figure 9.1 Numerical solution for the dynamics in Eq. (9.38) presenting the spatial distributions of: (a) quarantine and isolation ignored, and (b) quarantine and isolation applied, when $\alpha = 1, \beta = 2$, $\xi_c = 0.09$, and $\xi_{an} = 1.9$.

due to the possibility of non-zero initial conditions $(\mathscr{C}_1)_{t_f,x_f}(x,0), (\mathscr{C}_2)_{t_f,x_f}(x,0)$. But the Caputo derivative for the auxiliary functions $(\mathscr{E}_1)_{t_f,x_f}(x,t), (\mathscr{E}_2)_{t_f,x_f}(x,t)$ given in Eq. (9.39) is zero [28,29]. This implies that initial and boundary conditions for the well-posed system of linear discrete equations in Eq. (9.38) are zero.

Remark 9.1: *When $\alpha = 1$ and $\beta = 2$, the novel numerical method in Eq. (9.38) becomes the standard finite difference method [30].*

9.4 NUMERICAL RESULTS AND DISCUSSIONS

Here we present numerical simulations of the fractional derivative COVID-19 model in Eq. (9.4). We let $a = 0, b = 1$, and the number of days are 38. In the absence of scientific and/or experimental data, we set the spatial distribution $\mathscr{D} = [1 \times 10^{-9}, 1 \times 10^{-9}]^T$. The COVID-19 spread and immune proliferation effects are set to $\xi_c = 0.09, 0.9, 1.9$, and $\xi_{an} = 0.09, 0.9, 1.9$, respectively, and $\delta \in \mathbb{N} \setminus \{1\} \leq 38$ days.

In Figure 9.1, we present the respective spatial distributions of the spread for $\alpha = 1, \beta = 2$ and $\xi_c = 0.09$ and $\xi_{an} = 1.9$; the respective spatial distributions of the spread for $\alpha = 1, \beta = 2$ and $\xi_c = 0.9$ and $\xi_{an} = 1.9$ are presented in Figure 9.2; and we present the spatial distributions of the spread for $\alpha = 1, \beta = 2$ and $\xi_c = 1.9$ and $\xi_{an} = 1.9$ in Figure 9.3.

Similarly, in Figure 9.4 we present the respective spatial distributions of the spread for $\alpha = 0.5, \beta = 1.5$ and $\xi_c = 0.09$ and $\xi_{an} = 1.9$; the respective spatial distributions of the spread for $\alpha = 0.5, \beta = 1.5$ and $\xi_c = 0.9$ and $\xi_{an} = 1.9$ are presented in Figure 9.5; and we present the spatial distributions of the spread for $\alpha = 0.5, \beta = 1.5$ and $\xi_c = 1.9$ and $\xi_{an} = 1.9$ in Figure 9.6.

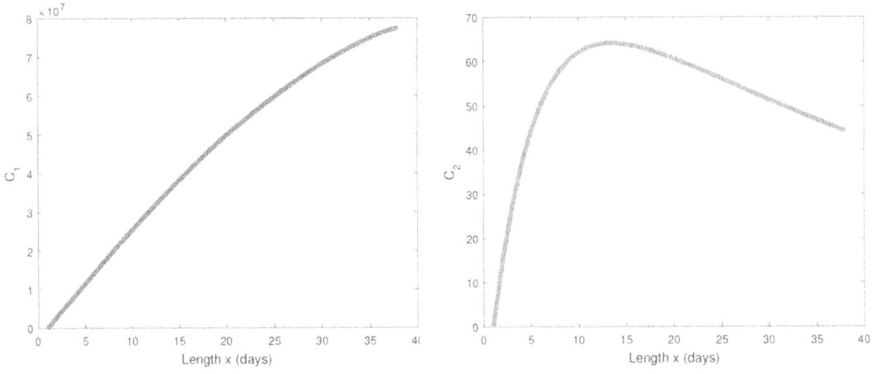

(a) (b)

Figure 9.2 Numerical solution for the dynamics in Eq. (9.38), presenting the spatial distributions of: (a) quarantine and isolation ignored, and (b) quarantine and isolation applied, when $\alpha = 1, \beta = 2, \xi_c = 0.9$, and $\xi_{an} = 1.9$.

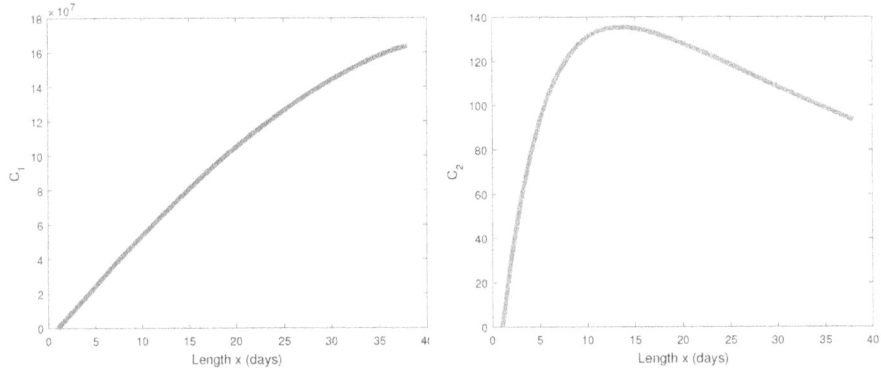

(a) (b)

Figure 9.3 Numerical solution for the dynamics in Eq. (9.38), presenting the spatial distributions of: (a) quarantine and isolation ignored, and (b) quarantine and isolation applied, when $\alpha = 1, \beta = 2, \xi_c = 1.9$, and $\xi_{an} = 1.9$.

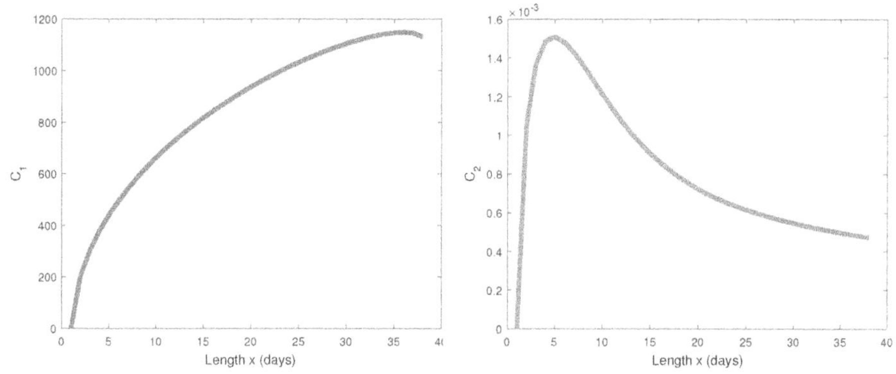

(a) (b)

Figure 9.4 Numerical solution for the dynamics in Eq. (9.38), presenting the spatial distributions of: (a) quarantine and isolation ignored, and (b) quarantine and isolation applied, when $\alpha = 0.5, \beta = 1.5$, $\xi_c = 0.09$, and $\xi_{an} = 1.9$.

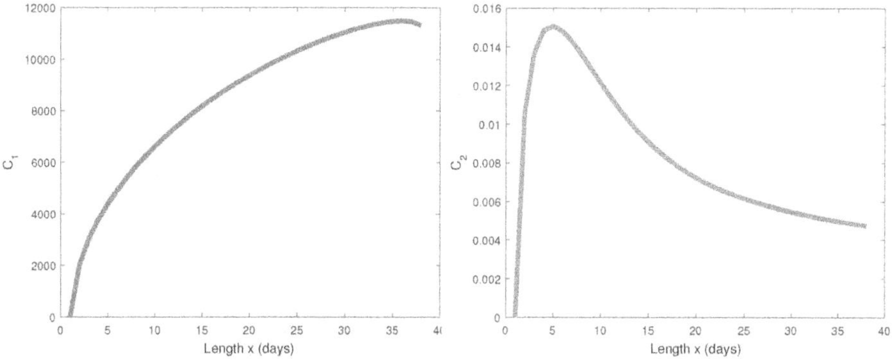

(a) (b)

Figure 9.5 Numerical solution for the dynamics in Eq. (9.38), presenting the spatial distributions of: (a) quarantine and isolation ignored, and (b) quarantine and isolation applied, when $\alpha = 0.5, \beta = 1.5$, $\xi_c = 0.9$, and $\xi_{an} = 1.9$.

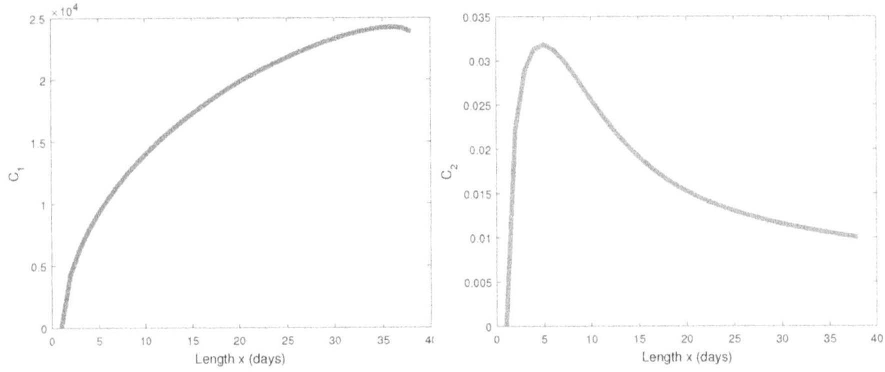

(a) (b)

Figure 9.6 Numerical solution for the dynamics in Eq. (9.38), presenting the spatial distributions of: (a) quarantine and isolation ignored, and (b) quarantine and isolation applied, when $\alpha = 0.5, \beta = 1.5, \xi_c = 1.9$, and $\xi_{an} = 1.9$.

In Figure 9.1, we see COVID-19 spreading at an exponential rate immediately from day one due to the fact that the principle of quarantine and isolation is not applied, whereas when the principle is applied we see the spread initially increasing steadily due to those tested positive, then within one or two days the COVID-19 spread is drastically reducing to a low level.

In Figure 9.2, when the proliferation effect of the COVID-19 spread is slightly increased and the effect on the immune system is kept the same, relative to the effects in Figure 9.1, we see similar behaviour for both dynamics, but only a slight increase for the dynamics corresponding to the principle of quarantine and isolation not applied but double increase in the dynamics for the principle of quarantine and isolation applied.

Similarly, in Figure 9.3, when both proliferation effects are equal, we see that both dynamics not applying and applying the quarantine and isolation have doubled their profiles, compared to their profiles presented in Figure 9.2.

In Figure 9.4, we see similar behaviours to the profiles presented in Figure 9.1, whereas in Figures 9.5 and 9.6, we again see similar profiles but the dynamics corresponding to the principle of quarantine and isolation ignored are doubling at a low profiles contrary to what we have presented in Figures 9.1–9.3. However, we see the profiles corresponding to the principle applied decreasing drastically, irrespective the increase in the effects of the proliferation effects of COVID-19 in a host and those tested positive.

9.5 CONCLUSION

In this work, we have considered the dynamics modelling COVID-19 spread in a given host and/or community. Since the spread is curtailed mainly by the dynamics of quarantine and isolation, the initial value problem in the form of an ordinary differential equation is derived to model such principle. However, the impact of COVID-19 spread has been huge worldwide. Thus, in an effort to present desirable results, we have extended the dynamics to include spatial effects of the COVID-19 spread. The dynamics are then modelled by a system of parabolic partial differential equations. The system of PPDEs is classical, while the deadly effect of COVID-19 has enormous across all human activities worldwide. For this reason, we have extended the dynamics from the classical PPDE to the system of fractional parabolic partial differential equations, in order to ensure that we are able to capture all the features of phenomena. Consequently, we have presented unique solutions, continuously dependent on the data, and have analyzed the phenomena, established the equilibrium, and designed a novel numerical method which in turn enabled us to capture the behaviours of the phenomena with respect to the parameter values presented. Since, the four cases presented have revealed differences in the influences of prevention strategies implemented towards the epidemic, we believe that our results indeed demonstrate that the phenomena modelled by the system of fractional parabolic partial differential equations outperform the classical phenomena. This can be viewed in the profile of quarantine and isolation as a prevention method, with COVID-19 spread decreasing as the number of people getting infected is reduced as compared to the profiles corresponding to the classical system. Thus, we believe that we have achieved our main goal towards helping society at large in understanding the principle of quarantine and isolation when a host tests positive for COVID-19. Therefore, we agree that policymakers and society must take quarantine and isolation as preventive measures to reduce the spread of the virus in their communities. Our future direction is to consider the system of fractional fractal phenomena in higher-dimensional space with a varying spatial distribution term.

AUTHORS' CONTRIBUTIONS

All authors contributed equally to the production of the manuscript.

ACKNOWLEDGEMENT*

This work was not funded.

REFERENCES

1. Chinazzi, M., Davis, J.T., Ajelli, M., Gioannini, C., Litvinova, M., Merler, S., Piontti, A.P., Mu, K., Rossi, L., Sun, K., Viboud, C., Xiong, X., Yu, H., Halloran, M.E., Longini, I.M. & Vespignani, A. (2020). The effect of travel restrictions on the spread of the 2019 novel coronavirus (COVID-19) outbreak. Science, 368(6489), 395–400.

2. Zhao, S. & Chen, H. (2020). Modeling the epidemic dynamics and control of COVID-19 outbreak in China. Quantitative Biology, 8(1), 11–19.

3. Hassan, S.A., Sheikh, F.N., Jamal, S., Ezeh, J.K., & Akhtar, A. (2020). Corona virus (COVID-19): A review of clinical features, diagnosis, and treatment. Cureus, 12(3), e7355.

4. Rustan, R. & Handayani, L. (2020). The Outbreak's modeling of coronavirus (COVID-19) using the modified SEIR model in Indonesia. Jurnal Fisika dan Aplikasinya, 5(1), 61–68.

5. Edwards, C.H. & Penney, D.E., Elementary Differential Equations with Boundary Value Problems, Pearson, Athens, 2008.

6. Kreutz, C., (2020). A new approximation approach for transient differential equation models. Frontiers in Physics, 8, 20.

7. Chen, T., Rui, J., Wang, Q., Zhao, Z., Cui, J. & Yin, L. (2020). A mathematical model for simulating the phase-based transmissibility of a novel corona virus. Infectious Diseases of Poverty, 9–24.

8. Liu, Z., Magal, P., Seydi, O. & Webb, A. (2020). A COVID-19 epidemic model with latency period. Infectious Disease Modelling, 5, 323–337.

9. Fanelli, D. & Piazza, F. (2020). Analysis and forecast of COVID-19 spreading in China, Italy and France. Chaos, Solitons & Fractals, 134, 109761.

10. Mishra, A.M., Purohit, S.D., Owolabi, K.M. & Sharma, Y.D. (2020). A nonlinear epidemiological model considering asymptotic and quarantine classes for SARS CoV-2 virus. Chaos, Solitons & Fractals, 109953.

11. Roques, L., Klein, E.K., Papaïx, J., Sar, A. & Soubeyrand, S. (2020). Using early data to estimate the actual infection fatality ratio from COVID-19 in France. Biology, 9, 97.

12. Caputo, M., Elasticitàe dissipazione, Zanichelli, Bologna, 1969.

13. Saichev, A. & Zaslavsky, G. (1997). Fractional kinetic equations: solutions and applications. Chaos, 7(4), 753–764.

14. Khan, M.A. & Atangana, A. (2020). Modeling the dynamics of novel coronavirus (2019-nCov) with fractional derivative, Alexandria Engineering Journal, https://doi.org/10.1016/j.aej.2020.02.03.

15. Singh, H. (2020). Analysis for fractional dynamics of Ebola virus model. Chaos, Solitons & Fractals, 138, 109992.

16. Podlubny, I. (2000). Matrix approach to discrete fractional calculus. Fractional Calculus and Applied Analysis, 3(4), 359–386.

17. Owolabi, K.M. & Shikongo, A. (2020). Mathematical modelling of multi-mutation and drug resistance model with fractional derivative. Alexandria Engineering Journal, https://doi.org/10.1016/j.aej.2020.02.014.

18. Owolabi, K.M. & Shikongo, A, (2020). Fractional operator method on a multi-mutation and intrinsic resistance model. Alexandria Engineering Journal, https://doi.org/10.1016/j.aej.2019.12.033.

19. Al-Refai, M. (2020). On weighted Atangana-Baleanu fractional operators. Advances in Difference Equations, 2020(3).

20. Singh, H., Srivastava, H.M. & Kumar, D. (2017). A reliable numerical algorithm for the fractional vibration equation. Chaos, Solitons & Fractals, 103, 131–138.

21. Singh, H. (2018). Approximate solution of fractional vibration equation using Jacobi polynomials. Applied Mathematics and Computation, 317, 85–100.

22. Singh, H., Kumar, V. & Baleanu, D. Methods of Mathematical Modelling: Fractional Differential Equations, CRC Press, Boca Raton, FL, 2019.

23. Luchko, Y. (2009). Maximum principle for the generalized time-fractional diffusion equation. Journal of Mathematical Analysis and Applications, 351, 218–223.

24. Mohammed, A. (2012). On the fractional derivatives at extreme points. Electronic Journal of Qualitative Theory of Differential Equations, 55, 1–5.

25. Podlubny, I., Chechkin, A.V., Skovranek, T., Chen, Y. & Jara, B.M.V. (2009). Matrix approach to discrete fractional calculus II: partial fractional differential equations. Journal of Computational Physics, 228(8), 3137–3153.

26. Ortigueira, M.D., (2006). Riesz potential operators and inverses via fractional centred derivatives. International Journal of Mathematics and Mathematical Sciences, 2006(48391), 1–12.

27. Ortigueira, M.D. & Batista, A.G. (2008). On the relation between the fractional Brownian motion and the fractional derivatives. Physics Letters A, 372, 958–968.

28. Meerschaert, M. & Tadjeran, C. (2004). Finite difference approximations for fractional advection-dispersion equations. Journal of Computational and Applied Mathematics, 172(1), 65–77.

29. Metzler, R., Barkai, E. & Klafter, J. (1999). Deriving fractional Fokker-Planck equations from a generalized master equation. Europhysics Letters, 46(4), 431–436.

30. Burden, R.L. & Faires, J.D., Numerical Analysis, Brooks/Cole, Boston, MA, 2011.

10 Superabundant Explicit Wave and Numerical Solutions of the Fractional Isotropic Extension Model of the KdV Model

Mostafa M.A. Khater[1,2] *and Raghda A.M. Attia*[3]
[1]Department of Mathematics, Faculty of Science, Jiangsu University, Zhenjiang, China
[2]Department of Mathematics, Obour High Institute For Engineering and Technology, Cairo, Egypt
[3]Department of Basic Science, Higher Technological Institute 10th of Ramadan City, El Sharqia, Egypt

CONTENTS

10.1 INTRODUCTION

Nowadays, fractional calculus has attracted many researchers in distinct fields such as mechanical engineering, quasi-chaotic dynamical systems, physics, electric control theory, finance, economics, solid-state biology, chemistry, food supplements, signal processing, fluid mechanics, economics, applied mathematics, hydrodynamics, system identification, statistics, dynamical systems with chaotic dynamical behavior, optical fibers, and so on [1–5]. Many complex phenomena cannot formulate in the nonlinear partial differential equation with the integer order because of their nonlocal property [6, 7]. In the other hand, these phenomena have been being formulated in fractional nonlinear partial differential equations [8–13] based on the historical and current states of the problem in the contract of the classical calculus [14–18]. However, despite the importance of these models, it is very difficult to evaluate their exact traveling wave solutions. Consequently, mathematicians and physicists have been devoting their valuable time to formulate many accurate analytical and numerical solutions of these phenomena [19–29]. Additionally, many fractional operators have been constructed to convert the fractional nonlinear partial differential equations into nonlinear ordinary differential equations with integer order, such as Riemann-Liouville derivatives, Caputo, Caputo-Fabrizio definition, and so on [30–33].

This chapter investigates the fractional (2+1)-dimensional generalized Nizhnik-Novikov-Veselov (GNNV) equations through seven recent computational methods (the exp $(-\phi(\Xi)$-expansion method, the extended Fan-expansion method, the extended $\left(\frac{G'(\Xi)}{G(\Xi)}\right)$-expansion method, the extended simplest equation method, the extended Tanh (Ξ)-expansion method, the modified Khater method, and the Adomian decomposition method), along with the Atangana conformable derivative operator. Additionally, we use the obtained analytical solutions to evaluate the boundary and initial conditions that allow application of the Adomian decomposition method and B-spline collection family for constructing the semi-analytical and numerical solutions where this part of chapter aims to show the accuracy of the obtained solutions. The (2+1)-dimensional GNNV equations are given by [34–38]

$$\begin{cases} \mathscr{U}_t + r_1\,\mathscr{U}_{xxx} + r_2\,\mathscr{U}_{yyy} + r_3\,\mathscr{U}_x + r_4\,\mathscr{U}_y - 3\,r_1\,(\mathscr{U}\,\mathscr{V})_x - 3\,r_2\,(\mathscr{U}\,\mathscr{W})_y = 0, \\ \mathscr{U}_x = \mathscr{V}_y, \\ \mathscr{U}_y = \mathscr{W}_x, \end{cases} \tag{10.1}$$

where a, b, c, d are arbitrary constants. This model is also known by isotropic extension of the well-known (1+1)-dimensional KdV equation. For $c = d = 0$, Eq. (10.1) reduces to the (2+1)-dimensional Nizhnik-Novikov-Veselov equations (NNVEs) [39–43].

Using the next, the Atangana conformable fractional transformation [44–46] $\left[\mathscr{U} = \mathscr{U}(x,y,t) = \mathscr{S}(\Xi), \mathscr{V} = \mathscr{V}(x,y,t) = \mathscr{F}(\Xi), \mathscr{W} = \mathscr{W}(x,y,t) = \mathscr{G}(\Xi), \Xi = x + y + \frac{\lambda}{\alpha}\left(t + \frac{1}{\Gamma(\alpha)}\right)^{\alpha}\right]$ and integrating the obtained equations once with zero constant of integration leads to

$$\begin{cases} (\lambda + r_3 + r_4)\,\mathscr{S} + (r_1 + r_2)\,\mathscr{S}'' - 3r_1\,(\mathscr{S}\,\mathscr{F}) - 3r_2\,(\mathscr{S}\,\mathscr{G}) = 0, \\ \mathscr{S} = \mathscr{F}, \\ \mathscr{S} = \mathscr{G}. \end{cases} \tag{10.2}$$

Substituting the second and third equations into the first equation of the same system yields

$$\mathscr{L}_1\,\mathscr{S} - 3\,\mathscr{L}_2\,\mathscr{S}^2 + \mathscr{L}_2\,\mathscr{S}'' = 0, \tag{10.3}$$

where $\mathscr{L}_1 = \lambda + r_3 + r_2$, $\mathscr{L}_2 = r_1 + r_2$. Using the balance principle in Eq. (10.3) results in $n = 2$.

The remaining sections of this chapter are organized as follows: Sections 10.2 and 10.4 apply four recent schemes and one semi-analytical method, and B-spline collection schemes, to the fractional nonlinear (2+1)-dimensional GNNV equations to obtain distinct formulas of analytical and semi-analytical wave solutions, and numerical solutions. Additionally, the obtained solutions have been provided in 2D, 3D, and contour plots to show novel properties of the incompressible fluid. Section 10.3 checks the stability property of the obtained solutions. Section 10.5 illustrates the interpretation of the shown figures and tables. Section 10.6 provides the conclusion.

10.2 ANALYTICAL EXPLICIT WAVE SOLUTIONS

This section applies some recent computational schemes to the fractional isotropic extension model of the KdV model ((2+1)-dimensional GNNV equations). The aim of this section is to investigate novel explicit wave solutions to explain more physical properties of this fractional model.

10.2.1 EXP(−$\phi(\Xi)$)-EXPANSION METHOD

Applying the exp($-\phi(\Xi)$)-expansion method along the value of calculated balance to the (2+1)-dimensional GNNV equations Eq. (10.3) leads us to formulate its general solution in the following form:

$$\mathscr{S}(\Xi) = \sum_{i=-n}^{n} a_i e^{-i\phi(\Xi)} = a_{-2}\,e^{2\phi(\Xi)} + a_{-1}\,e^{\phi(\Xi)} + a_1\,e^{-\phi(\Xi)} + a_2\,e^{-2\phi(\Xi)} + a_0, \tag{10.4}$$

where $a_i, (i = -2, -1, 0, 1, 2)$ are arbitrary constants, and $\phi(\Xi)$ satisfies the following ODE:

$$\phi'(\Xi) = \sigma\,e^{\phi(\Xi)} + \frac{1}{e^{\phi(\Xi)}} + \rho, \tag{10.5}$$

where σ, ρ are arbitrary constants to be determined later. Using Eq. (10.4) along (10.5) in the framework of the suggested technique yields the following values of the above-mentioned arbitrary constants.

Family I

$$a_{-2} \to 2\sigma^2, a_{-1} \to 2\rho\sigma, a_0 \to 2\sigma, a_1 \to 0, a_2 \to 0, \mathscr{L}_1 \to 4\sigma\mathscr{L}_2 - \rho^2\mathscr{L}_2.$$

Family II

$$a_{-2} \to 2\sigma^2, a_{-1} \to 2\rho\sigma, a_0 \to \frac{1}{3}\left(\rho^2 + 2\sigma\right), a_1 \to 0, a_2 \to 0, \mathscr{L}_1 \to \mathscr{L}_2\left(\rho^2 - 4\sigma\right).$$

Family III

$$a_{-2} \to 0, a_{-1} \to 0, a_0 \to 2\sigma, a_1 \to 2\rho, a_2 \to 2, \mathscr{L}_1 \to 4\sigma\mathscr{L}_2 - \rho^2\mathscr{L}_2.$$

Family IV

$$a_{-2} \to 0, a_{-1} \to 0, a_0 \to \frac{1}{3}\left(\rho^2 + 2\sigma\right), a_1 \to 2\rho, a_2 \to 2, \mathscr{L}_1 \to \mathscr{L}_2\left(\rho^2 - 4\sigma\right).$$

Thus, the explicit wave solutions of the (2+1)-dimensional GNNV equations (10.1) are given in the following order:
For $\rho^2 - 4\sigma > 0$, $\sigma \neq 0$, we get

$$\mathscr{U}_{I,1}(x,t) = \frac{4\sigma - \rho^2}{\cosh\left(\sqrt{\rho^2 - 4\sigma}(\mathscr{H} + \vartheta)\right) + 1}, \tag{10.6}$$

$$\mathscr{U}_{I,2}(x,t) = \frac{\rho^2 - 4\sigma}{\cosh\left(\sqrt{\rho^2 - 4\sigma}(\mathscr{H} + \vartheta)\right) - 1}, \tag{10.7}$$

$$\mathscr{U}_{II,1}(x,t) = -\frac{1}{6}\left(\rho^2 - 4\sigma\right)\left(3\operatorname{sech}^2\left(\frac{1}{2}\sqrt{\rho^2 - 4\sigma}(\mathscr{H} + \vartheta)\right) - 2\right), \tag{10.8}$$

$$\mathscr{U}_{II,2}(x,t) = \frac{1}{6}\left(\rho^2 - 4\sigma\right)\left(3\operatorname{csch}^2\left(\frac{1}{2}\sqrt{\rho^2 - 4\sigma}(\mathscr{H} + \vartheta)\right) + 2\right), \tag{10.9}$$

$$\mathscr{U}_{III,1}(x,t) = -\frac{2\sigma\left(\rho^2 - 4\sigma\right)}{\left(\sqrt{\rho^2 - 4\sigma}\sinh\left(\frac{1}{2}\sqrt{\rho^2 - 4\sigma}(\mathscr{H} + \vartheta)\right) + \rho\cosh\left(\frac{1}{2}\sqrt{\rho^2 - 4\sigma}(\mathscr{H} + \vartheta)\right)\right)^2}, \tag{10.10}$$

$$\mathscr{U}_{\text{III},2}(x,t) = \frac{2\sigma\left(\rho^2-4\sigma\right)}{\left(\rho\sinh\left(\frac{1}{2}\sqrt{\rho^2-4\sigma}(\mathscr{H}+\vartheta)\right)+\sqrt{\rho^2-4\sigma}\cosh\left(\frac{1}{2}\sqrt{\rho^2-4\sigma}(\mathscr{H}+\vartheta)\right)\right)^2}, \tag{10.11}$$

$$\mathscr{U}_{\text{IV},1}(x,t) = \frac{1}{3}\left(\rho^2+2\sigma\right)+\frac{8\sigma^2}{\left(\rho+\sqrt{\rho^2-4\sigma}\tanh\left(\frac{1}{2}\sqrt{\rho^2-4\sigma}(\mathscr{H}+\vartheta)\right)\right)^2} - \frac{4\rho\sigma}{\rho+\sqrt{\rho^2-4\sigma}\tanh\left(\frac{1}{2}\sqrt{\rho^2-4\sigma}(\mathscr{H}+\vartheta)\right)}, \tag{10.12}$$

$$\mathscr{U}_{\text{IV},2}(x,t) = \frac{1}{3}\left(\rho^2+2\sigma\right)+\frac{8\sigma^2}{\left(\rho+\sqrt{\rho^2-4\sigma}\coth\left(\frac{1}{2}\sqrt{\rho^2-4\sigma}(\mathscr{H}+\vartheta)\right)\right)^2} - \frac{4\rho\sigma}{\rho+\sqrt{\rho^2-4\sigma}\coth\left(\frac{1}{2}\sqrt{\rho^2-4\sigma}(\mathscr{H}+\vartheta)\right)}. \tag{10.13}$$

For $\rho^2-4\sigma>0$, $\sigma=0$, we get

$$\mathscr{U}_{\text{III},3}(x,t) = \frac{1}{2}\rho^2\text{csch}^2\left(\frac{1}{2}\rho(\mathscr{H}+\vartheta)\right), \tag{10.14}$$

$$\mathscr{U}_{\text{IV},3}(x,t) = \frac{1}{3}\rho^2\left(\frac{3}{\cosh(\rho(\mathscr{H}+\vartheta))-1}+1\right). \tag{10.15}$$

For $\rho^2-4\sigma=0$, $\sigma\neq 0, \rho\neq 0$, we get

$$\mathscr{U}_{\text{I},4}(x,t) = 2\sigma\left(\frac{4\sigma(\rho(\mathscr{H}+\vartheta)+2)^2-4\rho^3(\mathscr{H}+\vartheta)}{\rho^4(\mathscr{H}+\vartheta)^2}-1\right), \tag{10.16}$$

$$\mathscr{U}_{\text{II},4}(x,t) = \frac{1}{3}\left(\rho^2-10\sigma+\frac{24\sigma^2(\rho(\mathscr{H}+\vartheta)+2)^2}{\rho^4(\mathscr{H}+\vartheta)^2}-\frac{24\sigma}{\rho(\mathscr{H}+\vartheta)}\right), \tag{10.17}$$

$$\mathscr{U}_{\text{III},4}(x,t) = 2\sigma-\frac{\rho^3(\mathscr{H}+\vartheta)(\rho(\mathscr{H}+\vartheta)+4)}{2(\rho(\mathscr{H}+\vartheta)+2)^2}, \tag{10.18}$$

$$\mathscr{U}_{\text{IV},4}(x,t) = \frac{1}{6}\left(4\sigma+\frac{\rho^2(8-\rho(\mathscr{H}+\vartheta)(\rho(\mathscr{H}+\vartheta)+4))}{(\rho(\mathscr{H}+\vartheta)+2)^2}\right). \tag{10.19}$$

For $\rho^2 - 4\sigma = 0$, $\sigma = 0, \rho = 0$, we get

$$\mathscr{U}_{\text{III}.5}(x,t) = \frac{2}{(\mathscr{H} + \vartheta)^2}, \tag{10.20}$$

$$\mathscr{U}_{\text{IV}.5}(x,t) = \frac{2}{(\mathscr{H} + \vartheta)^2}. \tag{10.21}$$

For $\rho^2 - 4\sigma < 0$, $\sigma \neq 0$, we get

$$\mathscr{U}_{\text{I}.6}(x,t) = \frac{4\sigma - \rho^2}{\cos\left(\sqrt{4\sigma - \rho^2}(\mathscr{H} + \vartheta)\right) + 1}, \tag{10.22}$$

$$\mathscr{U}_{\text{I}.7}(x,t) = \frac{\rho^2 - 4\sigma}{\cos\left(\sqrt{4\sigma - \rho^2}(\mathscr{H} + \vartheta)\right) - 1}, \tag{10.23}$$

$$\mathscr{U}_{\text{II}.6}(x,t) = -\frac{1}{6}\left(\rho^2 - 4\sigma\right)\left(3\sec^2\left(\frac{1}{2}\sqrt{4\sigma - \rho^2}(\mathscr{H} + \vartheta)\right) - 2\right), \tag{10.24}$$

$$\mathscr{U}_{\text{II}.7}(x,t) = -\frac{1}{6}\left(\rho^2 - 4\sigma\right)\left(3\csc^2\left(\frac{1}{2}\sqrt{4\sigma - \rho^2}(\mathscr{H} + \vartheta)\right) - 2\right), \tag{10.25}$$

$$\mathscr{U}_{\text{III}.6}(x,t) = \\ -\frac{2\sigma\left(\rho^2 - 4\sigma\right)}{\left(\rho\cos\left(\frac{1}{2}\sqrt{4\sigma - \rho^2}(\mathscr{H} + \vartheta)\right) - \sqrt{4\sigma - \rho^2}\sin\left(\frac{1}{2}\sqrt{4\sigma - \rho^2}(\mathscr{H} + \vartheta)\right)\right)^2}, \tag{10.26}$$

$$\mathscr{U}_{\text{III}.7}(x,t) = \\ -\frac{2\sigma\left(\rho^2 - 4\sigma\right)}{\left(\sqrt{4\sigma - \rho^2}\cos\left(\frac{1}{2}\sqrt{4\sigma - \rho^2}(\mathscr{H} + \vartheta)\right) - \rho\sin\left(\frac{1}{2}\sqrt{4\sigma - \rho^2}(\mathscr{H} + \vartheta)\right)\right)^2}, \tag{10.27}$$

$$\begin{aligned}\mathscr{U}_{\text{IV}.6}(x,t) =& \frac{1}{3}\left(\rho^2 + 2\sigma\right) + \frac{8\sigma^2}{\left(\rho - \sqrt{4\sigma - \rho^2}\tan\left(\frac{1}{2}\sqrt{4\sigma - \rho^2}(\mathscr{H} + \vartheta)\right)\right)^2} \\ & - \frac{4\rho\sigma}{\rho - \sqrt{4\sigma - \rho^2}\tan\left(\frac{1}{2}\sqrt{4\sigma - \rho^2}(\mathscr{H} + \vartheta)\right)},\end{aligned} \tag{10.28}$$

$$\mathscr{U}_{\mathrm{IV},7}(x,t) = \frac{1}{3}\left(\rho^2+2\sigma\right) + \frac{8\sigma^2}{\left(\rho-\sqrt{4\sigma-\rho^2}\cot\left(\frac{1}{2}\sqrt{4\sigma-\rho^2}(\mathscr{H}+\vartheta)\right)\right)^2} - \frac{4\rho\sigma}{\rho-\sqrt{4\sigma-\rho^2}\cot\left(\frac{1}{2}\sqrt{4\sigma-\rho^2}(\mathscr{H}+\vartheta)\right)}. \tag{10.29}$$

10.2.2 EXTENDED FAN-EXPANSION METHOD

Applying the extended Fan-expansion method along the value of calculated balance to the (2+1)-dimensional GNNV equations Eq. (10.3) leads us to formulate its general solution in the following form:

$$\mathscr{S}(\Xi) = \sum_{i=0}^{n} a_i\phi(\Xi)^i = a_2\,\phi(\Xi)^2 + a_1\,\phi(\Xi) + a_0, \tag{10.30}$$

where $a_i, (i = 0, 1, 2)$ are arbitrary constants, and $\phi(\Xi)$ satisfies the following ODE:

$$\phi'(\Xi)^2 = \phi(\Xi)^2\left(\beta^2+2\gamma\delta\right) + 2\beta\gamma\phi(\Xi)^3 + 2\beta\delta\phi(\Xi) + \gamma^2\phi(\Xi)^4 + \delta^2, \tag{10.31}$$

where β, γ, δ are arbitrary constants to be determined later. Using Eq. (10.30) along with Eq. (10.31) in the framework of the suggested technique yields the following values of the above-mentioned arbitrary constants:

Family I

$$a_0 \to 2\gamma\delta, a_1 \to 2\beta\gamma, a_2 \to 2\gamma^2, \mathscr{L}_1 \to 4\gamma\delta\mathscr{L}_2 - \beta^2\mathscr{L}_2.$$

Family II

$$a_0 \to \frac{1}{3}\left(\beta^2+2\gamma\delta\right), a_1 \to 2\beta\gamma, a_2 \to 2\gamma^2, \mathscr{L}_1 \to \mathscr{L}_2\left(\beta^2-4\gamma\delta\right).$$

Thus, the explicit wave solutions of the (2+1)-dimensional GNNV equations (10.1) are given in the following order:
For $\beta^2 - 4\gamma\delta > 0$, $\beta\gamma \neq 0 (\gamma\delta \neq 0)$, we get

$$\mathscr{U}_{1,1} = -\frac{\beta^2-4\gamma\delta}{\cosh\left(\mathscr{H}\sqrt{\beta^2-4\gamma\delta}\right)+1}, \tag{10.32}$$

$$\mathscr{U}_{1,2} = \frac{\beta^2-4\gamma\delta}{\cosh\left(\mathscr{H}\sqrt{\beta^2-4\gamma\delta}\right)-1}, \tag{10.33}$$

$$\mathscr{U}_{1,3} = \frac{1}{2}\left(\beta^2-4\gamma\delta\right)\left(-1+\left(\tanh\left(\mathscr{H}\sqrt{\beta^2-4\gamma\delta}\right)\pm i\operatorname{sech}\left(\mathscr{H}\sqrt{\beta^2-4\gamma\delta}\right)\right)^2\right), \tag{10.34}$$

$$\mathscr{U}_{I.4} = \frac{1}{2}\left(\beta^2 - 4\gamma\delta\right)\left(\left(\coth\left(\mathscr{H}\sqrt{\beta^2 - 4\gamma\delta}\right) \pm \operatorname{csch}\left(\mathscr{H}\sqrt{\beta^2 - 4\gamma\delta}\right)\right)^2 - 1\right), \tag{10.35}$$

$$\mathscr{U}_{I.5} = \frac{\beta^2 - 4\gamma\delta}{\cosh\left(\mathscr{H}\sqrt{\beta^2 - 4\gamma\delta}\right) - 1}, \tag{10.36}$$

$$\begin{aligned} \mathscr{U}_{I.6} = &\frac{1}{\left(A\sinh\left(\mathscr{H}\sqrt{\beta^2 - 4\gamma\delta}\right) + B\right)^2} \\ &\times \left[A\left(\left(\beta^2 - 4\gamma\delta\right)\left(A - B\sinh\left(\mathscr{H}\sqrt{\beta^2 - 4\gamma\delta}\right)\right)\right.\right. \\ &\left.\left. - \sqrt{\beta^2 - 4\gamma\delta}\sqrt{(A^2 + B^2)\left(\beta^2 - 4\gamma\delta\right)}\cosh\left(\mathscr{H}\sqrt{\beta^2 - 4\gamma\delta}\right)\right)\right], \end{aligned} \tag{10.37}$$

$$\begin{aligned} \mathscr{U}_{I.7} = &\frac{1}{\left(A\cosh\left(\mathscr{H}\sqrt{\beta^2 - 4\gamma\delta}\right) + B\right)^2} \\ &\times \left[A\left(\sqrt{\beta^2 - 4\gamma\delta}\sqrt{-(A - B)(A + B)\left(\beta^2 - 4\gamma\delta\right)}\sinh\left(\mathscr{H}\sqrt{\beta^2 - 4\gamma\delta}\right)\right.\right. \\ &\left.\left. - A\beta^2 + 4A\gamma\delta - B\left(\beta^2 - 4\gamma\delta\right)\cosh\left(\mathscr{H}\sqrt{\beta^2 - 4\gamma\delta}\right)\right)\right], \end{aligned} \tag{10.38}$$

$$\mathscr{U}_{II.1} = -\frac{1}{6}\left(\beta^2 - 4\gamma\delta\right)\left(3\operatorname{sech}^2\left(\frac{1}{2}\mathscr{H}\sqrt{\beta^2 - 4\gamma\delta}\right) - 2\right), \tag{10.39}$$

$$\mathscr{U}_{II.2} = \frac{1}{6}\left(\beta^2 - 4\gamma\delta\right)\left(3\operatorname{csch}^2\left(\frac{1}{2}\mathscr{H}\sqrt{\beta^2 - 4\gamma\delta}\right) + 2\right), \tag{10.40}$$

$$\mathscr{U}_{II.3} = \frac{1}{6}\left(\beta^2 - 4\gamma\delta\right)\left(-1 + 3\left(\tanh\left(\mathscr{H}\sqrt{\beta^2 - 4\gamma\delta}\right) \pm i\operatorname{sech}\left(\mathscr{H}\sqrt{\beta^2 - 4\gamma\delta}\right)\right)^2\right), \tag{10.41}$$

$$\mathscr{U}_{II.4} = \frac{1}{6}\left(\beta^2 - 4\gamma\delta\right)\left(3\left(\coth\left(\mathscr{H}\sqrt{\beta^2 - 4\gamma\delta}\right) \pm \operatorname{csch}\left(\mathscr{H}\sqrt{\beta^2 - 4\gamma\delta}\right)\right)^2 - 1\right), \tag{10.42}$$

$$\mathscr{U}_{II.5} = \frac{1}{6}\left(\beta^2 - 4\gamma\delta\right)\left(3\operatorname{csch}^2\left(\frac{1}{2}\mathscr{H}\sqrt{\beta^2 - 4\gamma\delta}\right) + 2\right), \tag{10.43}$$

$$
\begin{aligned}
\mathscr{U}_{\mathrm{II},6} = & \frac{1}{6\left(A\sinh\left(\mathscr{H}\sqrt{\beta^2-4\gamma\delta}\right)+B\right)^2}\left[(\beta^2-4\gamma\delta)\,(A^2\cosh\left(2\mathscr{H}\sqrt{\beta^2-4\gamma\delta}\right)\right. \\
& +5A^2-2AB\sinh\left(\mathscr{H}\sqrt{\beta^2-4\gamma\delta}\right)+2B^2) \\
& \left.-6A\sqrt{\beta^2-4\gamma\delta}\sqrt{(A^2+B^2)\,(\beta^2-4\gamma\delta)}\cosh\left(\mathscr{H}\sqrt{\beta^2-4\gamma\delta}\right)\right],
\end{aligned}
\tag{10.44}
$$

$$
\begin{aligned}
\mathscr{U}_{\mathrm{II},7} = & \frac{1}{6\left(A\cosh\left(\mathscr{H}\sqrt{\beta^2-4\gamma\delta}\right)+B\right)^2} \\
& \times\left[-(\beta^2-4\gamma\delta)\left(5A^2+2AB\cosh\left(\mathscr{H}\sqrt{\beta^2-4\gamma\delta}\right)-2B^2\right)+A^2\right. \\
& \times(\beta^2-4\gamma\delta)\cosh\left(2\mathscr{H}\sqrt{\beta^2-4\gamma\delta}\right) \\
& +6A\sqrt{\beta^2-4\gamma\delta}\sqrt{-(A-B)(A+B)\,(\beta^2-4\gamma\delta)} \\
& \left.\times\sinh\left(\mathscr{H}\sqrt{\beta^2-4\gamma\delta}\right)\right],
\end{aligned}
\tag{10.45}
$$

where the arbitrary constants A, B satisfy $B^2-A^2<0$.

$$
\mathscr{U}_{\mathrm{I},8} = \frac{2\gamma\delta\,(4\gamma\delta-\beta^2)}{\left(\beta\cosh\left(\frac{1}{2}\mathscr{H}\sqrt{\beta^2-4\gamma\delta}\right)-\sqrt{\beta^2-4\gamma\delta}\sinh\left(\frac{1}{2}\mathscr{H}\sqrt{\beta^2-4\gamma\delta}\right)\right)^2},
\tag{10.46}
$$

$$
\mathscr{U}_{\mathrm{I},9} = \frac{2\gamma\delta\,(\beta^2-4\gamma\delta)}{\left(\sqrt{\beta^2-4\gamma\delta}\cosh\left(\frac{1}{2}\mathscr{H}\sqrt{\beta^2-4\gamma\delta}\right)-\beta\sinh\left(\frac{1}{2}\mathscr{H}\sqrt{\beta^2-4\gamma\delta}\right)\right)^2},
\tag{10.47}
$$

$$
\begin{aligned}
\mathscr{U}_{\mathrm{I},10} = & 2\gamma\delta\left(1+\frac{1}{\left(-\sqrt{\beta^2-4\gamma\delta}\sinh\left(\mathscr{H}\sqrt{\beta^2-4\gamma\delta}\right)+\left(\beta\cosh\left(\mathscr{H}\sqrt{\beta^2-4\gamma\delta}\right)\pm i\sqrt{\beta^2-4\gamma\delta}\right)\right)^2}\right. \\
& \times\left[2\cosh\left(\mathscr{H}\sqrt{\beta^2-4\gamma\delta}\right)\left(\beta\sqrt{\beta^2-4\gamma\delta}\sinh\left(\mathscr{H}\sqrt{\beta^2-4\gamma\delta}\right)+2\gamma\delta\cosh\left(\mathscr{H}\sqrt{\beta^2-4\gamma\delta}\right)\right.\right. \\
& \left.\left.\left.-\beta\left(\beta\cosh\left(\mathscr{H}\sqrt{\beta^2-4\gamma\delta}\right)\pm i\sqrt{\beta^2-4\gamma\delta}\right)\right)\right]\right).
\end{aligned}
\tag{10.48}
$$

$$
\mathscr{U}_{\mathrm{I},11} = \frac{2\gamma\delta\left(\left(\sqrt{\beta^2-4\gamma\delta}\cosh\left(\mathscr{H}\sqrt{\beta^2-4\gamma\delta}\right)\pm\sqrt{\beta^2-4\gamma\delta}\right)^2-(\beta^2-4\gamma\delta)\sinh^2\left(\mathscr{H}\sqrt{\beta^2-4\gamma\delta}\right)\right)}{\left(\left(\sqrt{\beta^2-4\gamma\delta}\cosh\left(\mathscr{H}\sqrt{\beta^2-4\gamma\delta}\right)\pm\sqrt{\beta^2-4\gamma\delta}\right)-\beta\sinh\left(\mathscr{H}\sqrt{\beta^2-4\gamma\delta}\right)\right)^2}.
\tag{10.49}
$$

$$\mathscr{U}_{I,12}=\frac{\gamma\delta\left(4\gamma\delta-\beta^2\right)\left(\cosh\left(\mathscr{H}\sqrt{\beta^2-4\gamma\delta}\right)-\cosh\left(4\mathscr{H}\sqrt{\beta^2-4\gamma\delta}\right)-2\right)}{\left(\sqrt{\beta^2-4\gamma\delta}\cosh\left(2\mathscr{H}\sqrt{\beta^2-4\gamma\delta}\right)-\beta\sinh\left(\frac{1}{2}\mathscr{H}\sqrt{\beta^2-4\gamma\delta}\right)\right)^2}. \tag{10.50}$$

$$\mathscr{U}_{II,8}=\frac{\left(\beta^2-4\gamma\delta\right)\left(-\beta\sqrt{\beta^2-4\gamma\delta}\sinh\left(\mathscr{H}\sqrt{\beta^2-4\gamma\delta}\right)+\left(\beta^2-2\gamma\delta\right)\cosh\left(\mathscr{H}\sqrt{\beta^2-4\gamma\delta}\right)-4\gamma\delta\right)}{3\left(\beta\cosh\left(\frac{1}{2}\mathscr{H}\sqrt{\beta^2-4\gamma\delta}\right)-\sqrt{\beta^2-4\gamma\delta}\sinh\left(\frac{1}{2}\mathscr{H}\sqrt{\beta^2-4\gamma\delta}\right)\right)^2}. \tag{10.51}$$

$$\mathscr{U}_{II,9}=\frac{\left(\beta^2-4\gamma\delta\right)\left(-\beta\sqrt{\beta^2-4\gamma\delta}\sinh\left(\mathscr{H}\sqrt{\beta^2-4\gamma\delta}\right)+\left(\beta^2-2\gamma\delta\right)\cosh\left(\mathscr{H}\sqrt{\beta^2-4\gamma\delta}\right)+4\gamma\delta\right)}{3\left(\sqrt{\beta^2-4\gamma\delta}\cosh\left(\frac{1}{2}\mathscr{H}\sqrt{\beta^2-4\gamma\delta}\right)-\beta\sinh\left(\frac{1}{2}\mathscr{H}\sqrt{\beta^2-4\gamma\delta}\right)\right)^2}. \tag{10.52}$$

$$\begin{aligned}\mathscr{U}_{II,10}&=\frac{1}{3}\left(\beta^2+2\gamma\delta\right)\\&+\frac{1}{\left(-\sqrt{\beta^2-4\gamma\delta}\sinh\left(\mathscr{H}\sqrt{\beta^2-4\gamma\delta}\right)+\left(\beta\cosh\left(\mathscr{H}\sqrt{\beta^2-4\gamma\delta}\right)\pm i\sqrt{\beta^2-4\gamma\delta}\right)\right)^2}\\&\times\Bigg[4\gamma\delta\cosh\left(\mathscr{H}\sqrt{\beta^2-4\gamma\delta}\right)\left(\beta\sqrt{\beta^2-4\gamma\delta}\sinh\left(\mathscr{H}\sqrt{\beta^2-4\gamma\delta}\right)+2\gamma\delta\cosh\left(\mathscr{H}\sqrt{\beta^2-4\gamma\delta}\right)\right.\\&\left.-\beta\left(\beta\cosh\left(\mathscr{H}\sqrt{\beta^2-4\gamma\delta}\right)\pm i\sqrt{\beta^2-4\gamma\delta}\right)\right)\Bigg].\end{aligned} \tag{10.53}$$

$$\begin{aligned}\mathscr{U}_{II,11}&=\frac{1}{3}\left(\beta^2+2\gamma\delta\right)\\&+\frac{1}{\left(\left(\sqrt{\beta^2-4\gamma\delta}\cosh\left(\mathscr{H}\sqrt{\beta^2-4\gamma\delta}\right)\pm\sqrt{\beta^2-4\gamma\delta}\right)-\beta\sinh\left(\mathscr{H}\sqrt{\beta^2-4\gamma\delta}\right)\right)^2}\\&\times\Bigg[4\gamma\delta\sinh\left(\mathscr{H}\sqrt{\beta^2-4\gamma\delta}\right)\left(\beta\left(\sqrt{\beta^2-4\gamma\delta}\cosh\left(\mathscr{H}\sqrt{\beta^2-4\gamma\delta}\right)\pm\sqrt{\beta^2-4\gamma\delta}\right)\right.\\&\left.-\left(\beta^2-2\gamma\delta\right)\sinh\left(\mathscr{H}\sqrt{\beta^2-4\gamma\delta}\right)\right)\Bigg].\end{aligned} \tag{10.54}$$

$$\begin{aligned}\mathscr{U}_{II,12}&=\frac{1}{6\left(\sqrt{\beta^2-4\gamma\delta}\cosh\left(2\mathscr{H}\sqrt{\beta^2-4\gamma\delta}\right)-\beta\sinh\left(\frac{1}{2}\mathscr{H}\sqrt{\beta^2-4\gamma\delta}\right)\right)^2}\Bigg[\left(\beta^2-4\gamma\delta\right)\\&\times\left(2\beta\sqrt{\beta^2-4\gamma\delta}\left(\sinh\left(\frac{3}{2}\mathscr{H}\sqrt{\beta^2-4\gamma\delta}\right)-\sinh\left(\frac{5}{2}\mathscr{H}\sqrt{\beta^2-4\gamma\delta}\right)\right)+\left(\beta^2-6\gamma\delta\right)\right.\\&\left.\times\cosh\left(\mathscr{H}\sqrt{\beta^2-4\gamma\delta}\right)+\left(\beta^2+2\gamma\delta\right)\cosh\left(4\mathscr{H}\sqrt{\beta^2-4\gamma\delta}\right)+8\gamma\delta\right)\Bigg].\end{aligned} \tag{10.55}$$

For $\beta^2-4\gamma\delta<0$, $\beta\gamma\neq\gamma 0\,(\gamma\delta\neq\gamma 0)$, we get

$$\mathscr{U}_{I,13}=-\frac{\beta^2-4\gamma\delta}{\cos\left(\mathscr{H}\sqrt{4\gamma\delta-\beta^2}\right)+1}, \tag{10.56}$$

$$\mathscr{U}_{I,14}=\frac{\beta^2-4\gamma\delta}{\cos\left(\mathscr{H}\sqrt{4\gamma\delta-\beta^2}\right)-1}, \tag{10.57}$$

$$\mathscr{U}_{\mathrm{I},15} = \frac{1}{2}\left(\beta^2 - 4\gamma\delta\right)\left(\left(\tan\left(\mathscr{H}\sqrt{4\gamma\delta-\beta^2}\right) \pm \sec\left(\mathscr{H}\sqrt{4\gamma\delta-\beta^2}\right)\right)^2 - 1\right), \tag{10.58}$$

$$\mathscr{U}_{\mathrm{I},16} = \frac{1}{2}\left(\beta^2 - 4\gamma\delta\right)\left(\left(\cot\left(\mathscr{H}\sqrt{4\gamma\delta-\beta^2}\right) \pm \csc\left(\mathscr{H}\sqrt{4\gamma\delta-\beta^2}\right)\right)^2 - 1\right), \tag{10.59}$$

$$\mathscr{U}_{\mathrm{I},17} = \frac{1}{2}\left(\beta^2 - 4\gamma\delta\right)\left(\csc^2\left(\frac{1}{2}\mathscr{H}\sqrt{4\gamma\delta-\beta^2}\right) - 2\right), \tag{10.60}$$

$$\begin{aligned}\mathscr{U}_{\mathrm{I},18} = & \frac{A}{2\left(A\sin\left(\mathscr{H}\sqrt{\beta^2-4\gamma\delta}\right)+B\right)^2} \\ & \times\Bigg[-2\sqrt{4\gamma\delta-\beta^2}\sqrt{-(A-B)(A+B)\left(\beta^2-4\gamma\delta\right)}\cos\left(\mathscr{H}\sqrt{4\gamma\delta-\beta^2}\right) \\ & - A\left(\beta^2-4\gamma\delta\right)\left(\sin^2\left(\mathscr{H}\sqrt{\beta^2-4\gamma\delta}\right)+\cos^2\left(\mathscr{H}\sqrt{4\gamma\delta-\beta^2}\right)+1\right) \\ & - 2B\left(\beta^2-4\gamma\delta\right)\sin\left(\mathscr{H}\sqrt{\beta^2-4\gamma\delta}\right)\Bigg],\end{aligned} \tag{10.61}$$

$$\begin{aligned}\mathscr{U}_{\mathrm{I},19} = & -\frac{A}{\left(A\cos\left(\mathscr{H}\sqrt{4\gamma\delta-\beta^2}\right)+B\right)^2} \\ & \times\Bigg[\sqrt{4\gamma\delta-\beta^2}\sqrt{-(A-B)(A+B)\left(\beta^2-4\gamma\delta\right)}\sin\left(\mathscr{H}\sqrt{4\gamma\delta-\beta^2}\right) \\ & + A\left(\beta^2-4\gamma\delta\right) + B\left(\beta^2-4\gamma\delta\right)\cos\left(\mathscr{H}\sqrt{4\gamma\delta-\beta^2}\right)\Bigg],\end{aligned} \tag{10.62}$$

$$\mathscr{U}_{\mathrm{II},13} = -\frac{1}{6}\left(\beta^2 - 4\gamma\delta\right)\left(3\sec^2\left(\frac{1}{2}\mathscr{H}\sqrt{4\gamma\delta-\beta^2}\right) - 2\right), \tag{10.63}$$

$$\mathscr{U}_{\mathrm{II},14} = -\frac{1}{6}\left(\beta^2 - 4\gamma\delta\right)\left(3\csc^2\left(\frac{1}{2}\mathscr{H}\sqrt{4\gamma\delta-\beta^2}\right) - 2\right), \tag{10.64}$$

$$\mathscr{U}_{\mathrm{II},15} = \frac{1}{6}\left(\beta^2 - 4\gamma\delta\right)\left(3\left(\tan\left(\mathscr{H}\sqrt{4\gamma\delta-\beta^2}\right) \pm \sec\left(\mathscr{H}\sqrt{4\gamma\delta-\beta^2}\right)\right)^2 - 1\right), \tag{10.65}$$

$$\mathscr{U}_{II.16} = \frac{1}{6}\left(\beta^2 - 4\gamma\delta\right)\left(3\left(\cot\left(\mathscr{H}\sqrt{4\gamma\delta - \beta^2}\right) \pm \csc\left(\mathscr{H}\sqrt{4\gamma\delta - \beta^2}\right)\right)^2 - 1\right), \tag{10.66}$$

$$\mathscr{U}_{II.17} = \frac{1}{6}\left(\beta^2 - 4\gamma\delta\right)\left(3\csc^2\left(\frac{1}{2}\mathscr{H}\sqrt{4\gamma\delta - \beta^2}\right) - 4\right), \tag{10.67}$$

$$\begin{aligned}\mathscr{U}_{II.18} =& \frac{-1}{6\left(A\sin\left(\mathscr{H}\sqrt{\beta^2 - 4\gamma\delta}\right) + B\right)^2}\\ &\times\left[\left(\beta^2 - 4\gamma\delta\right)\left(3A^2 + A\sin\left(\mathscr{H}\sqrt{\beta^2 - 4\gamma\delta}\right)\left(A\sin\left(\mathscr{H}\sqrt{\beta^2 - 4\gamma\delta}\right)\right.\right.\right.\\ &\left.\left. + 2B\right) - 2B^2\right) + 6A\sqrt{4\gamma\delta - \beta^2}\sqrt{\left(B^2 - A^2\right)\left(\beta^2 - 4\gamma\delta\right)}\\ &\left.\times\cos\left(\mathscr{H}\sqrt{4\gamma\delta - \beta^2}\right) + 3A^2\left(\beta^2 - 4\gamma\delta\right)\cos^2\left(\mathscr{H}\sqrt{4\gamma\delta - \beta^2}\right)\right],\end{aligned} \tag{10.68}$$

$$\begin{aligned}\mathscr{U}_{II.19} =& \frac{1}{6\left(A\cos\left(\mathscr{H}\sqrt{4\gamma\delta - \beta^2}\right) + B\right)^2}\\ &\times\left[-\left(\beta^2 - 4\gamma\delta\right)\left(5A^2 + 2AB\cos\left(\mathscr{H}\sqrt{4\gamma\delta - \beta^2}\right) - 2B^2\right)\right.\\ &+ A^2\left(\beta^2 - 4\gamma\delta\right)\cos\left(2\mathscr{H}\sqrt{4\gamma\delta - \beta^2}\right)\\ &\left. - 6A\sqrt{4\gamma\delta - \beta^2}\sqrt{-(A - B)(A + B)\left(\beta^2 - 4\gamma\delta\right)}\sin\left(\mathscr{H}\sqrt{4\gamma\delta - \beta^2}\right)\right],\end{aligned} \tag{10.69}$$

where the arbitrary constants A, B satisfy $B^2 - A^2$.

$$\mathscr{U}_{I.20} = \frac{2\gamma\delta\left(4\gamma\delta - \beta^2\right)}{\left(\sqrt{4\gamma\delta - \beta^2}\sin\left(\frac{1}{2}\mathscr{H}\sqrt{4\gamma\delta - \beta^2}\right) + \beta\cos\left(\frac{1}{2}\mathscr{H}\sqrt{4\gamma\delta - \beta^2}\right)\right)^2}, \tag{10.70}$$

$$\begin{aligned}\mathscr{U}_{I.21} =& 2\gamma\delta\left(\frac{2\beta\sinh\left(\frac{1}{2}\mathscr{H}\sqrt{4\gamma\delta - \beta^2}\right)}{\sqrt{4\gamma\delta - \beta^2}\cos\left(\frac{1}{2}\mathscr{H}\sqrt{4\gamma\delta - \beta^2}\right) - \beta\sin\left(\frac{1}{2}\mathscr{H}\sqrt{4\gamma\delta - \beta^2}\right)}\right.\\ &\left. + \frac{2\gamma\delta\left(\cosh\left(\mathscr{H}\sqrt{4\gamma\delta - \beta^2}\right) - 1\right)}{\left(\sqrt{4\gamma\delta - \beta^2}\cos\left(\frac{1}{2}\mathscr{H}\sqrt{4\gamma\delta - \beta^2}\right) - \beta\sin\left(\frac{1}{2}\mathscr{H}\sqrt{4\gamma\delta - \beta^2}\right)\right)^2} + 1\right),\end{aligned} \tag{10.71}$$

$$\mathscr{U}_{I,22} = 2\gamma\delta\left(1 - \frac{2\cos\left(\mathscr{H}\sqrt{4\gamma\delta-\beta^2}\right)}{\left(\sqrt{4\gamma\delta-\beta^2}\sin\left(\mathscr{H}\sqrt{4\gamma\delta-\beta^2}\right)+\left(\beta\cos\left(\mathscr{H}\sqrt{4\gamma\delta-\beta^2}\right)\pm\sqrt{4\gamma\delta-\beta^2}\right)\right)^2}\right.$$
$$\left(\beta\sqrt{4\gamma\delta-\beta^2}\sin\left(\mathscr{H}\sqrt{4\gamma\delta-\beta^2}\right)-2\gamma\delta\cos\left(\mathscr{H}\sqrt{4\gamma\delta-\beta^2}\right)+\beta\left(\beta\cos\left(\mathscr{H}\sqrt{4\gamma\delta-\beta^2}\right)\right.\right.$$
$$\left.\left.\left.\pm\sqrt{4\gamma\delta-\beta^2}\right)\right)\right). \tag{10.72}$$

$$\mathscr{U}_{I,23} = \frac{2\gamma\delta\left(\left(\sqrt{4\gamma\delta-\beta^2}\cos\left(\mathscr{H}\sqrt{4\gamma\delta-\beta^2}\right)\pm\sqrt{4\gamma\delta-\beta^2}\right)^2-(\beta^2-4\gamma\delta)\sin^2\left(\mathscr{H}\sqrt{4\gamma\delta-\beta^2}\right)\right)}{\left(\left(\sqrt{4\gamma\delta-\beta^2}\cos\left(\mathscr{H}\sqrt{4\gamma\delta-\beta^2}\right)\pm\sqrt{4\gamma\delta-\beta^2}\right)-\beta\sin\left(\mathscr{H}\sqrt{4\gamma\delta-\beta^2}\right)\right)^2}, \tag{10.73}$$

$$\mathscr{U}_{I,24} = \frac{\gamma\delta(\beta^2-4\gamma\delta)\left(\cos\left(\mathscr{H}\sqrt{4\gamma\delta-\beta^2}\right)-\cos\left(4\mathscr{H}\sqrt{4\gamma\delta-\beta^2}\right)-2\right)}{\left(\sqrt{4\gamma\delta-\beta^2}\cos\left(2\mathscr{H}\sqrt{4\gamma\delta-\beta^2}\right)-\beta\sin\left(\frac{1}{2}\mathscr{H}\sqrt{4\gamma\delta-\beta^2}\right)\right)^2}, \tag{10.74}$$

$$\mathscr{U}_{II,20} = \frac{(\beta^2-4\gamma\delta)\left(\beta\sqrt{4\gamma\delta-\beta^2}\sin\left(\mathscr{H}\sqrt{4\gamma\delta-\beta^2}\right)+(\beta^2-2\gamma\delta)\cos\left(\mathscr{H}\sqrt{4\gamma\delta-\beta^2}\right)-4\gamma\delta\right)}{3\left(\sqrt{4\gamma\delta-\beta^2}\sin\left(\frac{1}{2}\mathscr{H}\sqrt{4\gamma\delta-\beta^2}\right)+\beta\cos\left(\frac{1}{2}\mathscr{H}\sqrt{4\gamma\delta-\beta^2}\right)\right)^2}, \tag{10.75}$$

$$\mathscr{U}_{II,21} = \frac{1}{3}(\beta^2+2\gamma\delta)+\frac{4\gamma^2\delta^2\left(\cosh\left(\mathscr{H}\sqrt{4\gamma\delta-\beta^2}\right)-1\right)}{\left(\sqrt{4\gamma\delta-\beta^2}\cos\left(\frac{1}{2}\mathscr{H}\sqrt{4\gamma\delta-\beta^2}\right)-\beta\sin\left(\frac{1}{2}\mathscr{H}\sqrt{4\gamma\delta-\beta^2}\right)\right)^2}$$
$$+\frac{4\beta\gamma\delta\sinh\left(\frac{1}{2}\mathscr{H}\sqrt{4\gamma\delta-\beta^2}\right)}{\sqrt{4\gamma\delta-\beta^2}\cos\left(\frac{1}{2}\mathscr{H}\sqrt{4\gamma\delta-\beta^2}\right)-\beta\sin\left(\frac{1}{2}\mathscr{H}\sqrt{4\gamma\delta-\beta^2}\right)}, \tag{10.76}$$

$$\mathscr{U}_{II,22} = \frac{1}{3}(\beta^2+2\gamma\delta)+\frac{1}{\left(\sqrt{4\gamma\delta-\beta^2}\sin\left(\mathscr{H}\sqrt{4\gamma\delta-\beta^2}\right)+\left(\beta\cos\left(\mathscr{H}\sqrt{4\gamma\delta-\beta^2}\right)\pm\sqrt{4\gamma\delta-\beta^2}\right)\right)^2}$$
$$\times\left[4\gamma\delta\cos\left(\mathscr{H}\sqrt{4\gamma\delta-\beta^2}\right)\left(2\gamma\delta\cos\left(\mathscr{H}\sqrt{4\gamma\delta-\beta^2}\right)-\beta\left(\sqrt{4\gamma\delta-\beta^2}\sin\left(\mathscr{H}\sqrt{4\gamma\delta-\beta^2}\right)\right.\right.\right.$$
$$\left.\left.\left.+\left(\beta\cos\left(\mathscr{H}\sqrt{4\gamma\delta-\beta^2}\right)\pm\sqrt{4\gamma\delta-\beta^2}\right)\right)\right)\right], \tag{10.77}$$

$$\mathscr{U}_{II,23} = \frac{1}{3}(\beta^2+2\gamma\delta)+\frac{1}{\left(\left(\sqrt{4\gamma\delta-\beta^2}\cos\left(\mathscr{H}\sqrt{4\gamma\delta-\beta^2}\right)\pm\sqrt{4\gamma\delta-\beta^2}\right)-\beta\sin\left(\mathscr{H}\sqrt{4\gamma\delta-\beta^2}\right)\right)^2}$$
$$\times\left[4\gamma\delta\sin\left(\mathscr{H}\sqrt{4\gamma\delta-\beta^2}\right)\left(\beta\left(\sqrt{4\gamma\delta-\beta^2}\cos\left(\mathscr{H}\sqrt{4\gamma\delta-\beta^2}\right)\pm\sqrt{4\gamma\delta-\beta^2}\right)-(\beta^2-2\gamma\delta)\right.\right.$$
$$\left.\left.\times\sin\left(\mathscr{H}\sqrt{4\gamma\delta-\beta^2}\right)\right)\right], \tag{10.78}$$

$$\mathscr{U}_{11.24} = -\frac{(\beta^2 - 4\gamma\delta)}{6\left(\sqrt{4\gamma\delta - \beta^2}\cos\left(2\mathscr{H}\sqrt{4\gamma\delta - \beta^2}\right) - \beta\sin\left(\frac{1}{2}\mathscr{H}\sqrt{4\gamma\delta - \beta^2}\right)\right)^2}\Bigg[2\beta\sqrt{4\gamma\delta - \beta^2}$$
$$\times\left(\sin\left(\frac{5}{2}\mathscr{H}\sqrt{4\gamma\delta - \beta^2}\right) - \sin\left(\frac{3}{2}\mathscr{H}\sqrt{4\gamma\delta - \beta^2}\right)\right) + (\beta^2 - 6\gamma\delta)\cos\left(\mathscr{H}\sqrt{4\gamma\delta - \beta^2}\right)$$
$$+ (\beta^2 + 2\gamma\delta)\cos\left(4\mathscr{H}\sqrt{4\gamma\delta - \beta^2}\right) + 8\gamma\delta\Bigg]. \tag{10.79}$$

10.2.3 EXTENDED ($\frac{G'}{G}$)-EXPANSION METHOD

Applying the extended $\left(\frac{G'}{G}\right)$-expansion method along the value of calculated balance to the (2+1)-dimensional GNNV equations Eq. (10.3) leads us to formulate its general solution in the following form:

$$\begin{aligned}\mathscr{S}(\Xi) &= \sum_{i=1}^{n}\left(a_i\left(\frac{G'(\Xi)}{G(\Xi)}\right)^i + b_i\left(\frac{G'(\Xi)}{G(\Xi)}\right)^{i-1}\sqrt{\beta\left(\frac{\left(\frac{G'(\Xi)}{G(\Xi)}\right)^2}{\delta} + 1\right)}\right) + a_0\\ &= \frac{a_2 G'(\Xi)^2}{G(\Xi)^2} + \frac{a_1 G'(\Xi)}{G(\Xi)} + a_0 + \frac{b_2 G'(\Xi)\sqrt{\beta\left(\frac{G'(\Xi)^2}{\delta G(\Xi)^2} + 1\right)}}{G(\Xi)}\\ &\quad + b_1\sqrt{\beta\left(\frac{G'(\Xi)^2}{\delta G(\Xi)^2} + 1\right)},\end{aligned} \tag{10.80}$$

where $a_i, (i = 0, 1, 2), b_j, (j = 1, 2)$ are arbitrary constants, and $G(\Xi)$ satisfies the following ODE:

$$\frac{G''(\xi)}{G(\xi)} = -\delta, \tag{10.81}$$

where δ is arbitrary constant to be determined later. Using Eq. (10.80) along with Eq. (10.81) in the framework of the suggested technique yields the following values of the above-mentioned arbitrary constants:

Family I

$$b_1 \to 0, b_2 \to -\frac{\sqrt{\delta}}{\sqrt{\beta}}, a_0 \to \frac{2\delta}{3}, a_1 \to 0, a_2 \to 1, \mathscr{L}_1 \to -\delta\mathscr{L}_2.$$

Family II

$$b_1 \to 0, b_2 \to -\frac{\sqrt{\delta}}{\sqrt{\beta}}, a_0 \to \delta, a_1 \to 0, a_2 \to 1, \mathscr{L}_1 \to \delta\mathscr{L}_2.$$

Family III

$$b_1 \to 0, b_2 \to \frac{\sqrt{\delta}}{\sqrt{\beta}}, a_0 \to \frac{2\delta}{3}, a_1 \to 0, a_2 \to 1, \mathscr{L}_1 \to -\delta\mathscr{L}_2.$$

Thus, the explicit wave solutions of the (2+1)-dimensional GNNV equations (10.1) are given in the following order:
For $\delta > 0$, we get

$$\begin{aligned}\mathscr{U}_{I,1} =& \frac{2\delta}{3} + \frac{\sqrt{\delta}\left(c_1 - c_2\tan\left(\sqrt{\delta}\mathscr{H}\right)\right)}{\left(c_1\tan\left(\sqrt{\delta}\mathscr{H}\right)+c_2\right)^2} \\ &\times \left[a_1\left(c_1\tan\left(\sqrt{\delta}\mathscr{H}\right)+c_2\right) + a_2\sqrt{\delta}\left(c_1 - c_2\tan\left(\sqrt{\delta}\mathscr{H}\right)\right)\right. \\ &\left. + b_2\left(c_1\tan\left(\sqrt{\delta}\mathscr{H}\right)+c_2\right)\sqrt{\frac{\beta\left(c_1{}^2+c_2{}^2\right)}{\left(c_1\sin\left(\sqrt{\delta}\mathscr{H}\right)+c_2\cos\left(\sqrt{\delta}\mathscr{H}\right)\right)^2}}\right],\end{aligned} \tag{10.82}$$

$$\begin{aligned}\mathscr{U}_{II,1} =& \delta + \frac{\sqrt{\delta}\left(c_1 - c_2\tan\left(\sqrt{\delta}\mathscr{H}\right)\right)}{\left(c_1\tan\left(\sqrt{\delta}\mathscr{H}\right)+c_2\right)^2} \\ &\times \left[a_1\left(c_1\tan\left(\sqrt{\delta}\mathscr{H}\right)+c_2\right) + a_2\sqrt{\delta}\left(c_1 - c_2\tan\left(\sqrt{\delta}\mathscr{H}\right)\right)\right. \\ &\left. + b_2\left(c_1\tan\left(\sqrt{\delta}\mathscr{H}\right)+c_2\right)\sqrt{\frac{\beta\left(c_1{}^2+c_2{}^2\right)}{\left(c_1\sin\left(\sqrt{\delta}\mathscr{H}\right)+c_2\cos\left(\sqrt{\delta}\mathscr{H}\right)\right)^2}}\right],\end{aligned} \tag{10.83}$$

$$\begin{aligned}\mathscr{U}_{III,1} =& \frac{2\delta}{3} + \frac{1}{\left(c_1\tan\left(\sqrt{\delta}\mathscr{H}\right)+c_2\right)^2} \\ &\times \left[\sqrt{\delta}\left(c_1 - c_2\tan\left(\sqrt{\delta}\mathscr{H}\right)\right)\left(a_1\left(c_1\tan\left(\sqrt{\delta}\mathscr{H}\right)+c_2\right) + a_2\sqrt{\delta}\right.\right. \\ &\times \left(c_1 - c_2\tan\left(\sqrt{\delta}\mathscr{H}\right)\right) + b_2\left(c_1\tan\left(\sqrt{\delta}\mathscr{H}\right)+c_2\right) \\ &\left.\left.\times\sqrt{\frac{\beta\left(c_1{}^2+c_2{}^2\right)}{\left(c_1\sin\left(\sqrt{\delta}\mathscr{H}\right)+c_2\cos\left(\sqrt{\delta}\mathscr{H}\right)\right)^2}}\right)\right].\end{aligned} \tag{10.84}$$

For $\delta > 0$, we get

$$\begin{aligned}
\mathscr{U}_{1,2} = & \frac{1}{6\left(c_1\cos\left(\sqrt{\delta}\mathscr{H}\right)+c_2\sinh\left(\sqrt{-\delta}\mathscr{H}\right)\right)^2} \\
& \times\left[-6a_2\delta\left(c_2\cos\left(\sqrt{\delta}\mathscr{H}\right)+c_1\sinh\left(\sqrt{-\delta}\mathscr{H}\right)\right)^2+3a_1\sqrt{-\delta}\right. \\
& \times\left(2c_1c_2\cos\left(2\sqrt{\delta}\mathscr{H}\right)+\left(c_1{}^2+c_2{}^2\right)\sinh\left(2\sqrt{-\delta}\mathscr{H}\right)\right) \\
& +6b_2\sqrt{-\delta}\left(c_1c_2\cos\left(2\sqrt{\delta}\mathscr{H}\right)+\left(c_1{}^2+c_2{}^2\right)\cos\left(\sqrt{\delta}\mathscr{H}\right)\sinh\left(\sqrt{-\delta}\mathscr{H}\right)\right) \\
& \times\sqrt{\frac{\beta\left(c_1{}^2-c_2{}^2\right)}{\left(c_1\cos\left(\sqrt{\delta}\mathscr{H}\right)+c_2\sinh\left(\sqrt{-\delta}\mathscr{H}\right)\right)^2}} \\
& \left.+4\delta\left(c_1\cos\left(\sqrt{\delta}\mathscr{H}\right)+c_2\sinh\left(\sqrt{-\delta}\mathscr{H}\right)\right)^2\right],
\end{aligned} \tag{10.85}$$

$$\begin{aligned}
\mathscr{U}_{II,2} = & \frac{1}{2\left(c_1\cos\left(\sqrt{\delta}\mathscr{H}\right)+c_2\sinh\left(\sqrt{-\delta}\mathscr{H}\right)\right)^2} \\
& \times\left[2\left(-\delta a_2\left(c_2\cos\left(\sqrt{\delta}\mathscr{H}\right)+c_1\sinh\left(\sqrt{-\delta}\mathscr{H}\right)\right)^2\right.\right. \\
& +b_2\sqrt{-\delta}\left(c_1c_2\cos\left(2\sqrt{\delta}\mathscr{H}\right)+\left(c_1{}^2+c_2{}^2\right)\cos\left(\sqrt{\delta}\mathscr{H}\right)\sinh\left(\sqrt{-\delta}\mathscr{H}\right)\right) \\
& \times\sqrt{\frac{\beta\left(c_1{}^2-c_2{}^2\right)}{\left(c_1\cos\left(\sqrt{\delta}\mathscr{H}\right)+c_2\sinh\left(\sqrt{-\delta}\mathscr{H}\right)\right)^2}} \\
& \left.+\delta\left(c_1\cos\left(\sqrt{\delta}\mathscr{H}\right)+c_2\sinh\left(\sqrt{-\delta}\mathscr{H}\right)\right)^2\right) \\
& \left.+a_1\sqrt{-\delta}\left(2c_1c_2\cos\left(2\sqrt{\delta}\mathscr{H}\right)+\left(c_1{}^2+c_2{}^2\right)\sinh\left(2\sqrt{-\delta}\mathscr{H}\right)\right)\right],
\end{aligned} \tag{10.86}$$

$$\begin{aligned}
\mathscr{U}_{III,2} = & \frac{1}{6\left(c_1\cos\left(\sqrt{\delta}\mathscr{H}\right)+c_2\sinh\left(\sqrt{-\delta}\mathscr{H}\right)\right)^2} \\
& \times\left[-6a_2\delta\left(c_2\cos\left(\sqrt{\delta}\mathscr{H}\right)+c_1\sinh\left(\sqrt{-\delta}\mathscr{H}\right)\right)^2\right. \\
& +3a_1\sqrt{-\delta}\left(2c_1c_2\cos\left(2\sqrt{\delta}\mathscr{H}\right)+\left(c_1{}^2+c_2{}^2\right)\sinh\left(2\sqrt{-\delta}\mathscr{H}\right)\right) \\
& +6b_2\sqrt{-\delta}\left(c_1c_2\cos\left(2\sqrt{\delta}\mathscr{H}\right)+\left(c_1{}^2+c_2{}^2\right)\cos\left(\sqrt{\delta}\mathscr{H}\right)\sinh\left(\sqrt{-\delta}\mathscr{H}\right)\right) \\
& \times\sqrt{\frac{\beta\left(c_1{}^2-c_2{}^2\right)}{\left(c_1\cos\left(\sqrt{\delta}\mathscr{H}\right)+c_2\sinh\left(\sqrt{-\delta}\mathscr{H}\right)\right)^2}} \\
& \left.+4\delta\left(c_1\cos\left(\sqrt{\delta}\mathscr{H}\right)+c_2\sinh\left(\sqrt{-\delta}\mathscr{H}\right)\right)^2\right].
\end{aligned} \tag{10.87}$$

10.2.4 EXTENDED SIMPLEST EQUATION METHOD

Applying the extended simplest equation method along the value of calculated balance to the (2+1)-dimensional GNNV equations Eq. (10.3) leads us to formulate its general solution in the following form:

$$\mathscr{S}(\Xi) = \sum_{i=-n}^{n} a_i \mathscr{F}(\Xi)^i = a_2 \mathscr{F}(\Xi)^2 + a_1 \mathscr{F}(\Xi) + \frac{a_{-2}}{\mathscr{F}(\Xi)^2} + \frac{a_{-1}}{\mathscr{F}(\Xi)} + a_0, \quad (10.88)$$

where $a_i, (i = -2, \cdots, 2)$ are arbitrary constants, and $\mathscr{F}(\Xi)$ satisfies the following ODE:

$$\mathscr{F}'(\Xi) = \alpha + \lambda \mathscr{F}(\Xi) + \mu \mathscr{F}(\Xi)^2, \quad (10.89)$$

where α, λ, μ are arbitrary constants to be determined later. Using Eq. (10.88) along (10.89) in the framework of the suggested technique yields the following values of the above-mentioned arbitrary constants:

Family I

$$a_{-2} \to 2\alpha^2, a_{-1} \to 2\alpha\lambda, a_0 \to \frac{1}{3}(2\alpha\mu + \lambda^2), a_1 \to 0, a_2 \to 0, \mathscr{L}_1 \to \lambda^2 \mathscr{L}_2 - 4\alpha\mu\mathscr{L}_2.$$

Family II

$$a_{-2} \to 0, a_{-1} \to 0, a_0 \to 2\alpha\mu, a_1 \to 2\lambda\mu, a_2 \to 2\mu^2, \mathscr{L}_1 \to \mathscr{L}_2\left(-\left(\lambda^2 - 4\alpha\mu\right)\right).$$

Thus, the explicit wave solutions of the (2+1)-dimensional GNNV equations (10.1) are given in the following order:
For $\lambda = 0$, $\alpha\mu > 0$, we get

$$\mathscr{U}_{\mathrm{I},1}(x,t) = \frac{2\alpha\mu}{3} + 2\alpha\mu \cot^2\left(\sqrt{\alpha\mu}(\mathscr{H} + \vartheta)\right), \quad (10.90)$$

$$\mathscr{U}_{\mathrm{I},2}(x,t) = \frac{2\alpha\mu}{3} + 2\alpha\mu \tan^2\left(\sqrt{\alpha\mu}(\mathscr{H} + \vartheta)\right), \quad (10.91)$$

$$\mathscr{U}_{\mathrm{II},1}(x,t) = 2\alpha\mu \sec^2\left(\sqrt{\alpha\mu}(\mathscr{H} + \vartheta)\right), \quad (10.92)$$

$$\mathscr{U}_{\mathrm{II},2}(x,t) = 2\alpha\mu \csc^2\left(\sqrt{\alpha\mu}(\mathscr{H} + \vartheta)\right). \quad (10.93)$$

For $\lambda = 0$, $\alpha\mu < 0$, we get

$$\mathscr{U}_{\mathrm{I},3}(x,t) = \frac{2\alpha\mu}{3} - 2\alpha\mu \coth^2\left(\mathscr{H}\sqrt{-\alpha\mu} \mp \frac{\log(\vartheta)}{2}\right), \quad (10.94)$$

$$\mathscr{U}_{\mathrm{I},4}(x,t) = \frac{2\alpha\mu}{3} - 2\alpha\mu \tanh^2\left(\mathscr{H}\sqrt{-\alpha\mu} \mp \frac{\log(\vartheta)}{2}\right), \quad (10.95)$$

$$\mathscr{U}_{II.3}(x,t) = 2\alpha\mu\,\text{sech}^2\left(\mathscr{H}\sqrt{-\alpha\mu} \mp \frac{\log(\vartheta)}{2}\right), \tag{10.96}$$

$$\mathscr{U}_{II.4}(x,t) = -2\alpha\mu\,\text{csch}^2\left(\mathscr{H}\sqrt{-\alpha\mu} \mp \frac{\log(\vartheta)}{2}\right). \tag{10.97}$$

For $\alpha = 0, \lambda > 0$, we get

$$\mathscr{U}_{II.5}(x,t) = \frac{2\lambda^2\mu e^{\lambda(\mathscr{H}+\vartheta)}}{\left(\mu e^{\lambda(\mathscr{H}+\vartheta)} - 1\right)^2}. \tag{10.98}$$

For $\alpha = 0, \lambda < 0$, we get

$$\mathscr{U}_{II.6}(x,t) = -2\lambda\mu + 2\mu^2 - \frac{4\mu^2}{\mu e^{\lambda(\mathscr{H}+\vartheta)} + 1} + \frac{2\mu^2}{\left(\mu e^{\lambda(\mathscr{H}+\vartheta)} + 1\right)^2} + \frac{2\lambda\mu}{\mu e^{\lambda(\mathscr{H}+\vartheta)} + 1}. \tag{10.99}$$

For $4\alpha\mu > \lambda^2$, $\alpha \neq 0$, $\mu \neq 0$, $\lambda \neq 0$, we get

$$\begin{aligned}\mathscr{U}_{I.7}(x,t) =& \frac{2\alpha\mu}{3} + \frac{\lambda^2}{3} + \frac{8\alpha^2\mu^2}{\left(\lambda - \sqrt{4\alpha\mu - \lambda^2}\tan\left(\frac{1}{2}(\mathscr{H}+\vartheta)\sqrt{4\alpha\mu-\lambda^2}\right)\right)^2} \\ &- \frac{4\alpha\lambda\mu}{\lambda - \sqrt{4\alpha\mu - \lambda^2}\tan\left(\frac{1}{2}(\mathscr{H}+\vartheta)\sqrt{4\alpha\mu-\lambda^2}\right)},\end{aligned} \tag{10.100}$$

$$\begin{aligned}\mathscr{U}_{I.8}(x,t) =& \frac{2\alpha\mu}{3} + \frac{\lambda^2}{3} + \frac{8\alpha^2\mu^2}{\left(\lambda - \sqrt{4\alpha\mu - \lambda^2}\cot\left(\frac{1}{2}(\mathscr{H}+\vartheta)\sqrt{4\alpha\mu-\lambda^2}\right)\right)^2} \\ &- \frac{4\alpha\lambda\mu}{\lambda - \sqrt{4\alpha\mu - \lambda^2}\cot\left(\frac{1}{2}(\mathscr{H}+\vartheta)\sqrt{4\alpha\mu-\lambda^2}\right)},\end{aligned} \tag{10.101}$$

$$\mathscr{U}_{II.7}(x,t) = \frac{4\alpha\mu}{\cos\left((\mathscr{H}+\vartheta)\sqrt{4\alpha\mu-\lambda^2}\right)+1} - \frac{\lambda^2}{\cos\left((\mathscr{H}+\vartheta)\sqrt{4\alpha\mu-\lambda^2}\right)+1}, \tag{10.102}$$

$$\mathscr{U}_{II.8}(x,t) = \frac{\lambda^2}{\cos\left((\mathscr{H}+\vartheta)\sqrt{4\alpha\mu-\lambda^2}\right)-1} - \frac{4\alpha\mu}{\cos\left((\mathscr{H}+\vartheta)\sqrt{4\alpha\mu-\lambda^2}\right)-1}, \tag{10.103}$$

10.2.5 EXTENDED TANH(Ξ)-EXPANSION METHOD

Applying the extended Fan-expansion method along the value of calculated balance to the (2+1)-dimensional GNNV equations Eq. (10.3) leads us to formulate its general solution in the following form:

$$\mathscr{S}(\Xi) = \sum_{i=-n}^{n} a_i\phi(\Xi)^i = a_2\phi(\Xi)^2 + a_1\phi(\Xi) + \frac{a_{-2}}{\phi(\Xi)^2} + \frac{a_{-1}}{\phi(\Xi)} + a_0, \quad (10.104)$$

where $a_i, (i = -2, \cdots, 2)$ are arbitrary constants, and $\phi(\Xi)$ satisfies the following ODE:

$$\phi'(\Xi) = d + \phi(\Xi)^2, \quad (10.105)$$

where d is arbitrary constant to be determined later. Using Eq. (10.104) along (10.105) in the framework of the suggested technique yields the following values of the above-mentioned arbitrary constants:
Family I

$$a_{-2} \to 2d^2, a_{-1} \to 0, a_0 \to -\frac{1}{3}(4d), a_1 \to 0, a_2 \to 2, \mathscr{L}_1 \to -16d\mathscr{L}_2.$$

Family II

$$a_{-2} \to 2d^2, a_{-1} \to 0, a_0 \to \frac{2d}{3}, a_1 \to 0, a_2 \to 0, \mathscr{L}_1 \to -4d\mathscr{L}_2.$$

Family III

$$a_{-2} \to 2d^2, a_{-1} \to 0, a_0 \to 2d, a_1 \to 0, a_2 \to 0, \mathscr{L}_1 \to 4d\mathscr{L}_2.$$

Family IV

$$a_{-2} \to 0, a_{-1} \to 0, a_0 \to 2d, a_1 \to 0, a_2 \to 2, \mathscr{L}_1 \to 4d\mathscr{L}_2.$$

Family V

$$a_{-2} \to 2d^2, a_{-1} \to 0, a_0 \to 4d, a_1 \to 0, a_2 \to 2, \mathscr{L}_1 \to 16d\mathscr{L}_2.$$

Thus, the explicit wave solutions of the (2+1)-dimensional GNNV equations (10.1) are given in the following order:
For $d < 0$, we get

$$\mathscr{U}_{\mathrm{I},1}(x,t) = \frac{2}{3}d\left(3\tan^2\left(\sqrt{d}\mathscr{H}\right) + 3\cot^2\left(\sqrt{d}\mathscr{H}\right) - 2\right), \quad (10.106)$$

$$\mathscr{U}_{\mathrm{II},1}(x,t) = \frac{2}{3}d\left(3\cot^2\left(\sqrt{d}\mathscr{H}\right) + 1\right), \quad (10.107)$$

$$\mathscr{U}_{\mathrm{II},2}(x,t) = \frac{2}{3}d\left(3\tan^2\left(\sqrt{d}\mathscr{H}\right)+1\right), \tag{10.108}$$

$$\mathscr{U}_{\mathrm{III},1}(x,t) = 2d\csc^2\left(\sqrt{d}\mathscr{H}\right), \tag{10.109}$$

$$\mathscr{U}_{\mathrm{III},2}(x,t) = 2d\sec^2\left(\sqrt{d}\mathscr{H}\right), \tag{10.110}$$

$$\mathscr{U}_{\mathrm{IV},1}(x,t) = 2d\sec^2\left(\sqrt{d}\mathscr{H}\right), \tag{10.111}$$

$$\mathscr{U}_{\mathrm{IV},2}(x,t) = 2d\csc^2\left(\sqrt{d}\mathscr{H}\right), \tag{10.112}$$

$$\mathscr{U}_{\mathrm{IIV},1}(x,t) = d\left(a_2\tan^2\left(\sqrt{d}\mathscr{H}\right)+2\cot^2\left(\sqrt{d}\mathscr{H}\right)+4\right), \tag{10.113}$$

$$\mathscr{U}_{\mathrm{IIV},2}(x,t) = d\left(a_2\cot^2\left(\sqrt{d}\mathscr{H}\right)+2\tan^2\left(\sqrt{d}\mathscr{H}\right)+4\right). \tag{10.114}$$

For $d > 0$, we get

$$\mathscr{U}_{\mathrm{I},3}(x,t) = \frac{2}{3}d\left(3\tan^2\left(\sqrt{d}\mathscr{H}\right)+3\cot^2\left(\sqrt{d}\mathscr{H}\right)-2\right), \tag{10.115}$$

$$\mathscr{U}_{\mathrm{II},3}(x,t) = \frac{2}{3}d\left(3\cot^2\left(\sqrt{d}\mathscr{H}\right)+1\right), \tag{10.116}$$

$$\mathscr{U}_{\mathrm{II},4}(x,t) = \frac{2}{3}d\left(3\tan^2\left(\sqrt{d}\mathscr{H}\right)+1\right), \tag{10.117}$$

$$\mathscr{U}_{\mathrm{III},3}(x,t) = 2d\csc^2\left(\sqrt{d}\mathscr{H}\right), \tag{10.118}$$

$$\mathscr{U}_{\mathrm{III},4}(x,t) = 2d\sec^2\left(\sqrt{d}\mathscr{H}\right), \tag{10.119}$$

$$\mathscr{U}_{\mathrm{IV},3}(x,t) = 2d\sec^2\left(\sqrt{d}\mathscr{H}\right), \tag{10.120}$$

$$\mathscr{U}_{\mathrm{IV},4}(x,t) = 2d\csc^2\left(\sqrt{d}\mathscr{H}\right), \tag{10.121}$$

$$\mathscr{U}_{\mathrm{IIV},3}(x,t) = d\left(a_2\tan^2\left(\sqrt{d}\mathscr{H}\right)+2\cot^2\left(\sqrt{d}\mathscr{H}\right)+4\right), \tag{10.122}$$

$$\mathscr{U}_{\mathrm{IIV},4}(x,t) = d\left(a_2\cot^2\left(\sqrt{d}\mathscr{H}\right)+2\tan^2\left(\sqrt{d}\mathscr{H}\right)+4\right), \tag{10.123}$$

10.2.6 MODIFIED KHATER METHOD

Applying the modified Khater method along the value of calculated balance to the (2+1)-dimensional GNNV equations Eq. (10.3) leads us to formulate its general solution in the following form:

$$\begin{aligned}\mathscr{S}(\Xi) &= \sum_{i=1}^{n} a_i k^{i.\mathscr{F}(\Xi)} + \sum_{i=1}^{n} b_i k^{-i.\mathscr{F}(\Xi)} + a_0 \\ &= a_1 k^{\mathscr{F}(\Xi)} + a_2 k^{2.\mathscr{F}(\Xi)} + a_0 + b_2 k^{-2.\mathscr{F}(\Xi)} + b_1 k^{-\mathscr{F}(\Xi)},\end{aligned} \tag{10.124}$$

where $a_i, b_j, (i = 0, 1, 2), (j = 1, 2)$ are arbitrary constants, and $\mathscr{F}(\Xi)$ satisfies the following ODE:

$$\mathscr{F}'(\Xi) = \frac{\delta + \rho k^{\mathscr{F}(\Xi)} + \varkappa k^{-\mathscr{F}(\Xi)}}{\ln(k)}, \tag{10.125}$$

where p, q, r are arbitrary constants to be determined later. Using Eq. (10.124) along (10.125) in the framework of the suggested technique yields the following values of the above-mentioned arbitrary constants:

Family I

$$b_1 \to 0, b_2 \to 0, a_0 \to 2\rho\varkappa, a_1 \to 2\delta\rho, a_2 \to 2\rho^2, \mathscr{L}_1 \to 4\rho\mathscr{L}_2\varkappa - \delta^2\mathscr{L}_2.$$

Family II

$$b_1 \to 0, b_2 \to 0, a_0 \to \frac{1}{3}\left(\delta^2 + 2\rho\varkappa\right), a_1 \to 2\delta\rho, a_2 \to 2\rho^2, \mathscr{L}_1 \to \mathscr{L}_2\left(\delta^2 - 4\rho\varkappa\right).$$

Family III

$$b_1 \to 2\delta\varkappa, b_2 \to 2\varkappa^2, a_0 \to 2\rho\varkappa, a_1 \to 0, a_2 \to 0, \mathscr{L}_1 \to 4\rho\mathscr{L}_2\varkappa - \delta^2\mathscr{L}_2.$$

Family IV

$$b_1 \to 2\delta\varkappa, b_2 \to 2\varkappa^2, a_0 \to \frac{1}{3}\left(\delta^2 + 2\rho\varkappa\right), a_1 \to 0, a_2 \to 0, \mathscr{L}_1 \to \mathscr{L}_2\left(\delta^2 - 4\rho\varkappa\right).$$

Thus, the explicit wave solutions of the (2+1)-dimensional GNNV equations (10.1) are given in the following order:
For $\delta^2 - 4\rho\varkappa < 0, \rho \neq 0$, we get

$$\mathscr{U}_{1,1} = -\frac{\delta^2 - 4\rho\varkappa}{\cos\left(\mathscr{H}\sqrt{4\rho\varkappa - \delta^2}\right) + 1}, \tag{10.126}$$

$$\mathscr{U}_{1,2} = \frac{\delta^2 - 4\rho\varkappa}{\cos\left(\mathscr{H}\sqrt{4\rho\varkappa - \delta^2}\right) - 1}, \tag{10.127}$$

$$\mathscr{U}_{\mathrm{II},1} = -\frac{1}{6}\left(\delta^2 - 4\rho\varkappa\right)\left(3\sec^2\left(\frac{1}{2}\mathscr{H}\sqrt{4\rho\varkappa - \delta^2}\right) - 2\right), \tag{10.128}$$

$$\mathscr{U}_{\mathrm{II},2} = -\frac{1}{6}\left(\delta^2 - 4\rho\varkappa\right)\left(3\csc^2\left(\frac{1}{2}\mathscr{H}\sqrt{4\rho\varkappa - \delta^2}\right) - 2\right), \tag{10.129}$$

$$\mathscr{U}_{\mathrm{III},1} = \frac{2\rho\varkappa\left(4\rho\varkappa - \delta^2\right)}{\left(\delta\cos\left(\frac{1}{2}\mathscr{H}\sqrt{4\rho\varkappa - \delta^2}\right) - \sqrt{4\rho\varkappa - \delta^2}\sin\left(\frac{1}{2}\mathscr{H}\sqrt{4\rho\varkappa - \delta^2}\right)\right)^2}, \tag{10.130}$$

$$\mathscr{U}_{\mathrm{III},2} = \frac{2\rho\varkappa\left(4\rho\varkappa - \delta^2\right)}{\left(\sqrt{4\rho\varkappa - \delta^2}\cos\left(\frac{1}{2}\mathscr{H}\sqrt{4\rho\varkappa - \delta^2}\right) - \delta\sin\left(\frac{1}{2}\mathscr{H}\sqrt{4\rho\varkappa - \delta^2}\right)\right)^2}, \tag{10.131}$$

$$\begin{aligned}\mathscr{U}_{\mathrm{IV},1} =& \frac{1}{3}\left(\delta^2 - \frac{12\delta\rho\varkappa}{\delta - \sqrt{4\rho\varkappa - \delta^2}\tan\left(\frac{1}{2}\mathscr{H}\sqrt{4\rho\varkappa - \delta^2}\right)}\right. \\ &\left. + 2\rho\varkappa\left(\frac{12\rho\varkappa}{\left(\delta - \sqrt{4\rho\varkappa - \delta^2}\tan\left(\frac{1}{2}\mathscr{H}\sqrt{4\rho\varkappa - \delta^2}\right)\right)^2} + 1\right)\right),\end{aligned} \tag{10.132}$$

$$\begin{aligned}\mathscr{U}_{\mathrm{IV},2} =& \frac{1}{3}\left(\delta^2 - \frac{12\delta\rho\varkappa}{\delta - \sqrt{4\rho\varkappa - \delta^2}\cot\left(\frac{1}{2}\mathscr{H}\sqrt{4\rho\varkappa - \delta^2}\right)}\right. \\ &\left. + 2\rho\varkappa\left(\frac{12\rho\varkappa}{\left(\delta - \sqrt{4\rho\varkappa - \delta^2}\cot\left(\frac{1}{2}\mathscr{H}\sqrt{4\rho\varkappa - \delta^2}\right)\right)^2} + 1\right)\right).\end{aligned} \tag{10.133}$$

For $\delta^4 - 4\rho\varkappa > 0,\ \rho \neq 0$, we get

$$\mathscr{U}_{\mathrm{I},3} = -\frac{\delta^2 - 4\rho\varkappa}{\cosh\left(\mathscr{H}\sqrt{\delta^2 - 4\rho\varkappa}\right) + 1}, \tag{10.134}$$

$$\mathscr{U}_{\mathrm{I},4} = \frac{\delta^2 - 4\rho\varkappa}{\cosh\left(\mathscr{H}\sqrt{\delta^2 - 4\rho\varkappa}\right) - 1}, \tag{10.135}$$

$$\mathscr{U}_{\mathrm{II},3} = -\frac{1}{6}\left(\delta^2 - 4\rho\varkappa\right)\left(3\mathrm{sech}^2\left(\frac{1}{2}\mathscr{H}\sqrt{\delta^2 - 4\rho\varkappa}\right) - 2\right), \tag{10.136}$$

$$\mathscr{U}_{II,4} = \frac{1}{6}\left(\delta^2 - 4\rho\varkappa\right)\left(3\operatorname{csch}^2\left(\frac{1}{2}\mathscr{H}\sqrt{\delta^2 - 4\rho\varkappa}\right) + 2\right), \tag{10.137}$$

$$\mathscr{U}_{III,3} = \frac{2\rho\varkappa\left(4\rho\varkappa - \delta^2\right)}{\left(\sqrt{\delta^2 - 4\rho\varkappa}\sinh\left(\frac{1}{2}\mathscr{H}\sqrt{\delta^2 - 4\rho\varkappa}\right) + \delta\cosh\left(\frac{1}{2}\mathscr{H}\sqrt{\delta^2 - 4\rho\varkappa}\right)\right)^2}, \tag{10.138}$$

$$\mathscr{U}_{III,4} = \frac{2\rho\varkappa\left(\delta^2 - 4\rho\varkappa\right)\operatorname{csch}^2\left(\frac{1}{2}\mathscr{H}\sqrt{\delta^2 - 4\rho\varkappa}\right)}{\left(\delta + \sqrt{\delta^2 - 4\rho\varkappa}\coth\left(\frac{1}{2}\mathscr{H}\sqrt{\delta^2 - 4\rho\varkappa}\right)\right)^2}, \tag{10.139}$$

$$\begin{aligned}\mathscr{U}_{IV,3} = &\frac{1}{3\left(\delta + \sqrt{\delta^2 - 4\rho\varkappa}\tanh\left(\frac{1}{2}\mathscr{H}\sqrt{\delta^2 - 4\rho\varkappa}\right)\right)^2}\\ &\times\left[\left(\delta^2 - 4\rho\varkappa\right)\left(\delta^2 + \left(\delta^2 + 2\rho\varkappa\right)\tanh^2\left(\frac{1}{2}\mathscr{H}\sqrt{\delta^2 - 4\rho\varkappa}\right)\right.\right.\\ &\left.\left.+ 2\delta\sqrt{\delta^2 - 4\rho\varkappa}\tanh\left(\frac{1}{2}\mathscr{H}\sqrt{\delta^2 - 4\rho\varkappa}\right) - 6\rho\varkappa\right)\right],\end{aligned} \tag{10.140}$$

$$\begin{aligned}\mathscr{U}_{IV,4} = &\frac{1}{3\left(\delta + \sqrt{\delta^2 - 4\rho\varkappa}\coth\left(\frac{1}{2}\mathscr{H}\sqrt{\delta^2 - 4\rho\varkappa}\right)\right)^2}\\ &\times\left[\left(\delta^2 - 4\rho\varkappa\right)\left(\delta^2 + \left(\delta^2 + 2\rho\varkappa\right)\coth^2\left(\frac{1}{2}\mathscr{H}\sqrt{\delta^2 - 4\rho\varkappa}\right)\right.\right.\\ &\left.\left.+ 2\delta\sqrt{\delta^2 - 4\rho\varkappa}\coth\left(\frac{1}{2}\mathscr{H}\sqrt{\delta^2 - 4\rho\varkappa}\right) - 6\rho\varkappa\right)\right].\end{aligned} \tag{10.141}$$

For $\rho\varkappa > 0$, $\varkappa \neq 0$, $\rho \neq 0$, $\delta = 0$, we get

$$\mathscr{U}_{I,5} = 2\rho\varkappa\sec^2\left(\mathscr{H}\sqrt{\rho\varkappa}\right), \tag{10.142}$$

$$\mathscr{U}_{I,6} = 2\rho\varkappa\csc^2\left(\mathscr{H}\sqrt{\rho\varkappa}\right), \tag{10.143}$$

$$\mathscr{U}_{II,5} = \frac{2}{3}\rho\varkappa\left(3\tan^2\left(\mathscr{H}\sqrt{\rho\varkappa}\right) + 1\right), \tag{10.144}$$

$$\mathscr{U}_{II,6} = \frac{2}{3}\rho\varkappa\left(3\cot^2\left(\mathscr{H}\sqrt{\rho\varkappa}\right) + 1\right), \tag{10.145}$$

$$\mathscr{U}_{III,5} = 2\rho\varkappa\csc^2\left(\mathscr{H}\sqrt{\rho\varkappa}\right), \tag{10.146}$$

$$\mathscr{U}_{III.6} = 2\rho\varkappa\sec^2\left(\mathscr{H}\sqrt{\rho\varkappa}\right), \quad (10.147)$$

$$\mathscr{U}_{IV.5} = \frac{2}{3}\rho\varkappa\left(3\cot^2\left(\mathscr{H}\sqrt{\rho\varkappa}\right)+1\right), \quad (10.148)$$

$$\mathscr{U}_{IV.6} = \frac{2}{3}\rho\varkappa\left(3\tan^2\left(\mathscr{H}\sqrt{\rho\varkappa}\right)+1\right). \quad (10.149)$$

For $\rho\varkappa < 0$, $\varkappa \neq 0$, $\rho \neq 0$, $\delta = 0$, we get

$$\mathscr{U}_{I.7} = 2\rho\varkappa\operatorname{sech}^2\left(\mathscr{H}\sqrt{\rho(-\varkappa)}\right), \quad (10.150)$$

$$\mathscr{U}_{I.8} = 2\rho\varkappa\csc^2\left(\sqrt{\rho}\mathscr{H}\sqrt{\varkappa}\right), \quad (10.151)$$

$$\mathscr{U}_{II.7} = \frac{2}{3}\rho\varkappa\left(3\tan^2\left(\sqrt{\rho}\mathscr{H}\sqrt{\varkappa}\right)+1\right), \quad (10.152)$$

$$\mathscr{U}_{II.8} = \frac{2}{3}\rho\varkappa\left(3\cot^2\left(\sqrt{\rho}\mathscr{H}\sqrt{\varkappa}\right)+1\right), \quad (10.153)$$

$$\mathscr{U}_{III.7} = 2\rho\varkappa\csc^2\left(\sqrt{\rho}\mathscr{H}\sqrt{\varkappa}\right), \quad (10.154)$$

$$\mathscr{U}_{III.8} = 2\rho\varkappa\operatorname{sech}^2\left(\mathscr{H}\sqrt{\rho(-\varkappa)}\right), \quad (10.155)$$

$$\mathscr{U}_{IV.7} = \frac{2}{3}\rho\varkappa\left(3\cot^2\left(\sqrt{\rho}\mathscr{H}\sqrt{\varkappa}\right)+1\right), \quad (10.156)$$

$$\mathscr{U}_{IV.8} = \frac{2}{3}\rho\varkappa\left(3\tan^2\left(\sqrt{\rho}\mathscr{H}\sqrt{\varkappa}\right)+1\right). \quad (10.157)$$

For $\delta = 0$, $\varkappa = -\rho$, we get

$$\mathscr{U}_{I.9} = 2\varkappa^2\operatorname{csch}^2(\mathscr{H}\varkappa), \quad (10.158)$$

$$\mathscr{U}_{II.9} = \frac{2}{3}\varkappa^2\left(3\operatorname{csch}^2(\mathscr{H}\varkappa)+2\right), \quad (10.159)$$

$$\mathscr{U}_{III.9} = -2\varkappa^2\operatorname{sech}^2(\mathscr{H}\varkappa), \quad (10.160)$$

$$\mathscr{U}_{\mathrm{IV},9} = \frac{2}{3}\varkappa^2\left(2 - 3\operatorname{sech}^2(\mathscr{H}\varkappa)\right). \tag{10.161}$$

For $\delta = \frac{\varkappa}{2} = \kappa$, $\rho = 0$, we get

$$\mathscr{U}_{\mathrm{III},10} = \frac{4\kappa^2 e^{\kappa\mathscr{H}}}{\left(e^{\kappa\mathscr{H}} - 2\right)^2}, \tag{10.162}$$

$$\mathscr{U}_{\mathrm{IV},10} = \frac{\kappa^2\left(8e^{\kappa\mathscr{H}} + e^{2\kappa\mathscr{H}} + 4\right)}{3\left(e^{\kappa\mathscr{H}} - 2\right)^2}. \tag{10.163}$$

For $\delta = \rho = \kappa$, $\varkappa = 0$, we get

$$\mathscr{U}_{\mathrm{I},11} = \frac{\kappa^2}{\cosh(\kappa\mathscr{H}) - 1}, \tag{10.164}$$

$$\mathscr{U}_{\mathrm{II},11} = \frac{1}{6}\kappa^2(\cosh(\kappa\mathscr{H}) + 2)\operatorname{csch}^2\left(\frac{\kappa\mathscr{H}}{2}\right). \tag{10.165}$$

For $\varkappa = 0$, $\delta \neq 0$, $\rho \neq 0$, we get

$$\mathscr{U}_{\mathrm{I},12} = \frac{4\delta^2\rho e^{\delta\mathscr{H}}}{\left(\rho e^{\delta\mathscr{H}} - 2\right)^2}, \tag{10.166}$$

$$\mathscr{U}_{\mathrm{II},12} = \frac{\delta^2\left(\rho e^{\delta\mathscr{H}}\left(\rho e^{\delta\mathscr{H}} + 8\right) + 4\right)}{3\left(\rho e^{\delta\mathscr{H}} - 2\right)^2}. \tag{10.167}$$

For $\delta = \rho = 0$, $\varkappa \neq 0$, we get

$$\mathscr{U}_{\mathrm{III},13} = \frac{2}{\mathscr{H}^2}, \tag{10.168}$$

$$\mathscr{U}_{\mathrm{IV},13} = \frac{2}{\mathscr{H}^2}. \tag{10.169}$$

For $\delta = \varkappa = 0$, $\rho \neq 0$, we get

$$\mathscr{U}_{\mathrm{I},14} = \frac{2}{\mathscr{H}^2}, \tag{10.170}$$

$$\mathscr{U}_{\mathrm{II},14} = \frac{2}{\mathscr{H}^2}. \tag{10.171}$$

For $\delta = 0$, $\varkappa = \rho$, we get

$$\mathscr{U}_{\mathrm{I},15} = 2\varkappa^2\sec^2(C + \mathscr{H}\varkappa), \tag{10.172}$$

$$\mathscr{U}_{\text{II},15} = \frac{2}{3}\varkappa^2\left(3\tan^2(C+\mathscr{H}\varkappa)+1\right), \tag{10.173}$$

$$\mathscr{U}_{\text{III},15} = 2\varkappa^2\csc^2(C+\mathscr{H}\varkappa), \tag{10.174}$$

$$\mathscr{U}_{\text{IV},15} = \frac{2}{3}\varkappa^2\left(3\cot^2(C+\mathscr{H}\varkappa)+1\right). \tag{10.175}$$

For $\rho = 0,\ \delta \neq 0,\ \varkappa \neq 0$, we get

$$\mathscr{U}_{\text{III},16} = \frac{2\delta^3\varkappa e^{\delta\mathscr{H}}}{\left(\varkappa-\delta e^{\delta\mathscr{H}}\right)^2}, \tag{10.176}$$

$$\mathscr{U}_{\text{IV},16} = \frac{\delta^2\left(\delta^2 e^{2\delta\mathscr{H}}+4\delta\varkappa e^{\delta\mathscr{H}}+\varkappa^2\right)}{3\left(\varkappa-\delta e^{\delta\mathscr{H}}\right)^2}. \tag{10.177}$$

For $\delta^2 - 4\rho\varkappa = 00$, we get

$$\mathscr{U}_{\text{I},17} = \frac{2\rho\varkappa\left(4\rho\varkappa(\delta\mathscr{H}+2)^2-\delta^3\mathscr{H}(\delta\mathscr{H}+4)\right)}{\delta^4\mathscr{H}^2}, \tag{10.178}$$

$$\mathscr{U}_{\text{II},17} = \frac{\delta^6\mathscr{H}^2-2\delta^3\rho\mathscr{H}\varkappa(5\delta\mathscr{H}+12)+24\rho^2\varkappa^2(\delta\mathscr{H}+2)^2}{3\delta^4\mathscr{H}^2}, \tag{10.179}$$

$$\mathscr{U}_{\text{III},17} = 2\rho\varkappa - \frac{\delta^3\mathscr{H}(\delta\mathscr{H}+4)}{2(\delta\mathscr{H}+2)^2}, \tag{10.180}$$

$$\mathscr{U}_{\text{IV},17} = \frac{2\rho\varkappa}{3} - \frac{\delta^2(\delta\mathscr{H}(\delta\mathscr{H}+4)-8)}{6(\delta\mathscr{H}+2)^2}. \tag{10.181}$$

10.3 STABILITY

This section aims to check the stability property of the obtained analytical solutions of the fractional isotropic extension model of the KdV model ((2+1)-dimensional GNNV equations) through the Hamiltonian system's property. According to the following steps, we can check the stability property of our obtained solutions:
Step I Calculating the momentum in the Hamiltonian system:

$$\mathscr{M} = \frac{1}{2}\int_{\mho}^{\Upsilon}\mathscr{U}^2(\Xi)\,d\Xi, \tag{10.182}$$

where $\mathscr{M}$ is the momentum in the Hamiltonian system, while $\mho, \Upsilon$ are arbitrary constants ($\mho < \Upsilon$).

Step II The fundamental conditions of the stability property is given by

$$\left.\frac{\partial \mathscr{M}}{\partial \lambda}\right|_{\lambda=\varphi} > 0. \tag{10.183}$$

where λ, φ are arbitrary constants.

Studying the stability of the solution of Eq. (10.1) by using Eq. (10.6) with the following values of the constants $\left[\alpha = 0.5, \rho = 3, \sigma = 2, y = 3, \vartheta = 4\right]$ yields:

$$\mathscr{M} = -\frac{2\left(\log\left(\cosh\left(\frac{\lambda+8}{2}\right)\right) - \log\left(\cosh\left(\frac{\lambda+17}{2}\right)\right) - \log(\cosh(5\lambda+4)) + \log\left(\cosh\left(5\lambda+\frac{17}{2}\right)\right)\right)}{\lambda}. \tag{10.184}$$

Consequently, we get

$$\left.\frac{\partial \mathscr{M}}{\partial \lambda}\right|_{\lambda=1} = -0.000246743 < 0. \tag{10.185}$$

Therefore, this solution is unstable. Applying same steps to other obtained analytical solutions examines their stability property.

10.4 NUMERICAL SOLUTIONS

In this section, we apply the Adomian decomposition method and B-spline collection schemes to Eqs. (10.8), (10.32), (10.95), and (10.134). The goals of this sections are to investigate the semi-analytical and numerical solutions of the fractional isotropic extention model of the KdV model ((2+1)-dimensional GNNV model).

10.4.1 SEMI-ANALYTICAL SOLUTIONS

Implement of the Adomian decomposition method enables us to rewrite Eq. (10.3) in the following form:

$$\mathfrak{L}\mathscr{S}(\Xi) + \mathscr{R}\mathscr{S}(\Xi) + \mathscr{N}\mathscr{S}(\Xi) = 0, \tag{10.186}$$

where $\mathfrak{L}, \mathscr{R}, \mathscr{N}$ represent a differential operator, a linear operator, and a nonlinear term, respectively. Using the inverse operator $\mathfrak{L}^{-1}$ on (10.186), we get

$$\sum_{i=0}^{\infty} \mathscr{S}_i(\Xi) = \mathscr{S}(0) + \mathscr{S}'(0)(\Xi) - \frac{\mathscr{L}_0}{\mathscr{L}_2}\mathfrak{L}^{-1}\left(\sum_{i=0}^{\infty} \pounds_i\right) + 3\,\mathfrak{L}^{-1}\left(\sum_{i=0}^{\infty} A_i\right). \tag{10.187}$$

Under the following condition $\left[\mathscr{L}_1 = 1, \rho = 3, \sigma = 2, \mathscr{L}_2 = 1, \vartheta = 0\right]$ on Eq. (10.8), we get:

$$\mathscr{S}_{\text{Exact}} = \frac{1}{6}\left(2 - 3\operatorname{sech}^2\left(\frac{\Xi}{2}\right)\right). \tag{10.188}$$

Thus, we get:

$$\mathscr{S}_0(\Xi) = -\frac{1}{6}, \tag{10.189}$$

$$\mathscr{S}_1(\Xi) = \frac{\Xi^2}{8}, \tag{10.190}$$

$$\mathscr{S}_2(\Xi) = -\frac{\Xi^4}{48}, \tag{10.191}$$

$$\mathscr{S}_3(\Xi) = \frac{13\Xi^6}{5760} - \frac{\Xi^4}{96}. \tag{10.192}$$

Equations (10.189)–(10.192) lead to the following form of an approximate solution of Eq. (10.3).

$$\mathscr{S}_{\text{Approximate}} = \frac{13\Xi^6}{5760} - \frac{\Xi^4}{32} + \frac{\Xi^2}{8} - \frac{1}{6} + \cdots. \tag{10.193}$$

Table 10.1

Exact and Semi-analytical solutions of Eq. (10.3) Through the Obtained Solution (10.8) via the exp(−$\phi(\Xi)$)-Expansion Method

Value of Ξ	Exact	Numerical	Absolute Error
0.001	0.166667	0.166667	1.03806×10^{-14}
0.002	0.166666	0.166666	1.66561×10^{-13}
0.003	0.166666	0.166666	8.43769×10^{-13}
0.004	0.166665	0.166665	2.66662×10^{-12}
0.005	0.166664	0.166664	6.51049×10^{-12}
0.006	0.166662	0.166662	1.35001×10^{-11}
0.007	0.166661	0.166661	2.50105×10^{-11}
0.008	0.166659	0.166659	4.26667×10^{-11}
0.009	0.166657	0.166657	6.83441×10^{-11}
0.01	0.166654	0.166654	1.04167×10^{-10}

Table 10.2
Exact and Numerical of Eq. (10.3) Through the Obtained Solution (10.32) via the Extended Fan-Expansion Method

Value of Ξ	Exact	Numerical	Absolute Error
0	−0.5	−0.5	0
0.001	−0.5	−0.500004	4.50014×10^{-6}
0.002	−0.5	−0.500008	8.00026×10^{-6}
0.003	−0.499999	−0.500009	1.05004E-05
0.004	−0.499998	−0.50001	1.20004E-05
0.005	−0.499997	−0.500009	1.25004E-05
0.006	−0.499996	−0.500008	1.20004E-05
0.007	−0.499994	−0.500004	1.05003E-05
0.008	−0.499992	−0.5	8.00022×10^{-6}
0.009	−0.49999	−0.499994	4.50011×10^{-6}
0.01	−0.499988	−0.499988	5.55112×10^{-17}

10.4.2 NUMERICAL SOLUTIONS

In this section, the B-spline schemes is applied to the fractional nonlinear KG equation to evaluate its numerical solution and also to show the accuracy of the obtained analytical solutions in section 10.2, by employing the modified Khater method under the following conditions on Eq. (10.137) as following:

$$\left[b = -2, c = 3, \delta = 6, \lambda = 1, \mu = 5\right].$$

These conditions allow us to apply the B-spline family in the following forms:

10.4.2.1 Cubic B-Spline

The cubic B-spline scheme formulates the general solution of Eq. (10.3) in the following form:

$$\pounds(\Xi) = \sum_{\jmath=-1}^{m+1} \mho_\jmath \eth_\jmath, \tag{10.194}$$

where $\mho_\jmath, \eth_\jmath$ are given in the next mathematical forms, respectively:

$$\mathscr{L}\pounds(\Xi) = f(\Xi_\jmath, \pounds(\Xi_\jmath)), (\jmath = 0, 1, ..., m)$$

Table 10.3
Exact and Numerical of Eq. (10.3) Through the Obtained Solution (10.95) via the Extended Simplest Equation Method

Value of Ξ	Exact	Numerical	Absolute Error
0	−6	−6	0
0.001	−5.99998	−5.99984	0.000143999
0.002	−5.99993	−5.99935	0.000575985
0.003	−5.99984	−5.99854	0.00129592
0.004	−5.99971	−5.99741	0.00230375
0.005	−5.99955	−5.99595	0.0035994
0.006	−5.99935	−5.99417	0.00518276
0.007	−5.99912	−5.99206	0.0070537
0.008	−5.99885	−5.98964	0.00921207
0.009	−5.99854	−5.98688	0.0116577
0.01	−5.9982	−5.98381	0.0143904

and

$$\eth_{\jmath}(\Xi)=\frac{1}{6\hbar^{3}}\begin{cases}(\Xi-\Xi_{\jmath-2})^{3}, & \Xi\in[\Xi_{\jmath-2},\Xi_{\jmath-1}],\\ -3(\Xi-\Xi_{\jmath-1})^{3}+3\hbar(\Xi-\Xi_{\jmath-1})^{2}+3\hbar^{2}(\Xi-\Xi_{\jmath-1})+\hbar^{3}, & \Xi\in[\Xi_{\jmath-1},\Xi_{\jmath}],\\ -3(\Xi_{\jmath+1}-\Xi)^{3}+3\hbar(\Xi_{\jmath+1}-\Xi)^{2}+3\hbar^{2}(\Xi_{i+1}-\Xi)+\hbar^{3}, & \Xi\in[\Xi_{i},\Xi_{i+1}],\\ (\Xi_{i+2}-\Xi)^{3}, & \Xi\in[\Xi_{i+1},\Xi_{i+2}],\\ 0, & \text{Otherwise},\end{cases} \tag{10.195}$$

where $\jmath\in[-2,m+2]$. Thus, we obtain

$$£_{\jmath}(\Xi)=\mho_{\jmath-1}+4\mho_{\jmath}+\mho_{\jmath+1}. \tag{10.196}$$

Substituting Eq. (10.196) into Eq. (10.3) yields $(m+3)$ equations. Using Mathematica 12 to solve this system to get the value of $\mho_{\jmath}$ leads to the following values of analytical, numerical under the different values of Ξ in Table 10.2.

10.4.2.2 Quantic B-Spline

The quantic B-spline scheme formulates the general solution of Eq. (10.3) in the following form:

$$£(\Xi)=\sum_{\jmath=-1}^{m+1}\mho_{\jmath}\eth_{\jmath}, \tag{10.197}$$

Table 10.4
Exact and Numerical of Eq. (10.3) Through the Obtained Solution (10.134) via the Modified Khater Method

Value of Ξ	Exact	Numerical	Absolute Error
0	−0.5	−0.5	0
0.001	−0.5	−0.5	5.55112×10^{-17}
0.002	−0.5	−0.5	1.11022×10^{-16}
0.003	−0.499999	−0.499999	5.55112×10^{-17}
0.004	−0.499998	−0.499998	1.11022×10^{-16}
0.005	−0.499997	−0.499997	0
0.006	−0.499996	−0.499996	0
0.007	−0.499994	−0.499994	5.55112×10^{-17}
0.008	−0.499992	−0.499992	5.55112×10^{-17}
0.009	−0.49999	−0.49999	0
0.01	−0.499988	−0.499988	5.55112×10^{-17}

where $\mho_\jmath$, $\eth_\jmath$ are given in the next mathematical forms, respectively:

$$\mathscr{L}\pounds(\Xi)=f(\Xi_\jmath,\pounds(\Xi_\jmath)),\ (\jmath=0,1,...,n)$$

and

$$\eth_\jmath(\Xi)=\frac{1}{\hbar^5}\begin{cases}(\Xi-\Xi_{\jmath-3})^5, & \Xi\in[\Xi_{\jmath-3},\Xi_{\jmath-2}],\\ (\Xi-\Xi_{\jmath-3})^5-6(\Xi-\Xi_{\jmath-2})^5, & \Xi\in[\Xi_{\jmath-2},\Xi_{\jmath-1}],\\ (\Xi-\Xi_{\jmath-3})^5-6(\Xi-\Xi_{\jmath-2})^5+15(\Xi-\Xi_{\jmath-1})^5, & \Xi\in[\Xi_{\jmath-1},\Xi_\jmath],\\ (\Xi_{\jmath+3}-\Xi)^5-6(\Xi_{\jmath+2}-\Xi)^5+15(\Xi_{\jmath+1}-\Xi)^5, & \Xi\in[\Xi_\jmath,\Xi_{\jmath+1}],\\ (\Xi_{\jmath+3}-\Xi)^5-6(\Xi_{\jmath+2}-\Xi)^5, & \Xi\in[\Xi_{\jmath+1},\Xi_{\jmath+2}],\\ (\Xi_{\jmath+3}-\Xi)^5, & x\in[\Xi_{\jmath+2},\Xi_{\jmath+3}],\\ 0, & \text{Otherwise},\end{cases} \tag{10.198}$$

where $\jmath\in[-2,m+2]$. Thus, we obtain

$$\pounds_\jmath(\Xi)=\mho_{\jmath-2}+26\mho_{\jmath-1}+66\mho_\jmath+26\mho_{\jmath+1}+\mho_{\jmath+2}. \tag{10.199}$$

Substituting Eq. (10.199) into Eq. (10.3) yields $(m+5)$ equations. Using Mathematica 12 to solve this system to get the value of $\mho_\jmath$ leads to the following values of analytical, numerical under the different values of Ξ in Table 10.3.

10.4.2.3 Septic B-Spline

The septic B-spline scheme formulates the general solution of Eq. (10.3) in the following form:

$$\pounds(\Xi) = \sum_{\jmath=-1}^{n+1} \mho_\jmath \eth_\jmath, \tag{10.200}$$

where $\mho_\jmath, \eth_\jmath$ are given in the next mathematical forms, respectively:

$$\mathscr{L}\pounds(\Xi) = \mathscr{F}(\Xi_\jmath, \pounds(\Xi_\jmath)), \ (\jmath = 0, 1, ..., m)$$

and

$$\eth_\jmath(\Xi) = \frac{1}{h^5}\begin{cases} (\Xi-\Xi_{\jmath-4})^7, & \Xi \in [\Xi_{\jmath-4}, \Xi_{\jmath-3}], \\ (\Xi-\Xi_{\jmath-4})^7 - 8(\Xi-\Xi_{\jmath-3})^7, & \Xi \in [\Xi_{\jmath-3}, \Xi_{i-2}], \\ (\Xi-\Xi_{\jmath-4})^7 - 8(\Xi-\Xi_{\jmath-3})^7 + 28(\Xi-\Xi_{\jmath-2})^7, & \Xi \in [\Xi_{\jmath-2}, \Xi_{\jmath-1}], \\ (\Xi-\Xi_{\jmath-4})^7 - 8(\Xi-\Xi_{\jmath-3})^7 + 28(\Xi-\Xi_{\jmath-2})^7 + 56(\Xi-\Xi_{\jmath-1})^7, & \Xi \in [\Xi_{\jmath-1}, \Xi_\jmath], \\ (\Xi_{\jmath+4}-\Xi)^7 - 8(\Xi_{\jmath+3}-\Xi)^7 + 28(\Xi_{\jmath+2}-\Xi)^7 + 56(\Xi_{\jmath+1}-\Xi)^7, & \Xi \in [\Xi_\jmath, \Xi_{\jmath+1}], \\ (\Xi_{\jmath+4}-\Xi)^7 - 8(\Xi_{\jmath+3}-\Xi)^7 + 28(\Xi_{\jmath+2}-\Xi)^7, & \Xi \in [\Xi_{\jmath+1}, \Xi_{\jmath+2}], \\ (\Xi_{\jmath+4}-\Xi)^7 - 8(\Xi_{\jmath+3}-\Xi)^7, & \Xi \in [\Xi_{\jmath+2}, \Xi_{i+3}], \\ (\Xi_{\jmath+4}-\Xi)^7, & \Xi \in [\Xi_{\jmath+3}, \Xi_{i+4}], \\ 0, & \text{Otherwise}, \end{cases} \tag{10.201}$$

where $\jmath \in [-3, m+3]$. Thus, we obtain

$$\pounds_\jmath(\Xi) = \mho_{\jmath-3} + 120\mho_{\jmath-2} + 1191\mho_{\jmath-1} + 2416\mho_\jmath + 1191\mho_{\jmath+1} + 120\mho_{\jmath+2} + \mho_{\jmath+3}. \tag{10.202}$$

Substituting Eq. (10.202) into Eq. (10.3), yields $(m+7)$ equations. Using Mathematica 12 to solve this system to get the value of $\mho_\jmath$ leads to the following values of analytical, numerical under the different values of Ξ in Table 10.4.

10.5 FIGURES AND TABLES REPRESENTATION

This section explains the physical representation of each figure, to show their dynamical behavior. Our goals are to show the breath and solitary waves of the solution, the propagation pattern of this kind of waves along the x-axis, and its overhead view.

1. Figure (10.1) shows periodic solitary wave solution of Eq. (10.1) according to Eq. (10.6) in three-, two-dimensional, and contour plot with the following values of the parameters $\left[\alpha = 0.5, \lambda = 1, \rho = 3, \sigma = 2, y = 3, \vartheta = 4\right]$.
2. Figure (10.2) shows breath wave solution of Eq. (10.1) according to Eq. (10.8) in three-, two-dimensional, and contour plot with the following values of the parameters $\left[\alpha = 0.5, \lambda = 6, \rho = 5, \sigma = 6, y = 2, \vartheta = 3\right]$.
3. Figure (10.3) shows breath wave solution of Eq. (10.3) according to Eq. (10.34) in three-, two-dimensional, and contour plot with the following values of the parameters $\left[\alpha = 0.5, \lambda = 10, \beta = 3, \gamma = 2, \delta = 1, y = 12\right]$.

4. Figure (10.4) shows breath wave solution of Eq. (10.1) according to Eq. (10.39) in three-, two-dimensional, and contour plot with the following values of the parameters $\left[\alpha=0.5, \lambda=10, \beta=3, \gamma=2, \delta=1, y=12\right]$.
5. Figure (10.5) shows solitary solution of Eq. (10.1) according to Eq. (10.83) in three-, two-dimensional, and contour plot with the following values of the parameters $\left[a_2=5, a_1=5, \alpha=0.5, b_2=2, \beta=4, c_1=1, c_2=3, \delta=-4, \lambda=10, y=12\right]$.
6. Figure (10.6) shows breath wave solution of Eq. (10.1) according to Eq. (10.95) in three-, two-dimensional, and contour plot with the following values of the parameters $\left[\alpha=-1, \alpha=0.5, \lambda=2, \mu=9, y=3, \vartheta=1\right]$.
7. Figure (10.7) shows solitary wave solution of Eq. (10.1) according to Eq. (10.97) in three-, two-dimensional, and contour plot with the following values of the parameters $\left[\alpha=-1, \alpha=0.5, \lambda=2, \mu=9, y=3, \vartheta=1\right]$.
8. Figure (10.8) shows breath wave solution of Eq. (10.1) according to Eq. (10.106) in three-, two-dimensional, and contour plot with the following values of the parameters $\left[\alpha=0.5, d=-4, \lambda=2, y=3\right]$.
9. Figure (10.9) shows solitary wave solution of Eq. (10.1) according to Eq. (10.108) in three-, two-dimensional, and contour plot with the following values of the parameters $\left[\alpha=0.5, d=-9, \lambda=5, y=2\right]$.
10. Figure (10.10) shows breath wave solution of Eq. (10.1) according to Eq. (10.134) in three-, two-dimensional, and contour plot with the following values of the parameters $\left[\alpha=0.5, \delta=3, \lambda=4, \rho=2, y=5, \varkappa=1\right]$.
11. Figure (10.11) shows solitary wave solution of Eq. (10.1) according to Eq. (10.137) in three-, two-dimensional, and contour plot with the following values of the parameters $\left[\alpha=0.5, \delta=5, \lambda=7, \rho=2, y=1, \varkappa=3\right]$.
12. Table 10.1 shows the values of analytical and semi-analytical solutions with different values of Ξ through the solution (10.8) along the exp $(-\phi(\Xi))$-expansion method and the Adomian decomposition method.
13. Figure (10.12) shows exact and semi-analytical wave solutions with different values Ξ to show the accuracy of obtained solutions of Eq. (10.1) according to Table 10.1 in three different forms.
14. Table 10.2 shows the values of analytical and numerical solutions with different values of Ξ through the solution (10.32) along the extended Fan-expansion method and the cubic B-spline scheme.
15. Figure (10.13) shows the accuracy between our obtained analytical and numerical solutions according to the shown values in Table 10.2 in three different sketches with respect to cubic B-spline.

16. Table 10.3 shows the values of analytical and numerical solutions with different values of Ξ through solution (10.95) along the extended simplest equation method and the quantic B-spline scheme.
17. Figure (10.14) shows the accuracy between our obtained analytical and numerical solutions according to the shown values in Table 10.3 in three different sketches with respect to quantic B-spline.
18. Table 10.4 shows the values of analytical and numerical solutions with different values of Ξ through the solution (10.134) along the modified Khater method and the cubic B-spline scheme.
19. Figure (10.15) shows the accuracy between our obtained analytical and numerical solutions according to the shown values in Table 10.4 in three different sketches with respect to septic B-spline.

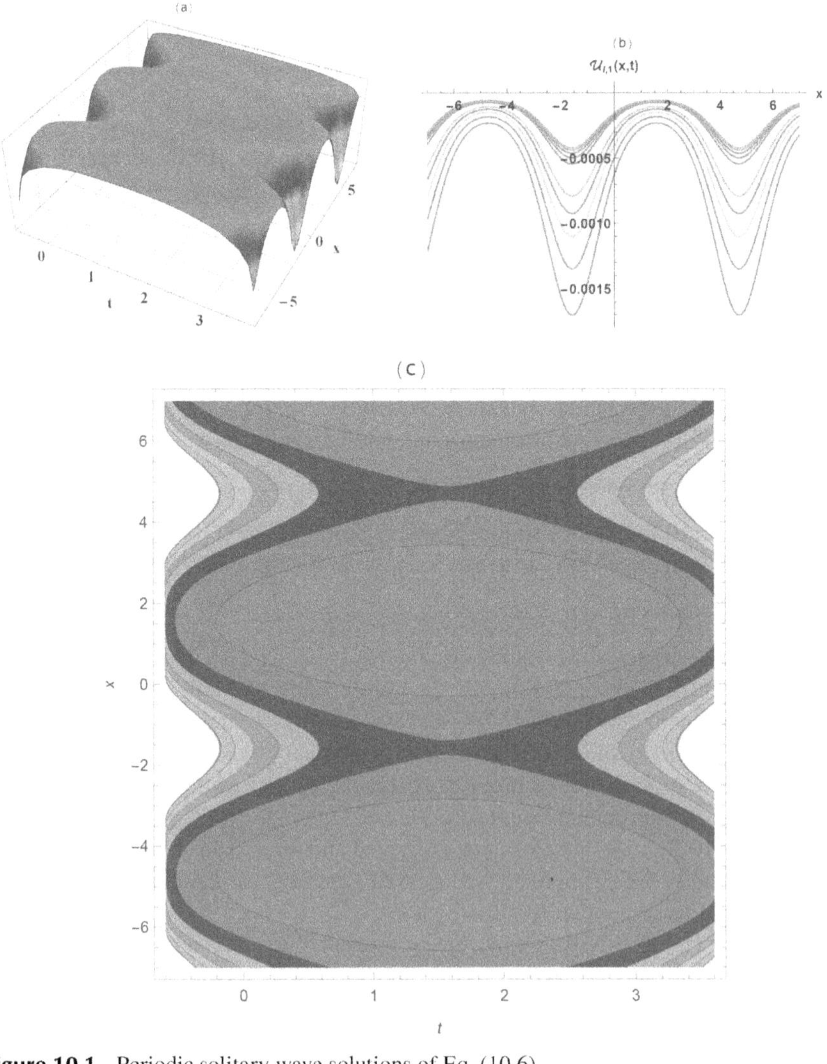

Figure 10.1 Periodic solitary wave solutions of Eq. (10.6).

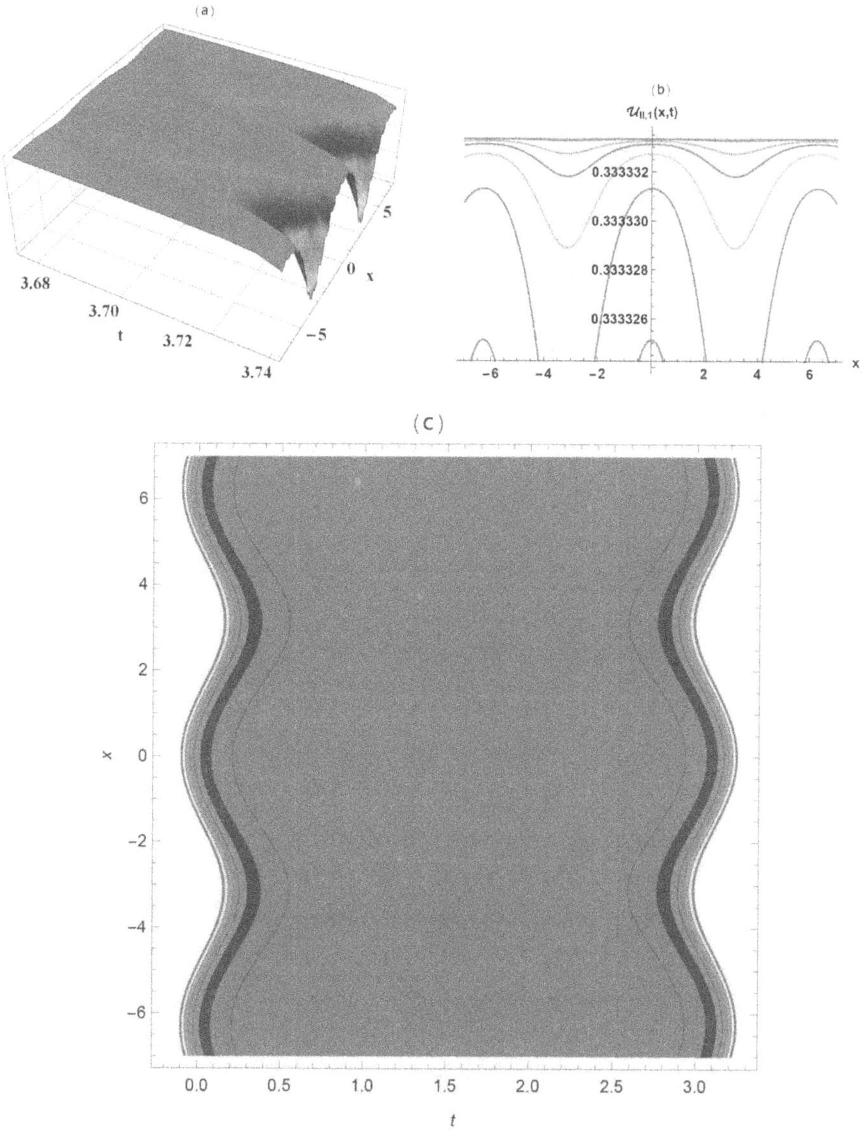

Figure 10.2 Breath solitary wave solutions of Eq. (10.8).

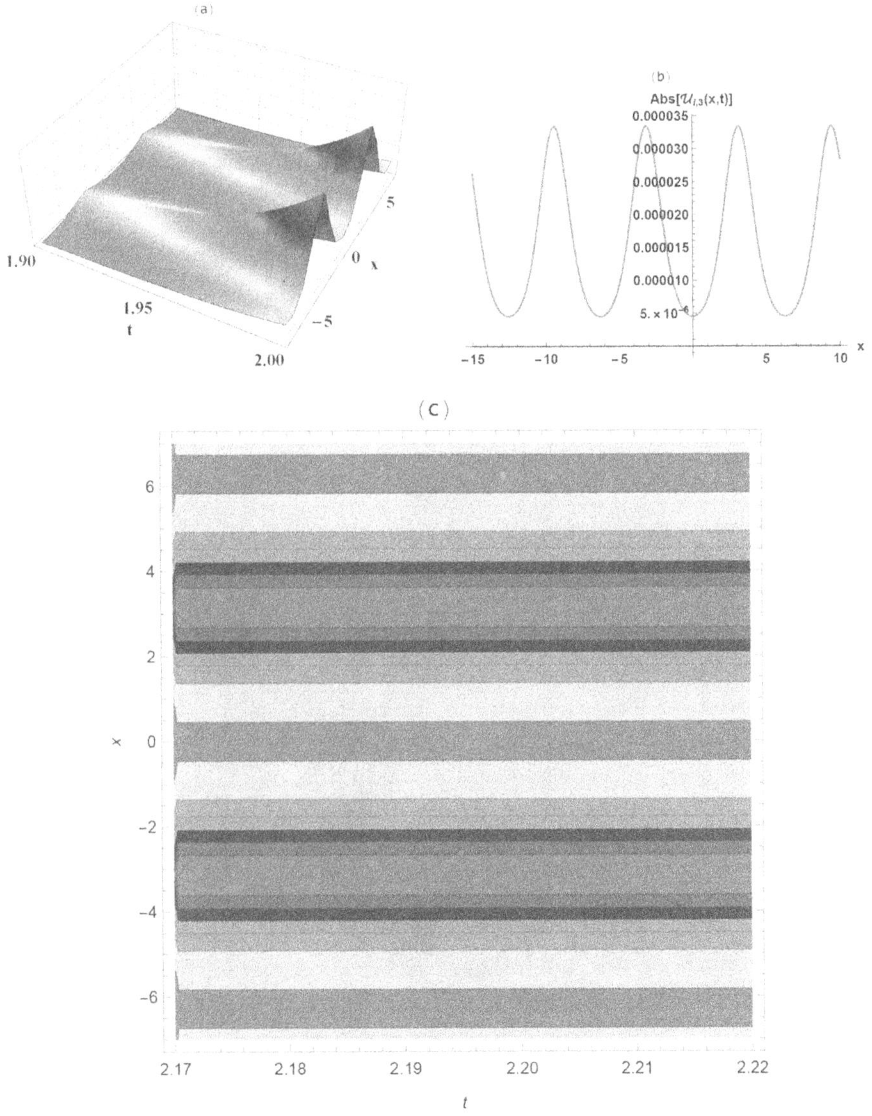

Figure 10.3 Breath solitary wave solutions of Eq. (10.34).

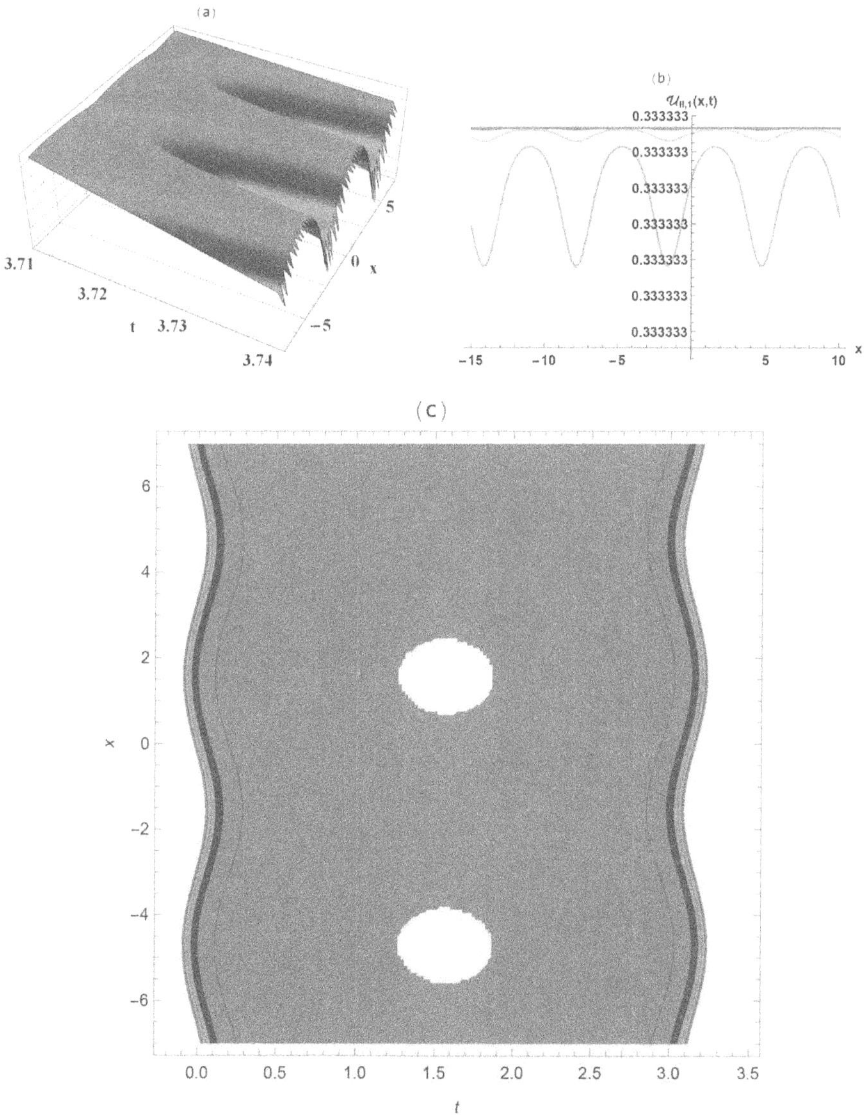

Figure 10.4 Breath solitary wave solutions of Eq. (10.39).

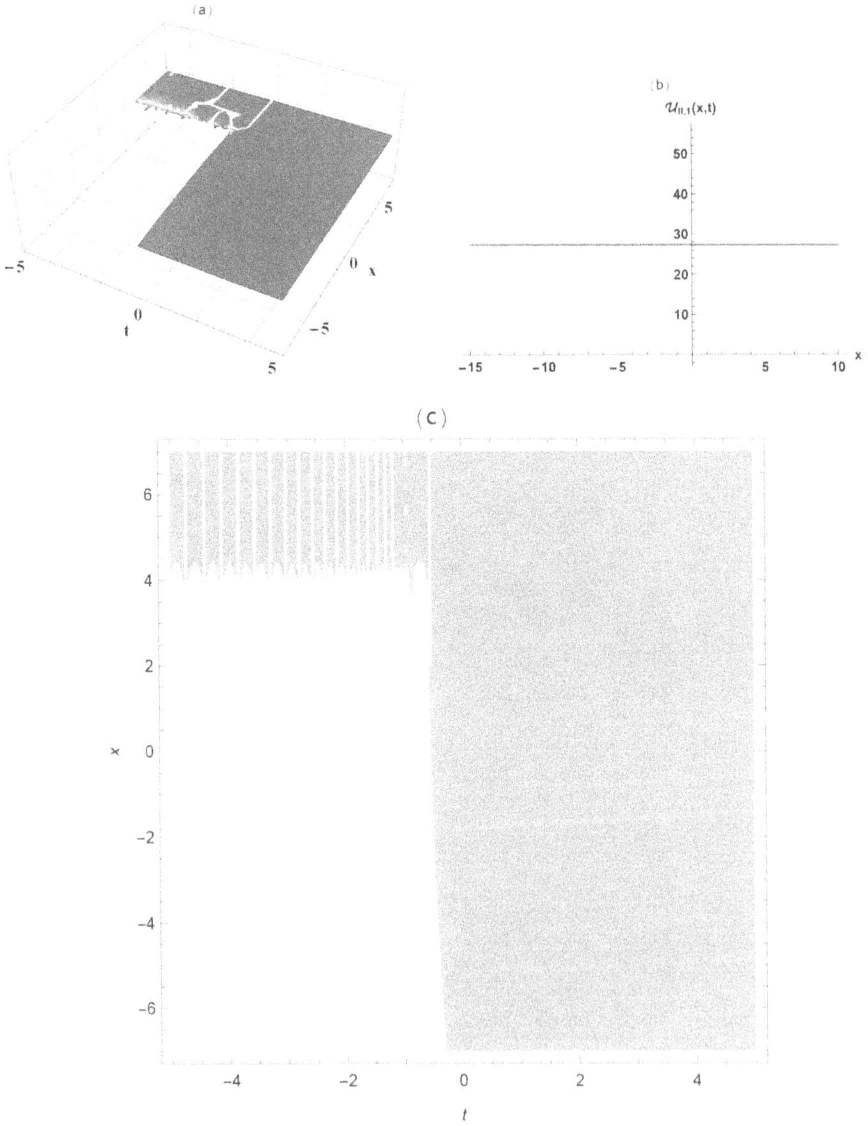

Figure 10.5 Solitary wave solutions of Eq. (10.83).

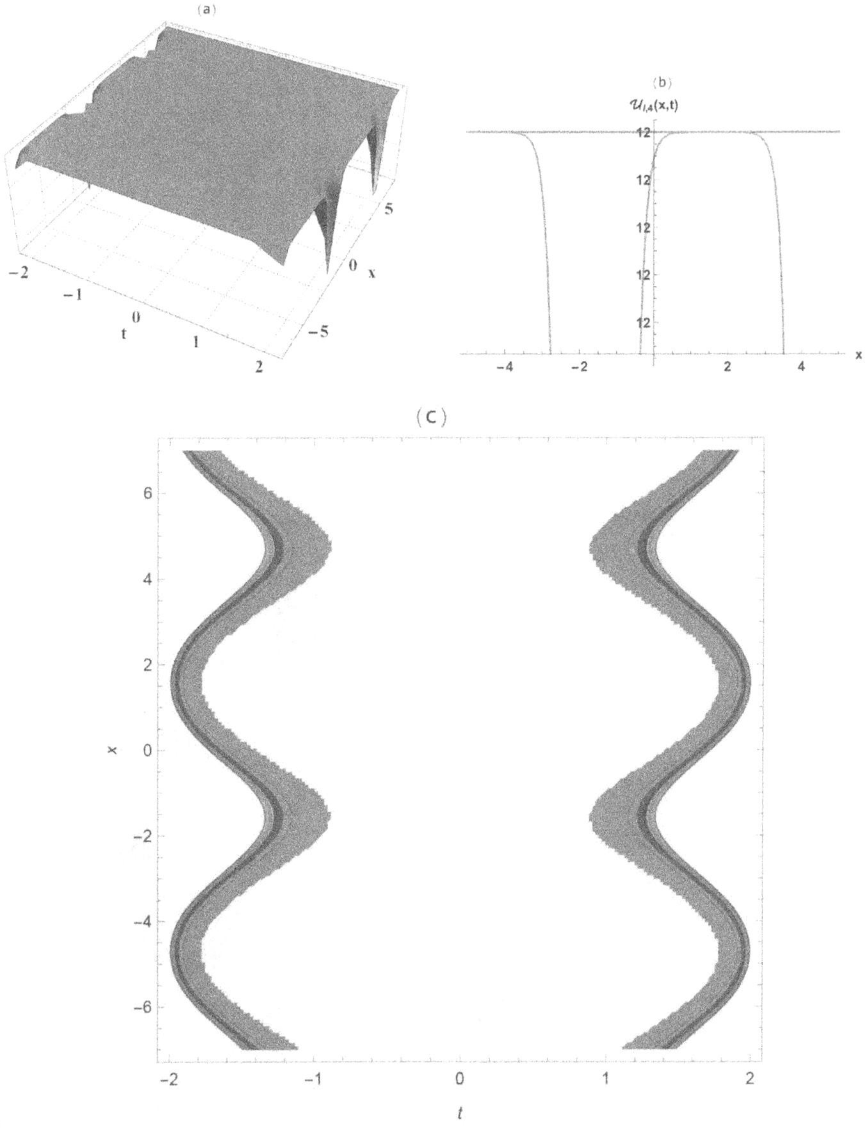

Figure 10.6 Breath solitary wave solutions of Eq. (10.95).

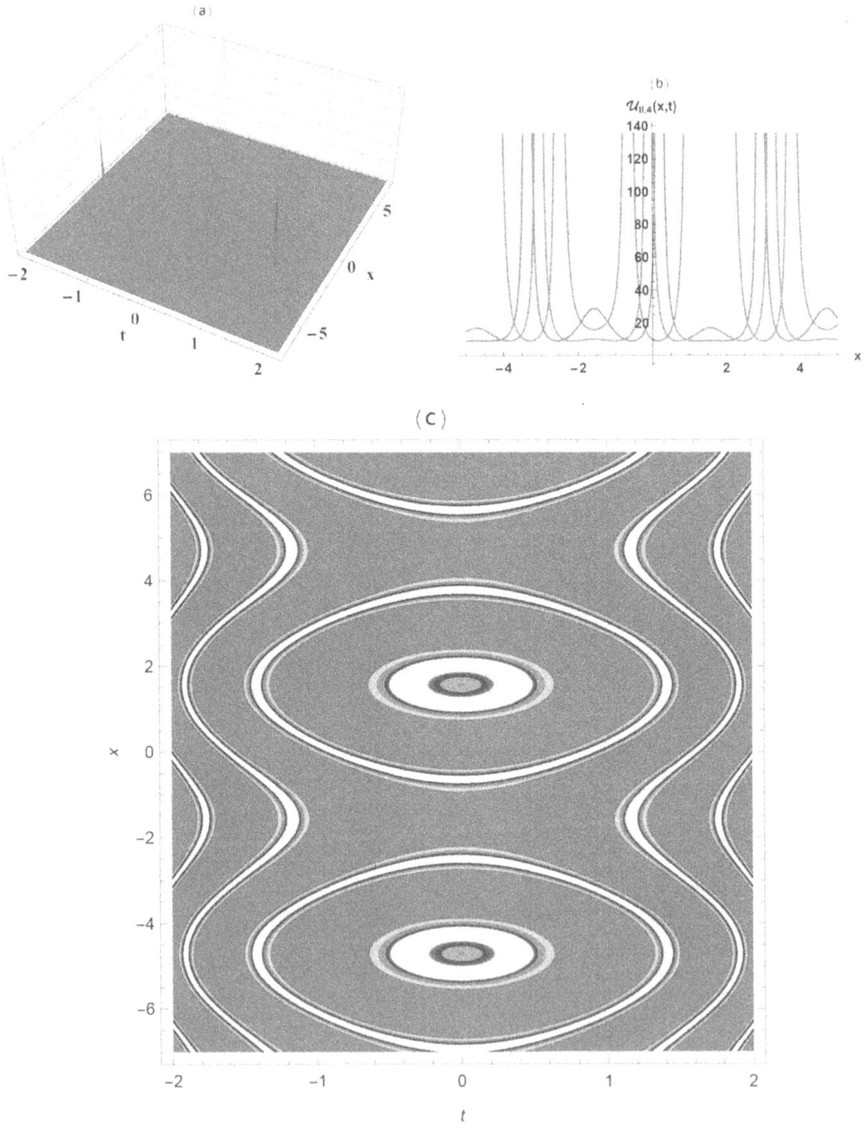

Figure 10.7 Solitary wave solutions of Eq. (10.97).

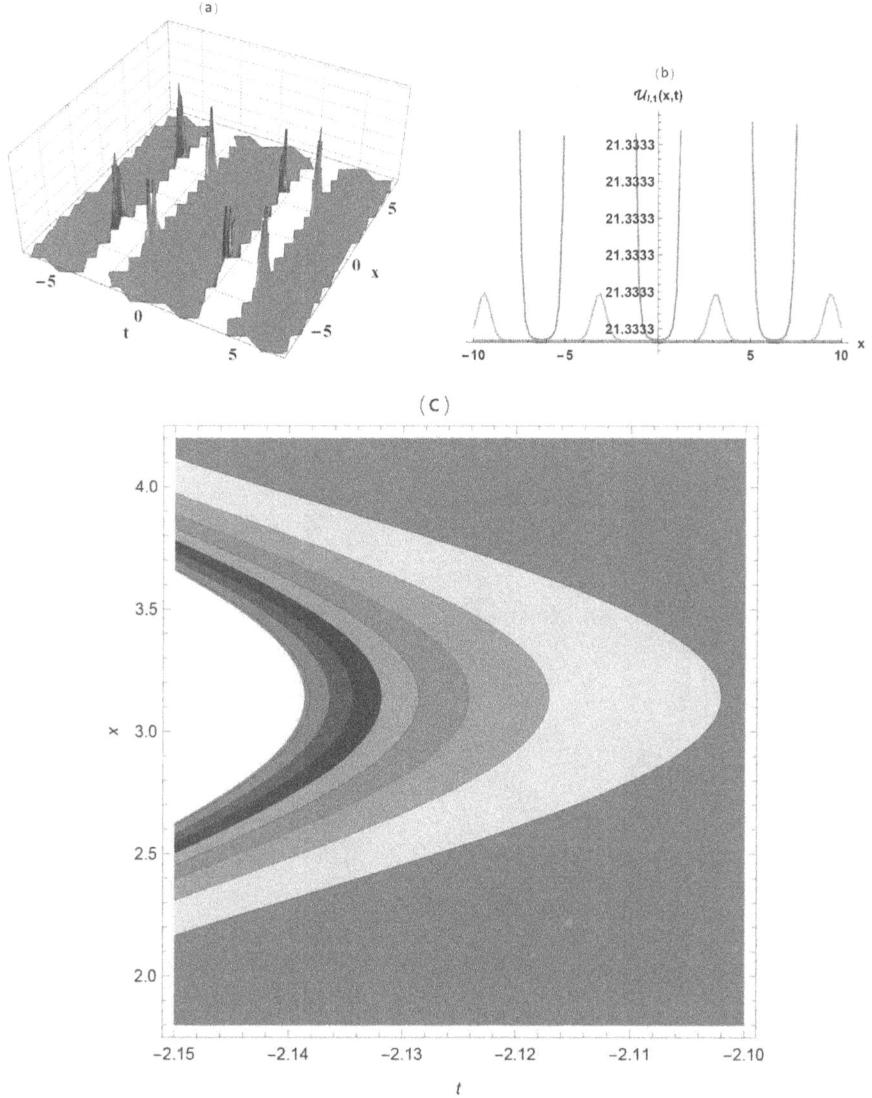

Figure 10.8 Breath solitary wave solutions of Eq. (10.106).

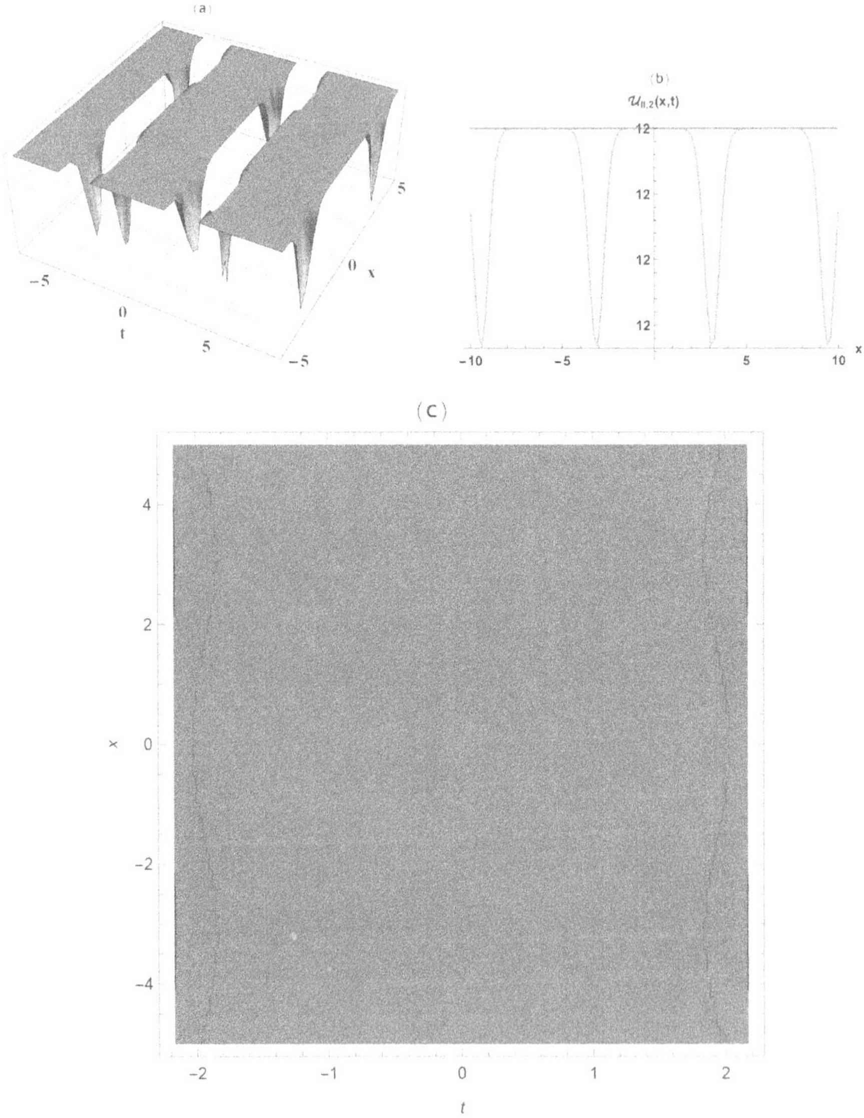

Figure 10.9 Solitary wave solutions of Eq. (10.108).

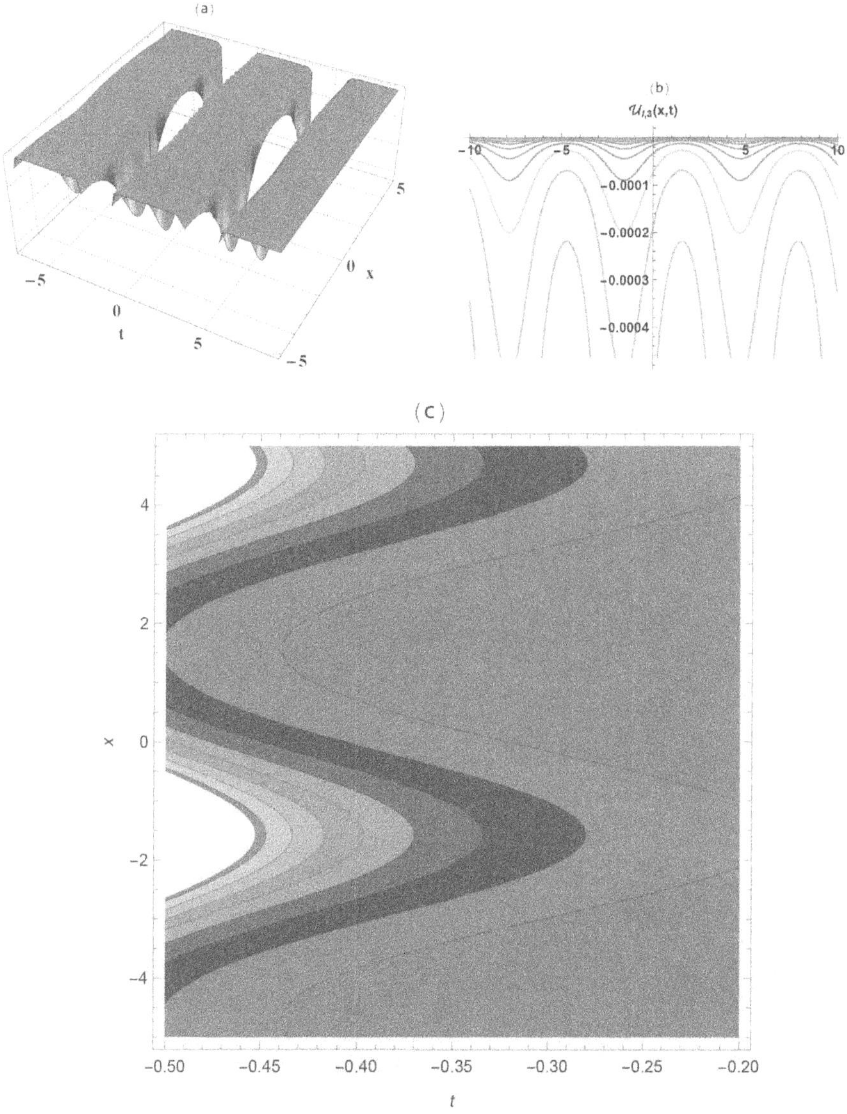

Figure 10.10 Breath solitary wave solutions of Eq. (10.134).

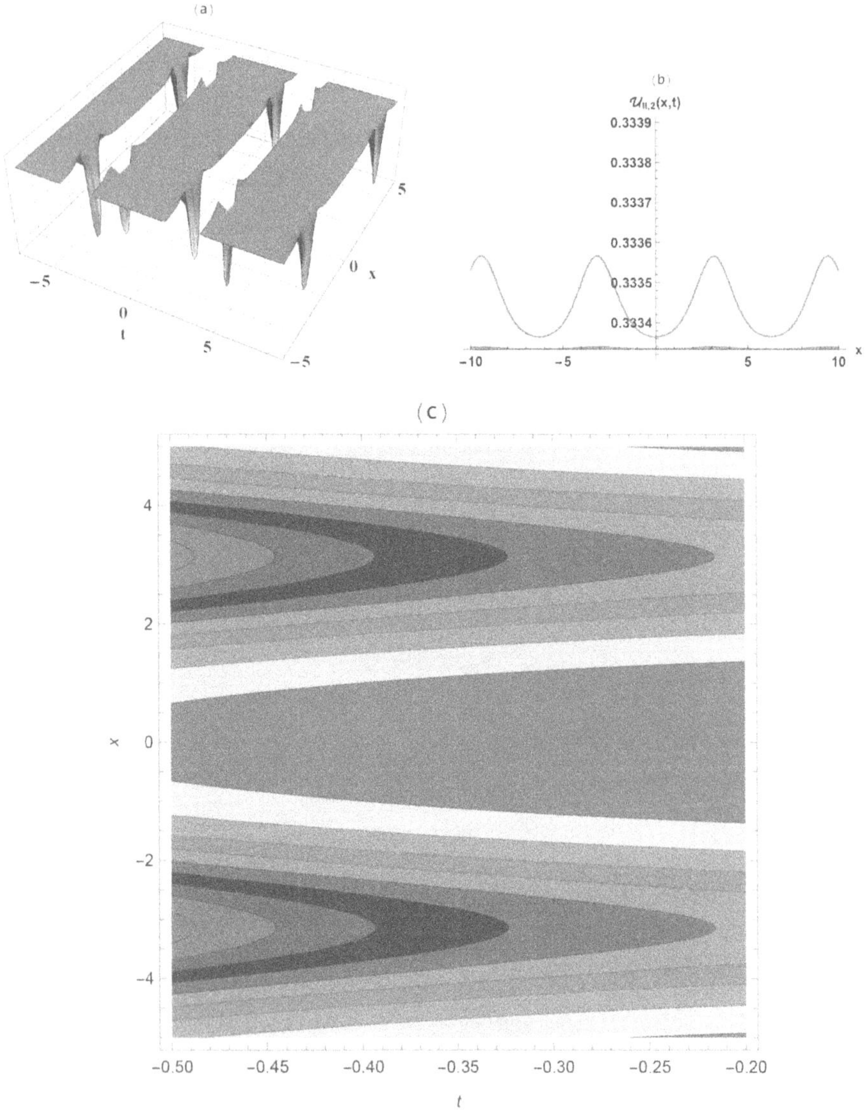

Figure 10.11 Solitary wave solutions of Eq. (10.137).

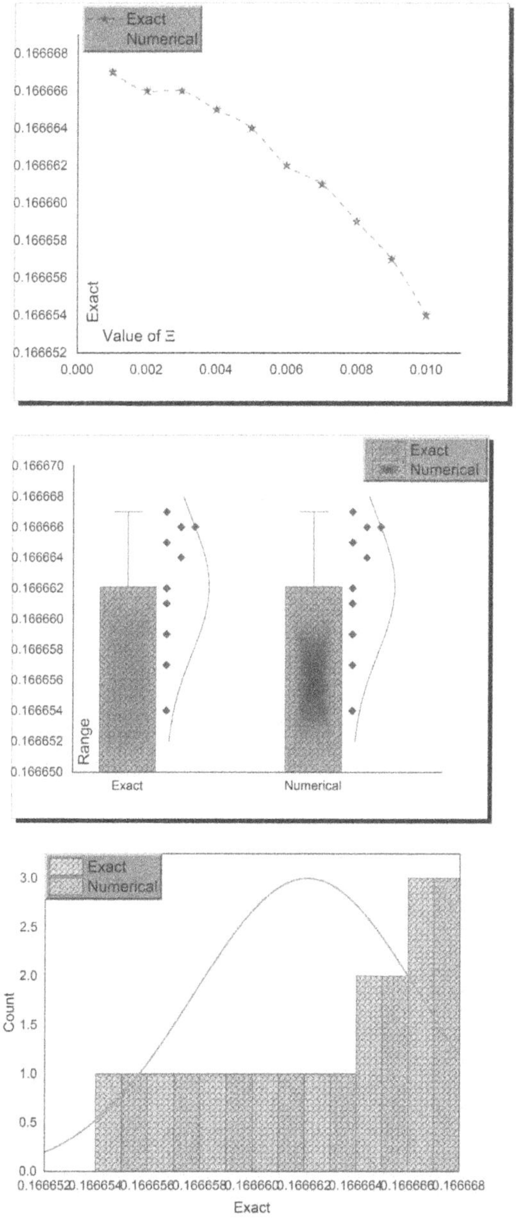

Figure 10.12 Exact and semi-analytical solutions of the (2+1)-dimensional generalized Nizhnik-Novikov-Veselov (GNNV) equations through the exp($-\phi(\Xi)$)-expansion method and the Adomian decomposition method.

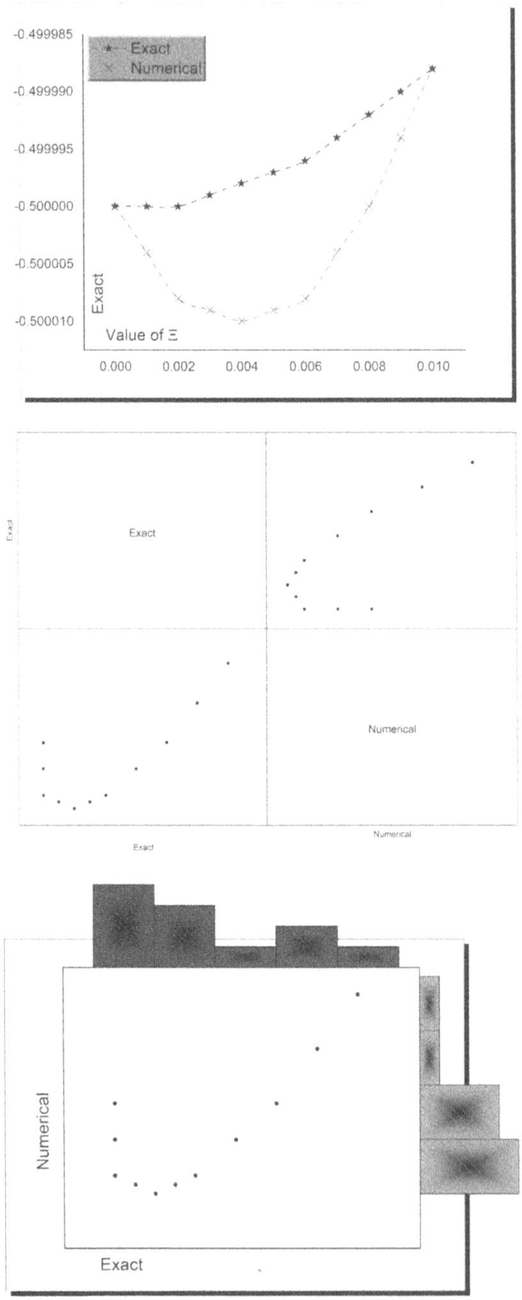

Figure 10.13 Exact and numerical solutions of the (2+1)-dimensional generalized Nizhnik-Novikov-Veselov (GNNV) equations through the extended Fan-expansion method and the cubic B-spline scheme.

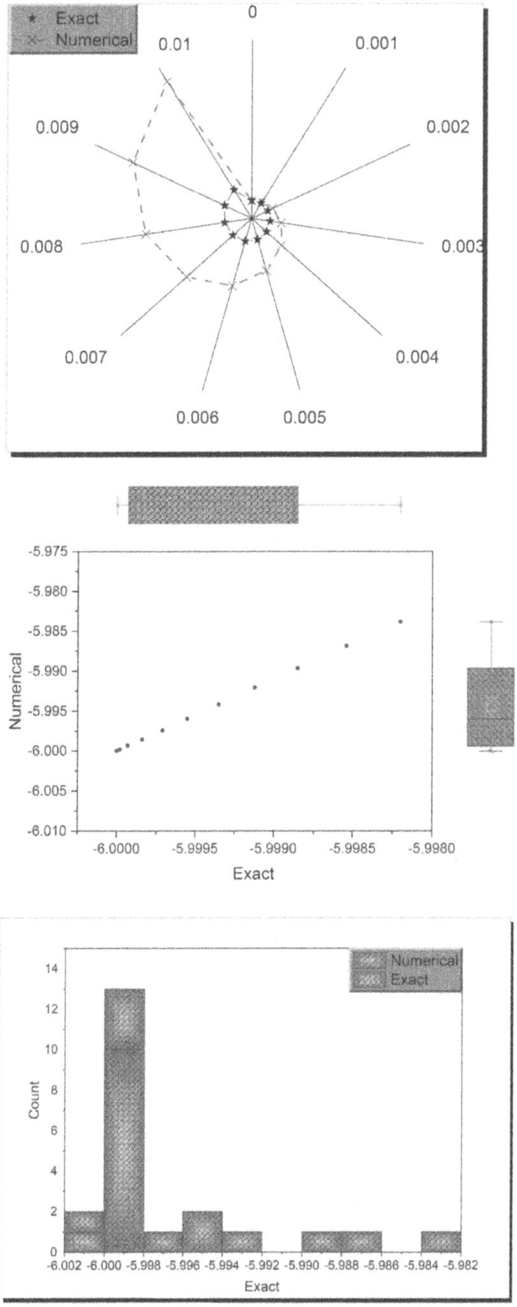

Figure 10.14 Exact and numerical solutions of the (2+1)-dimensional generalized Nizhnik-Novikov-Veselov (GNNV) equations through the extended simplest equation method and the quantic B-spline scheme.

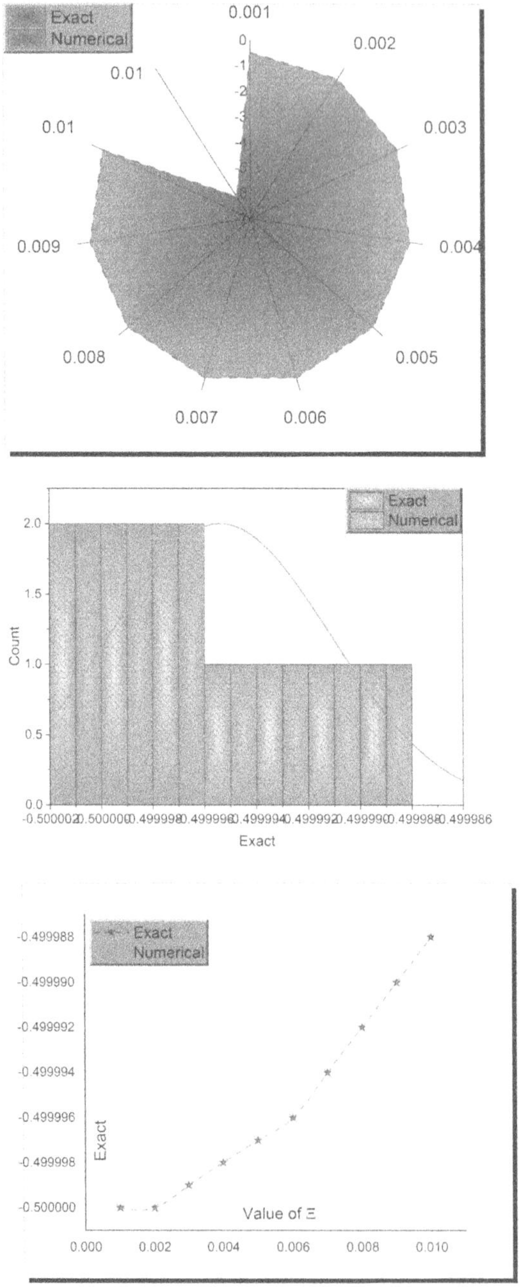

Figure 10.15 Exact and numerical solutions of the (2+1)-dimensional generalized Nizhnik-Novikov-Veselov (GNNV) equations through the modified Khater method and the septic B-spline scheme.

10.6 CONCLUSION

In this chapter, the exp $(-\phi(\Xi))$-expansion method, the extended Fan-expansion method, the extended $\left(\frac{G'(\Xi)}{G(\Xi)}\right)$-expansion method, the extended simplest equation method, the extended Tanh (Ξ)-expansion method, the modified Khater method, the Adomian decomposition method, and the B-spline collection of schemes (cubic, quantic, septic) have been successfully applied to the fractional isotropic extension model of the KdV model ((2+1)-dimensional generalized Nizhnik-Novikov-Veselov (GNNV) equations) for constructing analytical traveling wave solutions, semi-analytical, and numerical solutions. These solutions give more explanations of the dynamical behavior of incompressible fluid. Atangana conformable fractional derivatives, considered one of the most recent fractional operators in this field, have also been employed to convert the fractional form of the investigated model into the ordinary differential equation with an-integer order. Abundant wave solutions have been obtained via the above-mentioned schemes, and these solutions have then been used to calculate the initial and boundary conditions that allow application of the semi-analytical and numerical schemes that have been used to obtain the numerical solutions and to show the accuracy of the used analytical schemes. Additionally, this chapter has successfully shown the superiority of the modified Khater method over all the above schemes, as the number of its solutions is greater and the absolute error of its solutions is smaller. Many figures have been sketched in three- and two-dimensional, and contour plots to show the physical behavior of the obtained solutions.

Future goals are to discover more accurate fractional derivatives to avoid all the disadvantages of the current fractional operators. Moreover, to generalize the modified Khater method to give an infinite number of solutions for each suggested model. A final future aim is to evaluate the accuracy of the obtained solutions by employing the numerical schemes along the obtained computational solutions.

REFERENCES

1. M. Eslami, H. Rezazadeh, The first integral method for Wu–Zhang system with conformable time-fractional derivative, Calcolo 53 (3) (2016) 475–485.

2. M. Ekici, M. Mirzazadeh, M. Eslami, Q. Zhou, S. P. Moshokoa, A. Biswas, M. Belic, Optical soliton perturbation with fractional-temporal evolution by first integral method with conformable fractional derivatives, Optik 127 (22) (2016) 10659–10669.

3. H. Aminikhah, A. Refahi Sheikhani, H. Rezazadeh, Sub-equation method for the fractional regularized long-wave equations with conformable fractional derivatives, Scientia Iranica 23 (3) (2016) 1048–1054.

4. O. Ozkan, K. Ali, Conformable Double Laplace Transform For Fractional Partial Diferential Equations Arising in Mathematical Physics, Mathematical Studies and Applications 2018, 4–6 October 2018, (2018) 471.

5. A. Atangana, J. Gómez-Aguilar, Numerical approximation of Riemann-Liouville definition of fractional derivative: from Riemann-Liouville to Atangana-Baleanu, Numerical Methods for Partial Differential Equations 34 (5) (2018) 1502–1523.

6. K. M. Owolabi, A. Atangana, Numerical approximation of nonlinear fractional parabolic differential equations with Caputo–Fabrizio derivative in Riemann–Liouville sense, Chaos, Solitons & Fractals 99 (2017) 171–179.

7. S. Vong, P. Lyu, X. Chen, S.-L. Lei, High order finite difference method for time-space fractional differential equations with Caputo and Riemann-Liouville derivatives, Numerical Algorithms 72 (1) (2016) 195–210.

8. J. Hristov, Derivatives with non-singular kernels from the Caputo–Fabrizio definition and beyond: appraising analysis with emphasis on diffusion models, Frontiers in Fractional Calculus 1 (2017) 269–341.

9. H. Yépez-Martínez, J. F. Gómez-Aguilar, A new modified definition of Caputo–Fabrizio fractional-order derivative and their applications to the Multi Step Homotopy Analysis Method (MHAM), Journal of Computational and Applied Mathematics 346 (2019) 247–260.

10. M. A. Dokuyucu, E. Celik, H. Bulut, H. M. Baskonus, Cancer treatment model with the Caputo-Fabrizio fractional derivative, The European Physical Journal Plus 133 (3) (2018) 92.

11. H. Singh, H. Srivastava, D. Kumar, A reliable numerical algorithm for the fractional vibration equation, Chaos, Solitons & Fractals 103 (2017) 131–138.

12. H. Singh, D. Kumar, D. Baleanu, Methods of Mathematical Modelling: Fractional Differential Equations, CRC Press, Boca Raton, FL, 2019.

13. H. Singh, Approximate solution of fractional vibration equation using Jacobi polynomials, Applied Mathematics and Computation 317 (2018) 85–100.

14. H. Rezazadeh, D. Kumar, A. Neirameh, M. Eslami, M. Mirzazadeh, Applications of three methods for obtaining optical soliton solutions for the Lakshmanan–Porsezian–Daniel model with Kerr law nonlinearity, Pramana 94 (1) (2020) 39.

15. S. M. Mirhosseini-Alizamini, H. Rezazadeh, M. Eslami, M. Mirzazadeh, A. Korkmaz, New extended direct algebraic method for the Tzitzica type evolution equations arising in nonlinear optics, Computational Methods for Differential Equations 8 (1) (2020) 28–53.

16. H. Singh, H. Srivastava, Numerical simulation for fractional-order Bloch equation arising in nuclear magnetic resonance by using the Jacobi polynomials, Applied Sciences 10 (8) (2020) 2850.

17. H. Singh, Analysis for fractional dynamics of Ebola virus model, Chaos, Solitons & Fractals 138 (2020) 109992.

18. H. Singh, M. R. Sahoo, O. P. Singh, Numerical method based on Galerkin approximation for the fractional advection-dispersion equation, International Journal of Applied and Computational Mathematics 3 (3) (2017) 2171–2187.

19. M. A. Shallal, K. K. Ali, K. R. Raslan, H. Rezazadeh, A. Bekir, Exact solutions of the conformable fractional EW and MEW equations by a new generalized expansion method, Journal of Ocean Engineering and Science 5 (3) (2020) 223–229.

20. R. A. Attia, D. Lu, T. Ak, M. M. Khater, Optical wave solutions of the higher-order nonlinear Schrödinger equation with the non-Kerr nonlinear term via modified Khater method, Modern Physics Letters B (2020) 2050044.

21. M. M. Khater, C. Park, A.-H. Abdel-Aty, R. A. Attia, D. Lu, On new computational and numerical solutions of the modified Zakharov–Kuznetsov equation arising in electrical engineering, Alexandria Engineering Journal 59 (3) (2020) 1099–1105.

22. C. Yue, M. M. Khater, R. A. Attia, D. Lu, The plethora of explicit solutions of the fractional KS equation through liquid–gas bubbles mix under the thermodynamic conditions via Atangana–Baleanu derivative operator, Advances in Difference Equations 2020 (1) (2020) 1–12.

23. M. M. Khater, C. Park, D. Lu, R. A. Attia, Analytical, semi-analytical, and numerical solutions for the Cahn–Allen equation, Advances in Difference Equations 2020 (1) (2020) 1–12.

24. H. Qin, M. Khater, R. A. Attia, D. Lu, Approximate simulations for the non-linear long-short wave interaction system, Frontiers in Physics 7 (2020) 230.

25. H. Qin, R. A. Attia, M. Khater, D. Lu, Ample soliton waves for the crystal lattice formation of the conformable time-fractional (n+ 1) Sinh-Gordon equation by the modified Khater method and the Painlevé property, Journal of Intelligent & Fuzzy Systems 38 (3) (2020) 2745–s2752.

26. M. Al-Raeei, M. S. El-Daher, On: New optical soliton solutions for nonlinear complex fractional Schrödinger equation via new auxiliary equation method and novel (G'/G)-expansion method, Pramana 94 (1) (2020) 9.

27. H. Singh, R. K. Pandey, H. M. Srivastava, Solving non-linear fractional variational problems using Jacobi polynomials, Mathematics 7 (3) (2019) 224.

28. H. Singh, F. Akhavan Ghassabzadeh, E. Tohidi, C. Cattani, Legendre spectral method for the fractional Bratu problem, Mathematical Methods in the Applied Sciences 43 (9) (2020) 5941–5952.

29. H. Singh, C. Singh, A reliable method based on second kind Chebyshev polynomial for the fractional model of Bloch equation, Alexandria Engineering Journal 57 (3) (2018) 1425–1432.

30. M. Khater, R. A. Attia, H. Qin, H. Kadry, R. Kharabsheh, D. Lu, On the stable computational, semi-analytical, and numerical solutions of the langmuir waves in an ionized plasma, Journal of Intelligent & Fuzzy Systems 38 (3) (2020) 2833–2845.

31. E. Hosny Mohamad, E. Mostafa Hasan, F. Aly Shoukry, O. Ahmed El-Dydamoni, Assessment of the microscopic-observation-drug-susceptibility assay (MODS) and genexpert assay for diagnosis of pulmonary tuberculosis, Al-Azhar Medical Journal 49 (1) (2020) 91–102.

32. G. Akram, I. Zainab, Dark Peakon, Kink and periodic solutions of the nonlinear Biswas-Milovic equation with Kerr law nonlinearity, Optik 208 (2020) 164420.

33. R. M. Jena, S. Chakraverty, H. Rezazadeh, D. Domiri Ganji, On the solution of time-fractional dynamical model of Brusselator reaction-diffusion system arising in chemical reactions, Mathematical Methods in the Applied Sciences 8, (2020).

34. Z. Jie-Fang, Bäcklund transformation and variable separation solutions for the generalized Nozhnik-Novikov-Veselov equation, Chinese Physics 11 (7) (2002) 651.

35. C.-Q. Dai, Y.-Y. Wang, Combined wave solutions of the (2+1)-dimensional generalized Nizhnik–Novikov–Veselov system, Physics Letters A 372 (11) (2008) 1810–1815.

36. J.-F. Zhang, C.-L. Zheng, Abundant localized coherent structures of the (2+1)-dimensional generalized Nozhnik-Novikov-Veselov system, Chinese Journal of Physics 41 (3) (2003) 242–254.

37. C. Dai, J. Zhang, Variable separation solutions for the (2+1)-dimensional generalized Nizhnik–Novikov–Veselov equation, Chaos, Solitons & Fractals 33 (2) (2007) 564–571.

38. Y. Chen, Z. Dong, Symmetry reduction and exact solutions of the generalized Nizhnik–Novikov–Veselov equation, Nonlinear Analysis: Theory, Methods & Applications 71 (12) (2009) e810–e817.

39. Z. Zhao, Y. Chen, B. Han, Lump soliton, mixed lump stripe and periodic lump solutions of a (2+1)-dimensional asymmetrical Nizhnik–Novikov–Veselov equation, Modern Physics Letters B 31 (14) (2017) 1750157.

40. X.-B. Wang, B. Han, Characteristics of the breathers, rogue waves and soliton waves in a (2+1)-dimensional generalized Nizhnik–Novikov–Veselov equation, Modern Physics Letters B 33 (03) (2019) 1950014.

41. J.-G. Liu, Lump-type solutions and interaction solutions for the (2+1)-dimensional asymmetrical Nizhnik-Novikov-Veselov equation, The European Physical Journal Plus 134 (2) (2019) 56.

42. P.-X. Wu, Y.-F. Zhang, Lump, lumpoff and predictable rogue wave solutions to the (2+1)-dimensional asymmetrical Nizhnik-Novikov-Veselov equation, Physics Letters A 383 (15) (2019) 1755–1763.

43. Z. Li, Diverse oscillating soliton structures for the (2+1)-dimensional Nizhnik–Novikov–Veselov equation, The European Physical Journal Plus 135 (1) (2020) 8.

44. H. Yépez-Martínez, J. Gómez-Aguilar, Fractional sub-equation method for Hirota–Satsuma-coupled KdV equation and coupled mKdV equation using the Atangana's conformable derivative, Waves in Random and Complex Media 29 (4) (2019) 678–693.

45. N. Khan, O. Razzaq, M. Ayaz, Some properties and applications of conformable fractional laplace transform (CFLT), Journal of Fractional Calculus and Applications 9 (1) (2018) 72–81.

46. C. Park, M. M. Khater, R. A. Attia, W. Alharbi, S. S. Alodhaibi, An explicit plethora of solution for the fractional nonlinear model of the low–pass electrical transmission lines via Atangana–Baleanu derivative operator, Alexandria Engineering Journal 59 (3) (2020) 1205–1214.

11 A Modified Computational Scheme and Convergence Analysis for Fractional Order Hepatitis E Virus Model

Ved Prakash Dubey,[1] *Devendra Kumar,*[2]
Sarvesh Dubey[3]
[1]Faculty of Mathematical and Statistical Sciences, Shri Ramswaroop Memorial University, Lucknow, Uttar Pradesh, India
[2]Department of Mathematics, University of Rajasthan, Jaipur, Rajasthan, India
[3]Department of Physics, L.N.D. College (B.R. Ambedkar Bihar University, Muzaffarpur), Motihari, Bihar, India

CONTENTS

11.1 INTRODUCTION

The hepatitis E virus (HEV) is one among five known human hepatitis viruses and is unique in having multiple zoonotic reservoirs [1]. Hepatitis E disease spreads due to hepatitis E virus. The word hepatitis is made up of two parts: hepa (liver) and itis (inflammation). Hepatitis E disease generates inflammation in the liver [2]. Hepatitis E is a severe viral infectious disease transmitted mainly through polluted drinking water, food, infected blood transfusion, and cyclic outbreaks, the most common of these routes being contaminated drinking water. It usually develops with sharp hepatic symptoms. There are some usual symptoms of hepatitis disease such as fever, vomiting, nausea, and jaundice [1]. The risk of hepatitis E infection generally increases for livestock workers, especially pig farmers and employees working at meat processing plants and slaughterhouses. A mathematical model of hepatitis E infection can be helpful to anticipate the changes in its morbidity rate in controlled areas. It is remarkable that vertical transmission takes place in case of hepatitis E, that is, it is transmitted from pregnant mother to her foetus [3]. Actually, this is an infectious disease of the heart caused by HEV. There are four types of HEV, namely genotype 1, genotype 2, genotype 3, and genotype 4. Genotypes 1 and 2 are found in the human body, whereas genotypes 3 and 4 are found in animals, such as pigs, deer, etc. and can be transmitted to the human body [1].

In recent years, the dynamics of biological and ecological phenomena have been well expressed through mathematical models. Various authors have introduced mathematical models to describe HEV transmission dynamics. Some years ago, Mercera and Siddiqui [4] suggested a mathematical model to describe the HEV transmission dynamics utilizing Holling functional II response in order to assess the intervention strategies in Uganda. Backer et al. [5] formulated a mathematical model to show transmission dynamics of HEV in pigs. In 2018, Alzahrani and Khan [6] investigated the HEV model with optimal control in view of the Atangana-Baleanu fractional differential operator. Nannyonga and his co-workers [7] explored the dynamics, causes, and prevention of hepatitis E outbreaks in Uganda from 2007 to 2009. They also studied the interconnection of co-infection between malaria and hepatitis E disease. In 2013, Ren et al. [8] introduced a new HEV model to study the spread of HEV in Shanghai. In 2012, Rein et al. [9] estimated the global burden of HEV genotypes 1 and 2 in 2005. Gouttenoire and Moradpour [10] formulated a mouse model for HEV infection in 2016. Recently, Prakash et al. [11] and Khan et al. [12] used fractional modelling to study the dynamics of HEV in view of Atangana-Baleanu fractional derivative [13] and Caputo-Fabrizio fractional derivative [14], respectively. More recently, Sholikah et al. [15] explored the dynamics of HEV via a mathematical model to discuss optimal control of the spread of HEV.

Fractional differential equations are appropriate for the study of nonlinear physical phenomena. Fractional order models possess memory and hereditary properties, and so for this reason they become more useful and relevant to explore the dynamics of physical phenomena [16].This study employs the q-homotopy analysis Sumudu transform method (q-HASTM) proposed by Singh et al. [17] to handle the fractional model concerning interpersonal relationships. This method is a combination of the q-homotopy analysis method (q-HAM), homotopy polynomials, and the Sumudu transform (ST) algorithm. It is notable that the fusion of the q-HAM with the Sumudu transform provides a time-saving result. The q-HAM was developed by El-Tavil et al. [18, 19],and is a simple modification of HAM [20, 21] with the generalization of an embedding parameter occurring in HAM. It is notable that q-HASTM includes a certain auxiliary parameter $\hbar$ for adjustment and control of the convergence region, whereas q-HAM consists of the auxiliary parameter $\hbar$ as well as an asymptotic parameter ε in such a way that the q-HASTM reduces to HASTM for $\varepsilon = 1$. More recently, q-HASTM was utilized by Dubey et al. [22] to investigate the dynamics of a modified Degasperis-Procesi equation involving the Caputo fractional derivative.

This chapter focuses on the implementation of a semi-analytical scheme such as q-HASTM for a time-fractional nonlinear mathematical hepatitis E virus model with a perception to study the dynamics of the model. Using graphical presentations, we study the consequences of the variations of fractional order of a time derivative and time t on susceptible, exposed, infected, and recovered populations. The role of environment is also considered in disease dynamics of the hepatitis E model. This work strongly authenticates the computational strength of the employed scheme. Moreover, the uniqueness and convergence analysis of the method are also discussed with the help of fixed point theory. The remaining part of the chapter is organized as follows. Section 11.2 describes the basic definitions and formulae for the Caputo fractional derivative and ST operator. Section 11.3 presents the mathematical description of the HEV model. Section 11.4 describes the fundamental working procedure of the q-HASTM. Section 11.5 is devoted to the uniqueness and convergence analysis of the q-HASTM. In Section 11.6, q-HASTM is applied to the fractional HEV model. Section 11.7 discusses the numerical results through graphical presentations. Finally, Section 11.8 concludes the present work.

11.2 ELEMENTAL DEFINITIONS AND FORMULAE

This section presents formally the elemental concepts and formulae regarding the Caputo fractional derivative and the ST operator.

Definition 11.1: *(see [23]): The Riemann-Liouville (RL) fractional integral operator (J^{β}) of order β ($\beta > 0$) of a function $h \in C_{\eta}$, $\eta \geq -1$ is described by*

$$J^{\beta} h(\tau) = \frac{1}{\Gamma(\beta)} \int_0^{\tau} (\tau - \omega)^{\beta - 1} h(\omega)\, d\omega\,,\ \beta > 0,\ J^0 h(\tau) = h(\tau).$$

For the RL fractional integral operator, we have

$$J^{\beta} \tau^{\gamma} = \frac{\Gamma(\gamma+1)}{\Gamma(\gamma+\beta+1)} \tau^{\beta+\gamma}.$$

Definition 11.2 (23): *The definition of Caputo fractional derivative of $\wp(t)$ of order β ($\beta \geq 0$) is stated as*

$$D^{\beta} \wp(t) = J^{\ell-\beta} D^{\ell} \wp(t) = \frac{1}{\Gamma(\ell - \gamma)} \int_0^{t} (t - \tau)^{\ell - \beta - 1} \frac{d^{\ell}}{d\tau^{\ell}} \wp(\tau)\, d\tau,$$

$$\ell - 1 < \beta \leq \ell,\ t > 0,\ \ell \in \mathrm{N},$$

where D^{β} stands for the classical differential operator of order β.

Definition 11.3 (23): *The fractional derivative of $\wp(\varpi,t)$of order $\beta > 0$ in the Caputo sense is described by*

$$D_t^{\beta}\wp(\varpi,t) = \frac{\partial^{\beta}\wp(\varpi,t)}{\partial t^{\beta}} = \begin{cases} \frac{1}{\Gamma(\ell-\beta)}\int_0^t (t-\tau)^{\ell-\beta-1}\frac{\partial^{\ell}\wp(\varpi,\tau)}{\partial \tau^{\ell}}d\tau, & \ell-1<\beta<\ell \\ \frac{\partial^{\ell}\wp(\varpi,t)}{\partial t^{\ell}}, & \beta=\ell\in \mathrm{N}. \end{cases}$$

Definition 11.4: *(see [24]): The Sumudu transform is defined for the set of functions*

$$\Phi = \left\{\wp(t) \mid \exists K,\ \beta_1,\beta_2>0,\ |\wp(t)|<Ke^{|t|/\beta_l},\ if\, t\in(-1)^l\times[0,\infty)\right\},$$

by the following expression:

$$\bar{\wp}(u) = ST[\wp(t)] = \int_0^{\infty}\wp(ut)\,e^{-t}dt, u\in(-\beta_1,\beta_2).$$

Definition 11.5: *(see [25]): The Sumudu transform of the Caputo fractional derivative $D_t^{\beta}\wp(t)$ is given as*

$$ST\left[D_t^{\beta}\wp(t)\right] = u^{-\beta}ST[\wp(t)] - \sum_{\lambda=0}^{\kappa-1}u^{-\beta+\lambda}\wp^{(\lambda)}(0), \kappa-1<\beta\leq\kappa,$$

$$\beta>0,\ \beta\in \mathrm{N}.$$

The fundamental properties of Sumudu transform are provided by a number of notable researchers (see [26–28]).

Definition 11.6: *(see [29]): Let $0<\beta<1$ and $p=1-\beta$. Then the function space denoted by $C_p([t_0,T],R^n)$ is defined as*

$$C_p([t_0,T],R^n) = \left\{x\in C((t_0,T],R^n) \mid x(t)(t-t_0)^p \in C_p([t_0,T],R^n)\right\}.$$

Definition 11.7: *(see [30]): The Mittag-Leffler function is formulated as $E_{\beta}(z) = \sum_{k=0}^{\infty}\frac{z^k}{\Gamma(1+k\beta)}$, where $\beta>0$ and Γ signifies the gamma function [23]. The Mittag-Leffler function pertaining two parameters possesses the following form: $E_{\beta,\gamma}(z) = \sum_{k=0}^{\infty}\frac{z^k}{\Gamma(k\beta+\gamma)}$, where $\beta>0$, $\gamma>0$. For $\beta=1$, we have $E_{\beta}(z)=E_{\beta,1}(z)$. Also $E_{1,1}(z)=e^z$.*

11.3 MATHEMATICAL DESCRIPTION OF HEV MODEL

In this work, we have described the mathematical structure of the model related to the viral dynamics of hepatitis E recently formulated and studied by Khan et al. [12].

This nonlinear model system consists of five ODEs given as follows:

$$\begin{aligned}
\frac{dS}{dt} &= \wp(1-\ell I) - (\alpha I + \gamma_E P)\, S - \vartheta\, S, \\
\frac{dE}{dt} &= (\alpha I + \gamma_E P)\, S - (\vartheta + \rho)\, E + \wp \ell I, \\
\frac{dI}{dt} &= \rho\, E - (\vartheta + \sigma)\, I, \\
\frac{dR}{dt} &= \sigma I - \vartheta\, R, \\
\frac{dP}{dt} &= \varphi I - \lambda\, P,
\end{aligned} \tag{11.1}$$

subject to the initial conditions:

$$S(0) = S_0,\ E(0) = E_0, I(0) = I_0, R(0) = R_0, P(0) = P_0. \tag{11.2}$$

Here, the description about the variables and parameters of the model system (11.1) is as follows:

- $S(t), E(t), I(t)$, and $R(t)$, respectively, denote the population of susceptible, exposed, infected, and recovered individuals.
- $P(t)$ denotes the density of viral load in the environment.
- The constant $\wp$ specifies the recruitment rate.
- ℓ denotes the transfer rate of infection from an infected mother to her child.
- αdenotes the contact rate.
- γ_E signifies the rate of contact between susceptible individuals and the environment.
- ϑ signifies the natural death rate of humans.
- ρ and σ denote rate of infection and rate of recovery, respectively.
- φ signifies the rate of virus transfer from infected people to the environment.
- λ signifies the virus's decay in the environment.

Remark 11.1: *This model presumes positivity of all the above-mentioned parameters.*

Remark 11.2: *Here, the total population size of the individuals is distributed in four sub-compartments S, E, I, and R, so that $N = S + E + I + R$.*
In this chapter, the above-mentioned system (11.1) *is considered with the fractional order β as given below:*

$$\frac{d^\beta S}{dt^\beta} = \wp(1-\ell I) - (\alpha I + \gamma_E P)\, S - \vartheta\, S,$$

$$\frac{d^\beta E}{dt^\beta} = (\alpha I + \gamma_E P)\, S - (\vartheta + \rho)\, E + \wp \ell I,$$

$$\frac{d^{\beta}I}{dt^{\beta}} = \rho E - (\vartheta + \sigma) I,$$

$$\frac{d^{\beta}R}{dt^{\beta}} = \sigma I - \vartheta R,$$

$$\frac{d^{\beta}P}{dt^{\beta}} = \varphi I - \lambda P, \tag{11.3}$$

subject to the initial conditions:

$$S(0) = S_0,\ E(0) = E_0, I(0) = I_0, R(0) = R_0, P(0) = P_0. \tag{11.4}$$

Here, $\frac{d^{\beta}}{dt^{\beta}}$ *symbolizes the Caputo fractional differential operator and* $0 < \beta \leq 1$ *is a parameter specifying the fractional order of a time derivative.*

Remark 11.3: *Since population densities are non-negative, the solutions of system (11.3) will be restricted to the non-negative 5D orthant* $\{(S, E, I, R, P) \mid S \geq 0, E \geq 0, I \geq 0, R \geq 0, P \geq 0\} \subset R^5_+$*. Thus, the above described equations have to be solved according to the biologically meaningful conditions* $S \geq 0, E \geq 0, I \geq 0, R \geq 0, P \geq 0$.

Lemma 11.1: *[29]: If* $m \in C_p(R^+, R)$ *satisfies* ${}^{C}D^{q}_{t_0} m(t) \leq \lambda\, m(t) + d,\ m(t_0) = m_0$, $t \geq t_0 > 0$, *where* $\lambda, d \in R$. *Then* $m(t) \leq m(t_0)E_{\beta}(\lambda\,(t - t_0)^{\beta}) + d\,(t - t_0)^{\beta}\, E_{\beta,\beta+1}(\lambda\,(t - t_0)^{\beta}),\ t \geq t_0 > 0$.

Proof. See proof [29]. □

Lemma 11.2: *All solutions of system (11.3) which initiate in* R^5_+ *are uniformly bounded.*

Proof. To achieve the fact that solutions of system (11.3) is bounded, a function $W$is defined as follows:

$$W = S + E + I + R + P. \tag{11.5}$$

Clearly $W \geq 0$. Now, operating the Caputo fractional differential operator on both sides of the above Eq. (11.5) provides

$$D^{\beta}_t W = D^{\beta}_t S + D^{\beta}_t E + D^{\beta}_t I + D^{\beta}_t R + D^{\beta}_t P, D^{\beta}_t \equiv \frac{d^{\beta}}{dt^{\beta}}. \tag{11.6}$$

$$\begin{aligned} D^{\beta}_t W =& \wp(1 - \ell I) - (\alpha I + \gamma_E P)\, S - \vartheta S + (\alpha I + \gamma_E P) S \\ & - (\vartheta + \rho)\, E + \wp \ell I + \rho E - (\vartheta + \sigma)\, I \\ & + \sigma I - \vartheta R + \varphi I - \lambda P \\ =& \wp - \vartheta\,(S + E + I + R) + \varphi I - \lambda P \\ =& \wp + \varphi I - (\vartheta\,[S + E + I + R] + \lambda P) \\ \leq& \wp + \varphi I - k\,(S + E + I + R + P) \\ =& \wp + \varphi I - kW, \end{aligned} \tag{11.7}$$

where $k = \min\{\vartheta, \lambda\}$. Now by Lemma 11.1, we have

$$W(t) \leq W(0) E_{\beta}\left(-kt^{\beta}\right) + \varphi I(0) E_{\beta,\beta+1}\left(-kt^{\beta}\right) + \wp t^{\beta} E_{\beta,\beta+1}\left(-kt^{\beta}\right) = W_1, \tag{11.8}$$

where E_{β} signifies the Mittag-Leffler function. Thus, all the solutions of the model system (11.3) with initial conditions in $\Phi = \{(S, E, I, R, P) \in W : 0 \leq W \leq W_1\}$ remain in $\Phi \forall t > 0$. Thus, the region Φ is positively invariant in respect of the model (11.3). □

Table 11.1
Values of Parameters used in Numerical Simulation of the HEV Model

Parameter	Value	Unit
$\wp$	0.8	day^{-1}
ℓ	0.02	day^{-1}
α	0.0004	day^{-1}
γ_E	0.0005	day^{-1}
ϑ	1/67.7	day^{-1}
ρ	0.02	day^{-1}
σ	0.023801429	day^{-1}
φ	0.02	day^{-1}
λ	0.03	day^{-1}

Source: Khan et al. [12].

11.4 *q*-HASTM: BASIC METHODOLOGY

To explain the basic working steps of offered scheme q-HASTM for fractional PDEs, the ensuing general nonlinear non-homogeneous fractional order PDE is considered as follows:

$$D_{\tau}^{k\beta} g(\theta,\tau) + Mg(\theta,\tau) + Ng(\theta,\tau) = G(\theta,\tau), \tau > 0, \; \theta \in R, \; k-1 < k\beta \leq k, \tag{11.9}$$

where $D_{\tau}^{k\beta} = \frac{\partial^{k\beta}}{\partial \tau^{k\beta}}$ symbolizes the Caputo differential operator of order β. Here, $g(\theta,\tau)$ is a function of two variables θ and τ, M specifies the bounded linear differential operator in θ and τ, that is, $\exists$ a number $\delta > 0$ such that $\|Mg\| \leq \delta \|g\|$, N is a nonlinear differential operator of general nature in θ and τ and is also Lipschitz continuous with $\xi > 0$ satisfying the condition $|Ng - Nh| \leq \xi |g - h|$, and $G(\theta,\tau)$ denotes the source term.

Now the computational approach suggests the implementation of the Sumudu transform (ST) operator on Eq. (11.9), which delivers the following equation:

$$ST\left[D_{\tau}^{k\beta} g(\theta,\tau)\right] + ST\left[Mg(\theta,\tau) + Ng(\theta,\tau)\right] = ST\left[G(\theta,\tau)\right]. \tag{11.10}$$

Now, employing the formula of the Sumudu transform for the fractional derivative in Eq. (11.10) provides

$$u^{-k\beta} ST\left[g\left(\theta,\tau\right)\right] - \sum_{\gamma=0}^{k-1} u^{-k\beta+\gamma} g^{(\gamma)}\left(\theta,0\right) + ST\left[Mg\left(\theta,\tau\right) + Ng\left(\theta,\tau\right) - G\left(\theta,\tau\right)\right] = 0. \tag{11.11}$$

Simplification provides

$$ST\left[g\left(\theta,\tau\right)\right] - u^{k\beta} \sum_{\gamma=0}^{k-1} u^{-k\beta+\gamma} g^{\gamma}\left(\theta,0\right) + u^{k\beta} ST\left[Mg\left(\theta,\tau\right) + Ng\left(\theta,\tau\right) - G\left(\theta,\tau\right)\right] = 0. \tag{11.12}$$

Now in view of Eq. (11.12), the nonlinear operator is structured as follows:

$$\varsigma\left[\psi\left(\theta,\tau;q\right)\right] = ST\left[\psi\left(\theta,\tau;q\right)\right] - u^{k\beta} \sum_{\gamma=0}^{k-1} u^{-k\beta+\gamma} \psi^{\gamma}\left(\theta,0;q\right) + u^{k\beta}\left\{ST\left[M\psi\left(\theta,\tau;q\right) + N\psi\left(\theta,\tau;q\right)\right]\right\} - u^{k\beta} ST\left\{G\left(\theta,\tau;q\right)\right\}, \tag{11.13}$$

where $q \in \left[0, \frac{1}{\varepsilon}\right]$, $\varepsilon \geq 1$ signifies the embedding parameter, $\psi\left(\theta,\tau;q\right)$ denotes the real valued function of θ, τ and q, and the symbol ST stands for Sumudu transform operator.

Now, the linear operator ζ is selected as follows:

$\zeta\left[\psi\left(\theta,\tau;q\right)\right] = S\left[\psi\left(\theta,\tau;q\right)\right]$ with the characteristic $\zeta\left[a\right] = 0$ for a random constant a.

Now in view of q-HAM algorithm [18, 19], the homotopy equation can be constructed in this way:

$$\left(1 - \varepsilon q\right) ST\left[\psi\left(\theta,\tau;q\right) - g_0\left(\theta,\tau\right)\right] = \hbar q H\left(\theta,\tau\right) \varsigma\left[\psi\left(\theta,\tau;q\right)\right], \tag{11.14}$$

where $\hbar \neq 0$ denotes the auxiliary parameter but particularly negative in all practical situations, $H(\theta,\tau) \neq 0$ signifies the auxiliary function, $g_0(\theta,\tau)$ signifies the initial guess of $g(\theta,\tau)$, and $\psi(\theta,\tau;q)$ denotes an unknown function. Equation (11.14) is the zeroth-order deformation equation. Remarkably q-HASTM allows a greater degree of freedom for a proper choice of auxiliary parameters in a solution process. It can be easily seen that for $q = 0$ and $q = \frac{1}{\varepsilon}$, these subsequent expressions hold well as follows:

$$\psi\left(\theta,\tau;0\right) = g_0\left(\theta,\tau\right), \psi\left(\theta,\tau;\frac{1}{\varepsilon}\right) = g\left(\theta,\tau\right). \tag{11.15}$$

Hence, when q varies from 0 to $\frac{1}{\varepsilon}$, the solution $\psi(\theta,\tau;q)$ deviates from $g_0(\theta,\tau)$ to $g\left(\theta,\tau\right)$ Now, the Taylor's series expansion of the function $\psi\left(\theta,\tau;q\right)$ about q yields the series as follows:

$$\psi\left(\theta,\tau;q\right) = g_0\left(\theta,\tau\right) + \sum_{\mu=1}^{\infty} q^{\mu} g_{\mu}\left(\theta,\tau\right), \tag{11.16}$$

where

$$g_\mu(\theta,\tau) = \left[\frac{1}{\mu!}\frac{\partial^\mu \psi(\theta,\tau;q)}{\partial q^\mu}\right]_{q=0}. \tag{11.17}$$

The convergence-control parameter $\hbar \neq 0$ regulates the convergence domain of the series solution (11.16). Therefore the series represented by Eq. (11.16) converges at $q = \frac{1}{\varepsilon}$ by means of proper selection of the auxiliary linear operator, the initial guess $g_0(\theta,\tau)$, ε, $\hbar$, and $H(\theta,\tau)$. Thus, we have

$$g(\theta,\tau) = g_0(\theta,\tau) + \sum_{\mu=1}^{\infty} g_\mu(\theta,\tau)\left(\frac{1}{\varepsilon}\right)^\mu. \tag{11.18}$$

Equation (11.18) provides a relationship between $g_0(\theta,\tau)$ and the exact solution $g(\theta,\tau)$ through the term $g_\mu(\theta,\tau)$ ($\mu = 1,2,3,...$), which are still to be computed in forthcoming steps. Equation (11.18) provides the approximate solution of the concerned Eq. (11.9) in a series form. Now, the vectors are defined as follows:

$$\underset{\mu}{\overrightarrow{g}} = \left\{g_0(\theta,\tau), g_1(\theta,\tau), g_2(\theta,\tau), ...g_\mu(\theta,\tau)\right\}. \tag{11.19}$$

For computation of the μth-order deformation equation, the zeroth-order deformation Eq. (11.14) is differentiated μ-times with regard to q and then choosing $q = 0$, and at the end, dividing them finally by μ ! provides the μth-order deformation equation as follows:

$$ST\left[g_\mu(\theta,\tau) - \chi_\mu g_{\mu-1}(\theta,\tau)\right] = \hbar H(\theta,\tau)\,\Re_\mu\left(\vec{g}_{\mu-1}(\theta,\tau)\right). \tag{11.20}$$

Operating the inverse Sumudu transform ST^{-1} on above obtained μth-order deformation Eq. (11.20) and $H(\eta,\omega)$ is taken as 1, we have

$$g_\mu(\theta,\tau) = \chi_\mu g_{\mu-1}(\theta,\tau) + \hbar ST^{-1}\left[\Re_\mu\left(\vec{g}_{\mu-1}(\theta,\tau)\right)\right]. \tag{11.21}$$

In Eq. (11.21), the value of $\Re_\mu\left(g_{\mu-1}(\theta,\tau)\right)$ is expressed in a novel fashion as follows:

$$\begin{aligned}\Re_\mu\left(\vec{g}_{\mu-1}\right) &= ST\left[g_{\mu-1}(\theta,\tau)\right] - \left(1-\frac{\chi_\mu}{\varepsilon}\right)u^{k\beta} \\ &\quad\times\left(\sum_{\gamma=0}^{k-1} u^{-k\beta+\gamma} g^{(\gamma)}(\theta,0) + ST\left[G(\theta,\tau)\right]\right) \\ &\quad + u^{k\beta} ST\left[M g_{\mu-1}(\theta,\tau) + P_{\mu-1}\right],\end{aligned} \tag{11.22}$$

where the value of χ_μ is presented as

$$\chi_\mu = \begin{cases} 0, & \mu \leq 1 \\ \varepsilon, & \mu > 1. \end{cases} \tag{11.23}$$

In Eq. (11.22), P_μ stands for homotopy polynomial introduced by Odibat and Bataineh [31] in the computational procedure of HAM [20, 21], and is expressed as follows:

$$P_\mu = \frac{1}{\Gamma(\mu)}\left[\frac{\partial^\mu}{\partial q^\mu}\varsigma\psi(\theta,\tau;q)\right]_{q=0}, \tag{11.24}$$

where

$$\psi = \psi_0 + q\psi_1 + q^2\psi_2 + q^3\psi_3 + \cdots . \tag{11.25}$$

Putting the value of $\Re_\mu(g_{\mu-1})$ from Eq. (11.22) in Eq. (11.21) provides the equation as follows:

$$\begin{aligned} g_\mu(\theta,\tau) &= (\chi_\mu + \hbar)\, g_{\mu-1}(\theta,\tau) - \hbar\left(1 - \frac{\chi_\mu}{\varepsilon}\right) ST^{-1} \\ &\times \left(u^{k\beta}\sum_{\gamma=0}^{k-1} u^{-k\beta+\gamma} g^{(\gamma)}(\theta,0) + u^{k\beta} ST\left[G(\theta,\tau)\right]\right) \\ &+ \hbar S^{-1}\left(u^{k\beta} ST\left[M g_{\mu-1}(\theta,\tau) + P_{\mu-1}\right]\right). \end{aligned} \tag{11.26}$$

Now, the computation of $g_\mu(\theta,\tau)$ for $\mu \geq 1$ can be done in a very lucid way and the final structure of q-HASTM solution is expressed in the following way:

$$g(\theta,\tau) = \sum_{\mu=0}^{K} g_\mu(\theta,\tau)\left(\frac{1}{\varepsilon}\right)^\mu . \tag{11.27}$$

When $K \to \infty$, a precise approximation of the solution of given PDE (11.9) is acquired. It is worth noting that the q-HASTM reduces to HASTM for $\varepsilon = 1$.

Theorem 11.1: *If a constant* $0 < \omega < 1$ *can be obtained such that* $\|g_{\mu+1}(\theta,\tau)\| \leq \omega \|g_\mu(\theta,\tau)\|$ *for each value of* μ. *Moreover, if the truncated series* $\sum_{\mu=0}^{r} g_\mu(\theta,\tau)\left(\frac{1}{\varepsilon}\right)^\mu$ *is considered as an approximate solution* g, *then the maximum absolute truncated error is computed as*

$$\left\| g(\theta,\tau) - \sum_{\mu=0}^{r} g_\mu(\theta,\tau)\left(\frac{1}{\varepsilon}\right)^\mu \right\| \leq \frac{\omega^{r+1}}{\varepsilon^r(\varepsilon-\omega)} \|g_0(\theta,\tau)\| .$$

Proof: We proceed as follows:

$$\begin{aligned} \left\| g(\theta,\tau) - \sum_{\mu=0}^{r} g_\mu(\theta,\tau)\left(\frac{1}{\varepsilon}\right)^\mu \right\| &= \left\| \sum_{\mu=r+1}^{\infty} g_\mu(\theta,\tau)\left(\frac{1}{\varepsilon}\right)^\mu \right\| \\ &\leq \sum_{\mu=r+1}^{\infty} \|g_\mu(\theta,\tau)\| \left(\frac{1}{\varepsilon}\right)^\mu \\ &\leq \sum_{\mu=r+1}^{\infty} \omega^\mu \|g_0(\theta,\tau)\| \left(\frac{1}{\varepsilon}\right)^\mu \\ &\leq \left(\frac{\omega}{\varepsilon}\right)^{r+1}\left[1 + \left(\frac{\omega}{\varepsilon}\right) + \left(\frac{\omega}{\varepsilon}\right)^2 + \cdots\right] \|g_0(\theta,\tau)\| \\ &\leq \frac{\omega^{r+1}}{\varepsilon^r(\varepsilon-\omega)} \|g_0(\theta,\tau)\| . \end{aligned}$$

This completes the proof.

In the following section, we examine the uniqueness and convergence of the solution obtained via q-HASTM algorithm.

11.5 UNIQUENESS AND CONVERGENCE ANALYSIS FOR q-HASTM

In this section, uniqueness and convergence analysis of the q-HASTM scheme are discussed.

Theorem 11.2 (Uniqueness theorem): *The solution achieved through the implementation of q-HASTM for the fractional differential Eq.* (11.9) *is unique, wherever* $0<\rho<1$, *where*

$$\rho=(\chi_\mu+\hbar)+\hbar(\delta+\xi)\eta=(\varepsilon+\hbar)+\hbar(\delta+\xi)\eta.$$

Proof. The solution of nonlinear fractional differential Eq. (11.9) is obtained as

$$g(\theta,\tau)=\sum_{\mu=0}^{K} g_\mu(\theta,\tau)\left(\frac{1}{\varepsilon}\right)^{\mu}, \tag{11.28}$$

where

$$\begin{aligned} g_\mu(\theta,\tau)&=(\chi_\mu+\hbar)\,g_{\mu-1}(\theta,\tau)-\hbar\left(1-\frac{\chi_\mu}{\varepsilon}\right)ST^{-1}\\ &\quad\times\left(u^{k\beta}\sum_{\gamma=0}^{k-1}u^{-k\beta+\gamma}g^{(\gamma)}(\theta,0)+u^{k\beta}ST\left[G(\theta,\tau)\right]\right)\\ &\quad+\hbar ST^{-1}\left(u^{k\beta}ST\left[Mg_{\mu-1}(\theta,\tau)+P_{\mu-1}\right]\right). \end{aligned} \tag{11.29}$$

Within the bounds of possibility, we assume two different solutions $g(\theta,\tau)$ and $g^*(\theta,\tau)$ for Eq. (11.9), then making use of Eq. (11.29), we obtain

$$\begin{aligned} |g(\theta,\tau)-g^*(\theta,\tau)|&=\left|(\chi_\mu+\hbar)(g-g^*)+\hbar ST^{-1}\left\{u^{k\beta}ST\left(M(g-g^*)+N(g-g^*)\right)\right\}\right|\\ &\le(\chi_\mu+\hbar)\,|(g-g^*)|+\hbar ST^{-1}\left\{u^{k\beta}ST\left(|M(g-g^*)|+|N(g-g^*)|\right)\right\}. \end{aligned} \tag{11.30}$$

Using of convolution theorem for Sumudu transform in Eq. (11.30) yields

$$\begin{aligned} |g(\theta,\tau)-g^*(\theta,\tau)|&\le(\chi_\mu+\hbar)\,|g-g^*|+\hbar\int_0^\tau\left(|M(g-g^*)|+|N(g-g^*)|\right)\frac{(\tau-\varpi)^{k\beta-1}}{\Gamma(k\beta)}d\varpi\\ &\le(\chi_\mu+\hbar)\,|g-g^*|+\hbar\int_0^\tau\left(\delta|g-g^*|+\xi|g-g^*|\right)\frac{(\tau-\varpi)^{k\beta-1}}{\Gamma(k\beta)}d\varpi. \end{aligned} \tag{11.31}$$

Now in view of integral mean value theorem [32, 33], inequality (11.31) reduces to the following form:

$$\begin{aligned} |g(\theta,\tau)-g^*(\theta,\tau)|&\le(\chi_\mu+\hbar)\,|g-g^*|+\hbar\left(\delta|g-g^*|+\xi|g-g^*|\right)\eta\\ &=\left[(\chi_\mu+\hbar)+\hbar(\delta+\xi)\eta\right]|g-g^*|\\ &=\rho\,|g-g^*|.\\ \therefore\ (1-\rho)\,|g(\theta,\tau)-g^*(\theta,\tau)|&\le 0. \end{aligned} \tag{11.32}$$

where $\rho=(\chi_\mu+\hbar)+\hbar(\delta+\xi)\eta$. (11.33)
Since $0<\rho<1$, therefore $|g(\theta,\tau)-g^*(\theta,\tau)|=0$, which provides $g=g^*$. This confirms the uniqueness of the solution of Eq. (11.9). □

Theorem 11.3 (Convergence theorem): *Let Π be a Banach space and $H : \Pi \to \Pi$ be a nonlinear mapping. Assume that*

$$\|H(h) - H(g)\| \leq \rho \|h-g\|, \forall g, h \in \Pi.$$

Then in view of Banach's fixed point theory [32, 33], there exists a fixed point for H. Moreover, the sequence generated by the q-HASTM converges to the fixed points of $H$with random selections of $h_0, g_0 \in \Pi$ and

$$\|g_m - g_n\| \leq \frac{\rho^n}{1-\rho} \|g_1 - g_0\|, \forall h, g \in \Pi.$$

Proof. Here, we specially suppose that $(C[\Omega], \|.\|)$, where $C[\Omega]$ be a Banach space of all continuous real-valued functions on the interval Ω possessing the supremum norm. Now, we show that $\{g_n\}$ is a Cauchy sequence in the Banach space Π.
Now, consider

$$\begin{aligned} &\|g_m - g_n\| = \max_{t\in\Omega} |g_m - g_n| \\ &= \max_{t\in\Omega} \Big|(\varepsilon + \hbar)(g_{m-1} - g_{n-1}) \\ &+ \hbar ST^{-1}\left\{u^{k\beta} ST\left(M(g_{m-1} - g_{n-1}) + N(g_{m-1} - g_{n-1})\right)\right\}\Big|. \end{aligned} \tag{11.34}$$

Applying the convolution theorem for Sumudu transform, we obtain

$$\begin{aligned} \|g_m - g_n\| &= \max_{t\in\Omega}\left\{(\varepsilon+\hbar)|g_{m-1} - g_{n-1}| + \hbar\int_0^\tau \left(|M(g_{m-1} - g_{n-1})|\right.\right. \\ &\quad \left.\left. + |N(g_{m-1} - g_{n-1})|\right)\frac{(\tau-\varpi)^{k\beta-1}}{\Gamma(k\beta)} d\varpi\right\} \\ &\leq \max_{t\in\Omega}\left\{(\varepsilon+\hbar)|g_{m-1} - g_{n-1}| + \hbar\int_0^\tau \left(\delta|g_{m-1} - g_{n-1}|\right.\right. \\ &\quad \left.\left. + \xi|g_{m-1} - g_{n-1}|\right)\frac{(\tau-\varpi)^{k\beta-1}}{\Gamma(k\beta)} d\varpi\right\}. \end{aligned} \tag{11.35}$$

Now applying integral mean value theorem, inequality (11.35) reduces to the following form:

$$\begin{aligned} &\|g_m - g_n\| \leq \max_{t\in\Omega}\{(\varepsilon+\hbar)|g_{m-1} - g_{n-1}| \\ &+ \hbar(\delta|g_{m-1} - g_{n-1}| + \xi|g_{m-1} - g_{n-1}|)\eta\} \\ &= \max_{t\in\Omega}\{(\varepsilon+\hbar) + \hbar(\delta+\xi)\eta\}|g_{m-1} - g_{n-1}| \\ &= \rho\|g_{m-1} - g_{n-1}\| \\ &\therefore \|g_m - g_n\| \leq \rho\|g_{m-1} - g_{n-1}\|, \end{aligned} \tag{11.36}$$

where $\rho = (\varepsilon+\hbar) + \hbar(\delta+\xi)\eta$.

Assume $m = n+1$, then it produces

$$\|g_{n+1} - g_n\| \leq \rho \|g_n - g_{n-1}\| \leq \rho^2 \|g_{n-1} - g_{n-2}\| \cdots \leq \rho^n \|g_1 - g_0\|. \quad (11.37)$$

Utilizing the triangular inequality, we have

$$\begin{aligned}
\|g_m - g_n\| &\leq \|g_{n+1} - g_n\| + \|g_{n+2} - g_{n+1}\| + \cdots + \|g_m - g_{m-1}\| \\
&\leq \left(\rho^n + \rho^{n+1} + \rho^{n+2} + \cdots + \rho^{m-1}\right) \|g_1 - g_0\| \\
&= \rho^n \left(1 + \rho + \rho^2 + \cdots + \rho^{m-n-1}\right) \|g_1 - g_0\| \\
&= \rho^n \left[\frac{1 - \rho^{m-n-1}}{1 - \rho}\right] \|g_1 - g_0\|. \quad (11.38)
\end{aligned}$$

Since $0 < \rho < 1$, thus $1 - \rho^{m-n-1} < 1$, then we have

$$\|g_m - g_n\| \leq \frac{\rho^n}{1 - \rho} \|g_1 - g_0\|. \quad (11.39)$$

But $\|g_1 - g_0\| < \infty$, hence as $n \to \infty$ then $\|g_m - g_n\| \to 0$, therefore the sequence $\{g_n\}$ is a Cauchy sequence in $C[\Omega]$. Consequently $\{g_n\}$ is convergent. This ensures the convergence of the solution $g(\theta, \tau)$ of the fractional differential Eq. (11.9). Hence the theorem. □

Next, we apply the q-HASTM algorithm to the fractional order HEV model.

11.6 q-HASTM SOLUTION FOR THE FRACTIONAL HEV MODEL

In this section, the fundamental steps of the q-HASTM algorithm are applied for the fractional HEV model expressed in operator form as follows:

$$\begin{aligned}
D_t^\beta S &= \wp(1 - \ell I) - (\alpha I + \gamma_E P) S - \vartheta S, \\
D_t^\beta E &= (\alpha I + \gamma_E P) S - (\vartheta + \rho) E + \wp \ell I, \\
D_t^\beta I &= \rho E - (\vartheta + \sigma) I, \\
D_t^\beta R &= \sigma I - \vartheta R, \\
D_t^\beta P &= \varphi I - \lambda P, \quad (11.40)
\end{aligned}$$

subject to the initial conditions:

$$\begin{aligned}
&S(0) = S_0 = m_1,\ E(0) = E_0 = m_2, I(0) = I_0 = m_3, \\
&R(0) = R_0 = m_4, P(0) = P_0 = m_5. \quad (11.41)
\end{aligned}$$

Here, $D_t^\beta \equiv \frac{d^\beta}{dt^\beta}$ symbolizes the Caputo differential operator of order β $(0 < \beta \leq 1)$. It is remarkable that the functions $S(t)$, $E(t)$, $I(t)$, $R(t)$, and $P(t)$ are bounded in nature.

The set of initial approximations is prescribed as

$$S_0(t) = S(0) = S_0 = m_1, E_0(t) = E(0) = E_0 = m_2,$$

$$I_0(t) = I(0) = I_0 = m_3,$$
$$R_0(t) = R(0) = R_0 = m_4,$$
$$P_0(t) = P(0) = P_0 = m_5. \tag{11.42}$$

Operating the Sumudu transform ST on system of Eq. (11.40), the following system of equations is obtained as:

$$\begin{aligned}
ST\left[D_t^{\beta} S\right] &= \wp[ST(1) - \ell ST(I)] - [\alpha ST(IS) + \gamma_E ST(PS)] - \vartheta ST(S),\\
ST\left[D_t^{\beta} E\right] &= \alpha ST(IS) + \gamma_E ST(PS) - (\vartheta + \rho) ST(E) + \wp\ell ST(I),\\
ST\left[D_t^{\beta} I\right] &= \rho ST(E) - (\vartheta + \sigma) ST(I),\\
ST\left[D_t^{\beta} R\right] &= \sigma ST(I) - \vartheta ST(R),\\
ST\left[D_t^{\beta} P\right] &= \varphi ST(I) - \lambda ST(P).
\end{aligned} \tag{11.43}$$

After employing the formula of ST operator for fractional derivative in above obtained equations and further simplifying, we get

$$\begin{aligned}
u^{-\beta} ST[S(t)] - u^{-\beta} S(0) &= \wp[1 - \ell ST(I)] - \alpha ST(IS) - \gamma_E ST(PS) - \vartheta ST(S),\\
u^{-\beta} ST[E(t)] - u^{-\beta} E(0) &= \alpha ST(IS) + \gamma_E ST(PS) - (\vartheta + \rho) ST(E) + \wp\ell ST(I),\\
u^{-\beta} ST[I(t)] - u^{-\beta} I(0) &= \rho ST(E) - (\vartheta + \sigma) ST(I),\\
u^{-\beta} ST[R(t)] - u^{-\beta} R(0) &= \sigma ST(I) - \vartheta ST(R),\\
u^{-\beta} ST[P(t)] - u^{-\beta} P(0) &= \varphi ST(I) - \lambda ST(P).
\end{aligned} \tag{11.44}$$

Now, making use of the initial conditions provided by Eq. (11.42) in system of Eq. (11.44) with further simplification generates the following set of equations:

$$\begin{aligned}
&ST[S(t)] - S_0 - \wp u^{\beta} [1 - \ell ST(I)] + \alpha u^{\beta} ST(IS) + \gamma_E u^{\beta} ST(PS)\\
&+ \vartheta u^{\beta} ST(S) = 0,\\
&ST[E(t)] - E_0 - \alpha u^{\beta} ST(IS) - \gamma_E u^{\beta} ST(PS) + (\vartheta + \rho) u^{\beta} ST(E)\\
&- \wp\ell u^{\beta} ST(I) = 0,\\
&ST[I(t)] - I_0 - \rho u^{\beta} ST(E) + (\vartheta + \sigma) u^{\beta} ST(I) = 0,\\
&ST[R(t)] - R_0 - \sigma u^{\beta} ST(I) + \vartheta u^{\beta} ST(R) = 0,\\
&ST[P(t)] - P_0 - \varphi u^{\beta} ST(I) + \lambda u^{\beta} ST(P) = 0.
\end{aligned} \tag{11.45}$$

Now, in view of system of Eq. (11.45), the set of nonlinear operators is constructed in the following way:

$$\begin{aligned}
\eta_1[\phi_1(t;q)] &= ST[\phi_1(t;q)] - S_0 - \wp u^\beta + \wp \ell u^\beta ST(\phi_3) + \alpha u^\beta ST(\phi_1 \phi_3) \\
&\quad + \gamma_E u^\beta ST(\phi_1 \phi_5) + \vartheta u^\beta ST(\phi_1), \\
\eta_2[\phi_2(t;q)] &= ST[\phi_2(t;q)] - E_0 - \alpha u^\beta ST(\phi_1 \phi_3) - \gamma_E u^\beta ST(\phi_1 \phi_5) \\
&\quad + (\vartheta + \rho) u^\beta ST(\phi_2) - \wp \ell u^\beta ST(\phi_3), \\
\eta_3[\phi_3(t;q)] &= ST[\phi_3(t;q)] - I_0 - \rho u^\beta ST(\phi_2) + (\vartheta + \sigma) u^\beta ST(\phi_3), \\
\eta_4[\phi_4(t;q)] &= ST[\phi_4(t;q)] - R_0 - \sigma u^\beta ST(\phi_3) + \vartheta u^\beta ST(\phi_4), \\
\eta_5[\phi_5(t;q)] &= ST[\phi_5(t;q)] - P_0 - \varphi u^\beta ST(\phi_3) + \lambda u^\beta ST(\phi_5), \qquad (11.46)
\end{aligned}$$

where $q \in \left[0, \frac{1}{\varepsilon}\right]$ signifies the embedding parameter, $\phi_1(t;q)$, $\phi_2(t;q)$, $\phi_3(t;q)$, $\phi_4(t;q)$, and $\phi_5(t;q)$ denote the real-valued functions of q and t, and ST stands for Sumudu transform operator.

In the next step, we frame the system of zeroth-order homotopy equations in view of the q-HAM algorithm [18, 19] under the supposition $H(t) = 1$ in the following way:

$$\begin{aligned}
(1-\varepsilon q)\, ST\left[\phi_1(t;q) - S_0(t)\right] &= q\hbar\eta_1\left[\phi_1(t;q)\right], \\
(1-\varepsilon q)\, ST\left[\phi_2(t;q) - E_0(t)\right] &= q\hbar\eta_2\left[\phi_2(t;q)\right], \\
(1-\varepsilon q)\, ST\left[\phi_3(t;q) - I_0(t)\right] &= q\hbar\eta_3\left[\phi_3(t;q)\right], \\
(1-\varepsilon q)\, ST\left[\phi_4(t;q) - R_0(t)\right] &= q\hbar\eta_4\left[\phi_4(t;q)\right], \\
(1-\varepsilon q)\, ST\left[\phi_5(t;q) - P_0(t)\right] &= q\hbar\eta_5\left[\phi_5(t;q)\right], \qquad (11.47)
\end{aligned}$$

where $\hbar \neq 0$ symbolizes an auxiliary parameter, $H(t) \neq 0$ indicates an auxiliary function, and $\phi_1(t;q)$, $\phi_2(t;q)$, $\phi_3(t;q)$, $\phi_4(t;q)$, and $\phi_5(t;q)$ are unknown functions. It is remarkable that the q-HASTM algorithm provides a wide option to constitute auxiliary functions in a computational procedure.

For the embedding parameter $q = 0$ and $q = \frac{1}{\varepsilon}$, the following expressions hold good:

$$\begin{aligned}
\phi_1(t;0) &= S_0(t), \phi_1\left(t;\frac{1}{\varepsilon}\right) = S(t), \\
\phi_2(t;0) &= E_0(t), \phi_2\left(t;\frac{1}{\varepsilon}\right) = E(t), \\
\phi_3(t;0) &= I_0(t), \phi_3\left(t;\frac{1}{\varepsilon}\right) = I(t), \\
\phi_4(t;0) &= R_0(t), \phi_4\left(t;\frac{1}{\varepsilon}\right) = R(t), \\
\phi_5(t;0) &= P_0(t), \phi_5\left(t;\frac{1}{\varepsilon}\right) = P(t). \qquad (11.48)
\end{aligned}$$

Clearly, as q takes value from 0 to $\frac{1}{\varepsilon}$, the solutions $\phi_1(t;q)$, $\phi_2(t;q)$, $\phi_3(t;q)$, $\phi_4(t;q)$, and $\phi_5(t;q)$ vary from $S_0(t)$, $E_0(t)$, $I_0(t)$, $R_0(t)$, and $P_0(t)$, respectively. Now, the Taylor's series expansion of $\phi_1(t;q)$, $\phi_2(t;q)$, $\phi_3(t;q)$, $\phi_4(t;q)$, and $\phi_5(t;q)$ about q provides

$$\phi_1(t;q) = S_0(t) + \sum_{\mu=1}^{\infty} q^{\mu} S_{\mu}(t),$$

$$\phi_2(t;q) = E_0(t) + \sum_{\mu=1}^{\infty} q^{\mu} E_{\mu}(t),$$

$$\phi_3(t;q) = I_0(t) + \sum_{\mu=1}^{\infty} q^{\mu} I_{\mu}(t),$$

$$\phi_4(t;q) = R_0(t) + \sum_{\mu=1}^{\infty} q^{\mu} R_{\mu}(t),$$

$$\phi_5(t;q) = P_0(t) + \sum_{\mu=1}^{\infty} q^{\mu} P_{\mu}(t), \tag{11.49}$$

where

$$S_{\mu}(t) = \left[\frac{1}{\mu\,!}\frac{\partial^{\mu}\phi_1(t;q)}{\partial q^{\mu}}\right]_{q=0},$$

$$E_{\mu}(t) = \left[\frac{1}{\mu\,!}\frac{\partial^{\mu}\phi_2(t;q)}{\partial q^{\mu}}\right]_{q=0},$$

$$I_{\mu}(t) = \left[\frac{1}{\mu\,!}\frac{\partial^{\mu}\phi_3(t;q)}{\partial q^{\mu}}\right]_{q=0},$$

$$R_{\mu}(t) = \left[\frac{1}{\mu\,!}\frac{\partial^{\mu}\phi_4(t;q)}{\partial q^{\mu}}\right]_{q=0},$$

$$P_{\mu}(t) = \left[\frac{1}{\mu\,!}\frac{\partial^{\mu}\phi_5(t;q)}{\partial q^{\mu}}\right]_{q=0}. \tag{11.50}$$

Here, it is significant to observe that $\hbar$ regulates the convergence region of series solutions (11.49). The series solutions (11.49) converge for the appropriate selections of $\hbar$, $S_0(t)$, $E_0(t)$, $I_0(t)$, $R_0(t)$, and $P_0(t)$ at $q = 1$. Then it makes available the solutions for system of Eq. (11.40) as follows:

$$S(t) = S_0(t) + \sum_{\mu=1}^{\infty} S_{\mu}(t)\left(\frac{1}{\varepsilon}\right)^{\mu},$$

$$E(t) = E_0(t) + \sum_{\mu=1}^{\infty} E_{\mu}(t)\left(\frac{1}{\varepsilon}\right)^{\mu},$$

$$I(t) = I_0(t) + \sum_{\mu=1}^{\infty} I_{\mu}(t)\left(\frac{1}{\varepsilon}\right)^{\mu},$$

$$R(t) = R_0(t) + \sum_{\mu=1}^{\infty} R_\mu(t) \left(\frac{1}{\varepsilon}\right)^\mu,$$

$$R(t) = P_0(t) + \sum_{\mu=1}^{\infty} P_\mu(t) \left(\frac{1}{\varepsilon}\right)^\mu. \tag{11.51}$$

The above obtained system of Eq. (11.51) sets up a linkage between the exact solutions $S(t), E(t), I(t), R(t)$, and $P(t)$ and the initial guess $S_0(t), E_0(t), I_0(t), R_0(t)$, and $P_0(t)$ through the terms $S_\mu(t), E_\mu(t), I_\mu(t), R_\mu(t)$, and $P_\mu(t)(\mu = 1, 2, 3, ...)$ that are ready to be computed in forthcoming steps.

Now, differentiating the system of Eq. (11.47) with regard to q about μ-times and then multiplying by $\frac{1}{\mu!}$ and finally adopting $q = 0$, we compose the system of μth-order deformation equations in the following way:

$$ST\left[S_\mu(t) - \chi_\mu S_{\mu-1}(t)\right] = \hbar \Re_{1,\mu}\left(\vec{S}_{\mu-1}(t)\right),$$

$$ST\left[E_\mu(t) - \chi_\mu E_{\mu-1}(t)\right] = \hbar \Re_{2,\mu}\left(\vec{E}_{\mu-1}(t)\right),$$

$$ST\left[I_\mu(t) - \chi_\mu I_{\mu-1}(t)\right] = \hbar \Re_{3,\mu}\left(\vec{I}_{\mu-1}(t)\right),$$

$$ST\left[R_\mu(t) - \chi_\mu R_{\mu-1}(t)\right] = \hbar \Re_{4,\mu}\left(\vec{R}_{\mu-1}(t)\right),$$

$$ST\left[P_\mu(t) - \chi_\mu P_{\mu-1}(t)\right] = \hbar \Re_{5,\mu}\left(\vec{P}_{\mu-1}(t)\right), \tag{11.52}$$

where the vectors are stipulated as

$$\vec{S}_\mu(t) = \left\{S_0(t), S_1(t), S_2(t), ..., S_\mu(t)\right\},$$

$$\vec{E}_\mu(t) = \left\{E_0(t), E_1(t), E_2(t), ..., E_\mu(t)\right\},$$

$$\vec{I}_\mu(t) = \left\{I_0(t), I_1(t), I_2(t), ..., I_\mu(t)\right\},$$

$$\vec{R}_\mu(t) = \left\{R_0(t), R_1(t), R_2(t), ..., R_\mu(t)\right\},$$

$$\vec{P}_\mu(t) = \left\{P_0(t), P_1(t), P_2(t), ..., P_\mu(t)\right\}. \tag{11.53}$$

In the next step, the implementation of inverse ST operator ST^{-1} on system of Eq. (11.52) provides

$$S_\mu(t) = \chi_\mu S_{\mu-1}(t) + \hbar ST^{-1}\left[\Re_{1,\mu}\left(\vec{S}_{\mu-1}(t)\right)\right],$$

$$E_\mu(t) = \chi_\mu E_{\mu-1}(t) + \hbar ST^{-1}\left[\Re_{2,\mu}\left(\vec{E}_{\mu-1}(t)\right)\right],$$

$$I_\mu(t) = \chi_\mu I_{\mu-1}(t) + \hbar ST^{-1}\left[\Re_{3,\mu}\left(\vec{I}_{\mu-1}(t)\right)\right],$$

$$R_\mu(t) = \chi_\mu R_{\mu-1}(t) + \hbar ST^{-1}\left[\Re_{4,\mu}\left(\vec{R}_{\mu-1}(t)\right)\right],$$

$$P_\mu(t) = \chi_\mu P_{\mu-1}(t) + \hbar ST^{-1}\left[\Re_{5,\mu}\left(\vec{P}_{\mu-1}(t)\right)\right]. \tag{11.54}$$

Here, the terms $\Re_{1,\mu}\left(\vec{S}_{\mu-1}(t)\right)$, $\Re_{2,\mu}\left(\vec{E}_{\mu-1}(t)\right)$, $\Re_{3,\mu}\left(\vec{I}_{\mu-1}(t)\right)$, $\Re_{4,\mu}\left(\vec{R}_{\mu-1}(t)\right)$, and $\Re_{5,\mu}\left(\vec{P}_{\mu-1}(t)\right)$ inset of Eq. (11.54) are expressed in a new form as

$$\begin{aligned}
\Re_{1,\mu}\left(\vec{S}_{\mu-1}\right) &= ST\left(S_{\mu-1}\right) - S_0\left(1 - \frac{\chi_\mu}{\varepsilon}\right) - \wp u^\beta \\
&\quad + u^\beta ST(\wp\ell I_{\mu-1} + \alpha P_\mu + \gamma_E P'_\mu + \vartheta S_{\mu-1}), \mu \geq 1, \\
\Re_{2,\mu}\left(\vec{E}_{\mu-1}\right) &= ST\left(E_{\mu-1}\right) - E_0\left(1 - \frac{\chi_\mu}{\varepsilon}\right) \\
&\quad - u^\beta ST(\alpha P_\mu + \gamma_E P'_\mu - (\vartheta + \rho) E_{\mu-1} + \wp\ell I_{\mu-1}), \mu \geq 1, \\
\Re_{3,\mu}\left(\vec{I}_{\mu-1}\right) &= ST\left(I_{\mu-1}\right) - I_0\left(1 - \frac{\chi_\mu}{\varepsilon}\right) \\
&\quad - u^\beta ST\left(\rho E_{\mu-1} - (\vartheta + \sigma) I_{\mu-1}\right), \mu \geq 1, \\
\Re_{4,\mu}\left(\vec{R}_{\mu-1}\right) &= ST\left(R_{\mu-1}\right) - R_0\left(1 - \frac{\chi_\mu}{\varepsilon}\right) - u^\beta ST\left(\sigma I_{\mu-1} - \vartheta R_{\mu-1}\right), \mu \geq 1, \\
\Re_{5,\mu}\left(\vec{P}_{\mu-1}\right) &= ST\left(P_{\mu-1}\right) - P_0\left(1 - \frac{\chi_\mu}{\varepsilon}\right) - u^\beta ST\left(\varphi I_{\mu-1} - \lambda P_{\mu-1}\right), \mu \geq 1,
\end{aligned} \tag{11.55}$$

where the value of χ_μ is presented as

$$\chi_\mu = \begin{cases} 0, & \mu \leq 1 \\ \varepsilon, & \mu > 1. \end{cases} \tag{11.56}$$

In system of Eq. (11.55), P_μ and P'_μ represent the homotopy polynomials [31] and are expressed as

$$\begin{aligned}
P_\mu &= \frac{1}{\Gamma(\mu)}\left[\frac{\partial^\mu}{\partial q^\mu}\phi_1(t,q)\,\phi_3(t,q)\right]_{q=0}, \\
P'_\mu &= \frac{1}{\Gamma(\mu)}\left[\frac{\partial^\mu}{\partial q^\mu}\phi_1(t,q)\,\phi_5(t,q)\right]_{q=0},
\end{aligned} \tag{11.57}$$

and

$$\begin{aligned}
\phi_1(t,q) &= (\phi_1)_0 + q(\phi_1)_1 + q^2(\phi_1)_2 + \cdots, \\
\phi_2(t,q) &= (\phi_2)_0 + q(\phi_2)_1 + q^2(\phi_2)_2 + \cdots, \\
\phi_3(t,q) &= (\phi_3)_0 + q(\phi_3)_1 + q^2(\phi_3)_2 + \cdots, \\
\phi_4(t,q) &= (\phi_4)_0 + q(\phi_4)_1 + q^2(\phi_4)_2 + \cdots, \\
\phi_5(t,q) &= (\phi_5)_0 + q(\phi_5)_1 + q^2(\phi_5)_2 + \cdots.
\end{aligned} \tag{11.58}$$

In view of system of Eq. (11.55), the solutions of system of $\mu\,th$-order deformation Eq. (11.54) reduce to

$$\begin{aligned}
S_\mu(t) &= \chi_\mu S_{\mu-1}(t) + \hbar ST^{-1}\left[ST(S_{\mu-1}) - S_0\left(1-\frac{\chi_\mu}{\varepsilon}\right) - \wp u^\beta \right.\\
&\left. + u^\beta ST(\wp\ell I_{\mu-1} + \alpha P_\mu + \gamma_E P'_\mu + \vartheta S_{\mu-1})\right],\\
E_\mu(t) &= \chi_\mu E_{\mu-1}(t) + \hbar ST^{-1}\left[ST(E_{\mu-1}) - E_0\left(1-\frac{\chi_\mu}{\varepsilon}\right) - u^\beta ST(\alpha P_\mu \right.\\
&\left. + \gamma_E P'_\mu - (\vartheta+\rho) E_{\mu-1} + \wp\ell I_{\mu-1})\right],\\
I_\mu(t) &= \chi_\mu I_{\mu-1}(t) + \hbar ST^{-1}\left[ST(I_{\mu-1}) - I_0\left(1-\frac{\chi_\mu}{\varepsilon}\right)\right.\\
&\left. - u^\beta ST(\rho E_{\mu-1} - (\vartheta+\sigma) I_{\mu-1})\right],\\
R_\mu(t) &= \chi_\mu R_{\mu-1}(t) + \hbar ST^{-1}\left[ST(R_{\mu-1}) - R_0\left(1-\frac{\chi_\mu}{\varepsilon}\right)\right.\\
&\left. - u^\beta ST(\sigma I_{\mu-1} - \vartheta R_{\mu-1})\right],\\
P_\mu(t) &= \chi_\mu P_{\mu-1}(t) + \hbar ST^{-1}\left[ST(P_{\mu-1}) - P_0\left(1-\frac{\chi_\mu}{\varepsilon}\right)\right.\\
&\left. - u^\beta ST(\varphi I_{\mu-1} - \lambda P_{\mu-1})\right].
\end{aligned} \tag{11.59}$$

Simplification yields

$$\begin{aligned}
S_\mu(t) &= (\chi_\mu + \hbar) S_{\mu-1}(t) - \hbar S_0\left(1-\frac{\chi_\mu}{\varepsilon}\right) - \hbar\wp ST^{-1}\left(u^\beta\right)\\
&+ \hbar ST^{-1}\left[u^\beta ST(\wp\ell I_{\mu-1} + \alpha P_\mu + \gamma_E P'_\mu + \vartheta S_{\mu-1})\right], \mu \geq 1,\\
E_\mu(t) &= (\chi_\mu + \hbar) E_{\mu-1}(t) - \hbar E_0\left(1-\frac{\chi_\mu}{\varepsilon}\right)\\
&- \hbar ST^{-1}\left[u^\beta ST(\alpha P_\mu + \gamma_E P'_\mu - (\vartheta+\rho) E_{\mu-1} + \wp\ell I_{\mu-1})\right],\\
I_\mu(t) &= (\chi_\mu + \hbar) I_{\mu-1}(t) - \hbar I_0\left(1-\frac{\chi_\mu}{\varepsilon}\right)\\
&- \hbar ST^{-1}\left[u^\beta ST(\rho E_{\mu-1} - (\vartheta+\sigma) I_{\mu-1})\right],\\
R_\mu(t) &= (\chi_\mu + \hbar) R_{\mu-1}(t) - \hbar R_0\left(1-\frac{\chi_\mu}{\varepsilon}\right)\\
&- \hbar ST^{-1}\left[u^\beta ST(\sigma I_{\mu-1} - \vartheta R_{\mu-1})\right],\\
P_\mu(t) &= (\chi_\mu + \hbar) P_{\mu-1}(t) - \hbar P_0\left(1-\frac{\chi_\mu}{\varepsilon}\right)\\
&- \hbar ST^{-1}\left[u^\beta ST(\varphi I_{\mu-1} - \lambda P_{\mu-1})\right].
\end{aligned} \tag{11.60}$$

Utilizing the set of iterative schemes Eq. (11.60) along with the initial conditions provided by Eq. (11.41), we have computed the various iterative terms as follows:

$$S_1 = \hbar(-\wp + \ell\wp m_3 + \alpha m_1 m_3 + \gamma_E m_1 m_5 + \vartheta m_1)\frac{t^\beta}{\Gamma(\beta+1)},$$

$$E_1 = -\hbar(\alpha m_1 m_3 + \gamma_E m_1 m_5)\frac{t^\beta}{\Gamma(\beta+1)},$$

$$I_1 = \hbar[-\rho m_2 + (\vartheta+\sigma) m_3]\frac{t^\beta}{\Gamma(\beta+1)},$$

$$R_1 = \hbar(-\sigma m_3 + \vartheta m_4)\frac{t^\beta}{\Gamma(\beta+1)},$$

$$P_1 = \hbar(-\varphi m_3 + \lambda m_5)\frac{t^\beta}{\Gamma(\beta+1)},$$

$$S_2 = \hbar(\varepsilon+\hbar)(-\wp + \ell\wp m_3 + \alpha m_1 m_3 + \gamma_E m_1 m_5 + \vartheta m_1)\frac{t^\beta}{\Gamma(\beta+1)}$$
$$-\hbar\wp\frac{t^\beta}{\Gamma(\beta+1)} + \hbar^2$$
$$[(\ell\wp + \alpha m_1)(-\rho m_2 + (\vartheta+\sigma) m_3)$$
$$+(\alpha m_3 + \gamma_E m_5 + \vartheta)(-\wp + \ell\wp m_3 + \alpha m_1 m_3 + \gamma_E m_1 m_5 + \vartheta m_1)$$
$$+\gamma_E m_1(-\varphi m_3 + \lambda m_5)]\frac{t^{2\beta}}{\Gamma(2\beta+1)},$$

$$E_2 = -\hbar(\varepsilon+\hbar)(\alpha m_1 m_3 + \gamma_E m_1 m_5)\frac{t^\beta}{\Gamma(\beta+1)} - \hbar^2$$
$$[(\alpha m_3 + \gamma_E m_5)(-\wp + \ell\wp m_3 + \alpha m_1 m_3 + \gamma_E m_1 m_5 + \vartheta m_1)$$
$$+\alpha m_1(-\rho m_2 + (\vartheta+\sigma) m_3)$$
$$+\gamma_E m_1(-\varphi m_3 + \lambda m_5)]\frac{t^{2\beta}}{\Gamma(2\beta+1)},$$

$$I_2 = \hbar(\varepsilon+\hbar)(-\rho m_2 + (\vartheta+\sigma) m_3)\frac{t^\beta}{\Gamma(\beta+1)}$$
$$+\hbar^2[\rho(\alpha m_1 m_3 + \gamma_E m_1 m_5) + (\vartheta+\sigma)(-\rho m_2$$
$$+(\vartheta+\sigma) m_3)]\frac{t^{2\beta}}{\Gamma(2\beta+1)},$$

$$R_2 = \hbar(\varepsilon+\hbar)(-\sigma m_3 + \vartheta m_4)\frac{t^\beta}{\Gamma(\beta+1)}$$
$$+\hbar^2[-\sigma(-\rho m_2 + (\vartheta+\sigma) m_3) + \vartheta(-\sigma m_3 + \vartheta m_4)]\frac{t^{2\beta}}{\Gamma(2\beta+1)},$$

$$P_2 = \hbar(\varepsilon+\hbar)(-\varphi m_3 + \lambda m_5)\frac{t^\beta}{\Gamma(\beta+1)}$$
$$+\hbar^2[-\varphi(-\rho m_2 + (\vartheta+\sigma) m_3) + \lambda(-\varphi m_3 + \lambda m_5)]\frac{t^{2\beta}}{\Gamma(2\beta+1)}. \qquad (11.61)$$

Taking the convergence-control parameter $\hbar = -1$, we get

$$
\begin{aligned}
S_1 &= (\wp - \ell\wp m_3 - \alpha m_1 m_3 - \gamma_E m_1 m_5 - \vartheta m_1)\frac{t^\beta}{\Gamma(\beta+1)},\\
E_1 &= (\alpha m_1 m_3 + \gamma_E m_1 m_5)\frac{t^\beta}{\Gamma(\beta+1)},\\
I_1 &= [\rho m_2 - (\vartheta+\sigma) m_3]\frac{t^\beta}{\Gamma(\beta+1)},\\
R_1 &= (\sigma m_3 - \vartheta m_4)\frac{t^\beta}{\Gamma(\beta+1)},\\
P_1 &= (\varphi m_3 - \lambda m_5)\frac{t^\beta}{\Gamma(\beta+1)},\\
S_2 &= (\varepsilon-1)(\wp - \ell\wp m_3 - \alpha m_1 m_3 - \gamma_E m_1 m_5 - \vartheta m_1)\frac{t^\beta}{\Gamma(\beta+1)} + \wp\frac{t^\beta}{\Gamma(\beta+1)}\\
&\quad + [(\ell\wp + \alpha m_1)(-\rho m_2 + (\vartheta+\sigma) m_3)\\
&\quad + (\alpha m_3 + \gamma_E m_5 + \vartheta)(-\wp + \ell\wp m_3 + \alpha m_1 m_3 + \gamma_E m_1 m_5 + \vartheta m_1)\\
&\quad + \gamma_E m_1(-\varphi m_3 + \lambda m_5)]\frac{t^{2\beta}}{\Gamma(2\beta+1)},\\
E_2 &= (\varepsilon-1)(\alpha m_1 m_3 + \gamma_E m_1 m_5)\frac{t^\beta}{\Gamma(\beta+1)}\\
&\quad - [(\alpha m_3 + \gamma_E m_5)(-\wp + \ell\wp m_3 + \alpha m_1 m_3 + \gamma_E m_1 m_5 + \vartheta m_1)\\
&\quad + \alpha m_1(-\rho m_2 + (\vartheta+\sigma) m_3)\\
&\quad + \gamma_E m_1(-\varphi m_3 + \lambda m_5)]\frac{t^{2\beta}}{\Gamma(2\beta+1)},\\
I_2 &= (\varepsilon-1)[\rho m_2 - (\vartheta+\sigma) m_3]\frac{t^\beta}{\Gamma(\beta+1)}\\
&\quad + [\rho(\alpha m_1 m_3 + \gamma_E m_1 m_5) + (\vartheta+\sigma)(-\rho m_2 + (\vartheta+\sigma) m_3)]\\
&\quad \times \frac{t^{2\beta}}{\Gamma(2\beta+1)},\\
R_2 &= (\varepsilon-1)(\sigma m_3 - \vartheta m_4)\frac{t^\beta}{\Gamma(\beta+1)}\\
&\quad + [-\sigma(-\rho m_2 + (\vartheta+\sigma) m_3) + \vartheta(-\sigma m_3 + \vartheta m_4)]\frac{t^{2\beta}}{\Gamma(2\beta+1)},
\end{aligned}
$$

$$
\begin{aligned}
P_2 &= (\varepsilon-1)(\varphi m_3 - \lambda m_5)\frac{t^\beta}{\Gamma(\beta+1)}\\
&\quad + [-\varphi(-\rho m_2 + (\vartheta+\sigma) m_3) + \lambda(-\varphi m_3 + \lambda m_5)]\frac{t^{2\beta}}{\Gamma(2\beta+1)}.
\end{aligned}
\tag{11.62}
$$

Working in the same way, the remaining terms of S_μ, E_μ, I_μ, R_μ, and P_μ for $\mu \geq 3$ can be computed in a lucid manner, and consequently the series solution is fully obtained. Eventually, the q-HASTM solution for system of Eq. (11.40) is composed in this way:

$$S(t)=\sum_{\mu=o}^{K} S_\mu(t)\left(\frac{1}{\varepsilon}\right)^\mu, E(t)=\sum_{\mu=o}^{K} E_\mu(t)\left(\frac{1}{\varepsilon}\right)^\mu, I(t)=\sum_{\mu=o}^{K} I_\mu(t)\left(\frac{1}{\varepsilon}\right)^\mu,$$

$$R(t)=\sum_{\mu=o}^{K} R_\mu(t)\left(\frac{1}{\varepsilon}\right)^\mu, P(t)=\sum_{\mu=o}^{K} P_\mu(t)\left(\frac{1}{\varepsilon}\right)^\mu. \tag{11.63}$$

For $K \to \infty$, the desired solution for the fractional HEV model is acquired. The solutions are given by

$$\begin{aligned}
S(t) &= S_0(t)+\frac{1}{\varepsilon}S_1(t)+\frac{1}{\varepsilon^2}S_2(t)+\cdots,\\
E(t) &= E_0(t)+\frac{1}{\varepsilon}E_1(t)+\frac{1}{\varepsilon^2}E_2(t)+\cdots,\\
I(t) &= I_0(t)+\frac{1}{\varepsilon}I_1(t)+\frac{1}{\varepsilon^2}I_2(t)+\cdots,\\
R(t) &= R_0(t)+\frac{1}{\varepsilon}R_1(t)+\frac{1}{\varepsilon^2}R_2(t)+\cdots,\\
P(t) &= P_0(t)+\frac{1}{\varepsilon}P_1(t)+\frac{1}{\varepsilon^2}P_2(t)+\cdots.
\end{aligned} \tag{11.64}$$

Hence

$$\begin{aligned}
S(t) \;=\;& m_1+\frac{1}{\varepsilon}(\wp-\ell\wp m_3-\alpha m_1 m_3-\gamma_E m_1 m_5-\vartheta m_1)\frac{t^\beta}{\Gamma(\beta+1)}\\
&+\frac{(\varepsilon-1)}{\varepsilon^2}(\wp-\ell\wp m_3-\alpha m_1 m_3-\gamma_E m_1 m_5-\vartheta m_1)\frac{t^\beta}{\Gamma(\beta+1)}\\
&+\wp\frac{1}{\varepsilon^2}\frac{t^\beta}{\Gamma(\beta+1)}\\
&+\frac{1}{\varepsilon^2}[(\ell\wp+\alpha m_1)(-\rho m_2+(\vartheta+\sigma)m_3)+(\alpha m_3+\gamma_E m_5+\vartheta)\\
&\times(-\wp+\ell\wp m_3+\alpha m_1 m_3+\gamma_E m_1 m_5+\vartheta m_1)\\
&+\gamma_E m_1(-\varphi m_3+\lambda m_5)]\frac{t^{2\beta}}{\Gamma(2\beta+1)}+\cdots,
\end{aligned}$$

$$
\begin{aligned}
E(t) &= m_2 + \frac{1}{\varepsilon}\left(\alpha m_1 m_3 + \gamma_E m_1 m_5\right)\frac{t^{\beta}}{\Gamma(\beta+1)} + \frac{(\varepsilon-1)}{\varepsilon^2} \\
&\quad \times \left(\alpha m_1 m_3 + \gamma_E m_1 m_5\right)\frac{t^{\beta}}{\Gamma(\beta+1)} - \frac{1}{\varepsilon^2}\left[\left(\alpha m_3 + \gamma_E m_5\right)\right. \\
&\quad \times \left(-\wp + \ell\wp m_3 + \alpha m_1 m_3 + \gamma_E m_1 m_5 + \vartheta m_1\right) \\
&\quad + \alpha m_1\left(-\rho m_2 + (\vartheta+\sigma) m_3\right) \\
&\quad \left. + \gamma_E m_1\left(-\varphi m_3 + \lambda m_5\right)\right]\frac{t^{2\beta}}{\Gamma(2\beta+1)} + \cdots, \\
I(t) &= m_3 + \frac{1}{\varepsilon}\left[\rho m_2 - (\vartheta+\sigma) m_3\right]\frac{t^{\beta}}{\Gamma(\beta+1)} \\
&\quad + \frac{(\varepsilon-1)}{\varepsilon^2}\left[\rho m_2 - (\vartheta+\sigma) m_3\right]\frac{t^{\beta}}{\Gamma(\beta+1)} \\
&\quad + \frac{1}{\varepsilon^2}\left[\rho\left(\alpha m_1 m_3 + \gamma_E m_1 m_5\right) + (\vartheta+\sigma)\right. \\
&\quad \left. \times\left(-\rho m_2 + (\vartheta+\sigma) m_3\right)\right]\frac{t^{2\beta}}{\Gamma(2\beta+1)} + \cdots,
\end{aligned}
$$

$$
\begin{aligned}
R(t) &= m_4 + \frac{1}{\varepsilon}\left(\sigma m_3 - \vartheta m_4\right)\frac{t^{\beta}}{\Gamma(\beta+1)} \\
&\quad + \frac{(\varepsilon-1)}{\varepsilon^2}\left(\sigma m_3 - \vartheta m_4\right)\frac{t^{\beta}}{\Gamma(\beta+1)} \\
&\quad + \frac{1}{\varepsilon^2}\left[-\sigma\left(-\rho m_2 + (\vartheta+\sigma) m_3\right)\right. \\
&\quad \left. + \vartheta\left(-\sigma m_3 + \vartheta m_4\right)\right]\frac{t^{2\beta}}{\Gamma(2\beta+1)} + \cdots, \\
P(t) &= m_5 + \frac{1}{\varepsilon}\left(\varphi m_3 - \lambda m_5\right)\frac{t^{\beta}}{\Gamma(\beta+1)} \\
&\quad + \frac{(\varepsilon-1)}{\varepsilon^2}\left(\varphi m_3 - \lambda m_5\right)\frac{t^{\beta}}{\Gamma(\beta+1)} \\
&\quad + \frac{1}{\varepsilon^2}\left[-\varphi\left(-\rho m_2 + (\vartheta+\sigma) m_3\right)\right. \\
&\quad \left. + \lambda\left(-\varphi m_3 + \lambda m_5\right)\right]\frac{t^{2\beta}}{\Gamma(2\beta+1)} + \cdots.
\end{aligned}
$$

In the previous section, the fractional HEV model was properly solvedvia q-HASTM. We were motivated to get the solutions of this model by the q-HASTM algorithm. The q-HASTM is simpler to employ than other methods. The combination of the semi-analytical scheme q-HAM with a useful Sumudu transform algorithm lessens the time of computational procedure for a solution to a nonlinear model. In addition, the q-HASTM does not require any other restrictive assumption. It depicts small perturbations, lessens mathematical computations, generates non-local effects, and does not involve complicated polynomials or integrations. The q-HASTM shows an improvement of $q \in [0, 1]$ in the HASTM algorithm to $q \in \left[0, \frac{1}{\varepsilon}\right]$, $\varepsilon \geq 1$. The occurrence of the term $\left(\frac{1}{\varepsilon}\right)^{\mu}$ in the obtained solution provides faster convergence as compared to the standard HASTM. The convergence of the q-HASTM solution can be adjusted and controlled through the auxiliary parameter $\hbar$ and the asymptotic parameter ε.The numerical results depicted in the form of graphs authenticate the efficiency and accuracy of the q-HASTM.

11.7 NUMERICAL SIMULATIONS

In this section, the varying behaviour of populations of susceptible (S), exposed (E), infected (I), recovered (R), and density of viral load in the environment (P) are investigated in respect of change in time t and the fractional parameter β. The numerical scheme q-HASTM is utilized to prepare the graphical results. The numerical treatment of the fractional HEV model takes different values of the fractional order $\beta = 1, 0.9, 0.8, 0.7$. The initial conditions in the HEV model are considered in view of the original nature of the model and its dependent variables. The values of the parameters described in the model and used in this simulation are provided in Table 11.1. Here, the time range is taken up to 30(in days). Figures 11.1–11.6 present the graphical results for parameter $\varepsilon = 1$ and for different values of the fractional order $\beta = 1, 0.9, 0.8, 0.7$. Figures 11.7–11.12 depict the graphical results for parameter $\varepsilon = 2$ and for different values of the fractional order β. Figures 11.1–11.5, respectively, present the variation of individual populations of S, E, I, R, and P in respect of change in t and β. Figure 11.6 shows the mutual variation of S, E, I, R, and P with regard to t for $\varepsilon = 1$ and $\beta = 1$. Similarly, Figures 11.7–11.11, respectively, present the variation of individual populations of S, E, I, R, and P in respect of change in t and β. Figure 11.12 shows mutual variations of S, E, I, R, and P with regard to t for $\varepsilon = 2$ and $\beta = 1$.

Figures 11.1–11.5 depict that the population of susceptible individuals S increases with decreasing β and increasing t, whereas the population of exposed, infected, recovered, and density of virus load decreases with decreasing β for $\varepsilon = 1$. Similarly, Figures 11.7–11.11 depict that an increase occurs in susceptible individuals S with decreasing β and increasing t, whereas the population of exposed, infected, recovered, and density of virus load decreases with decreasing value of β for $\varepsilon = 2$. It is clearly observed from the graphical results that values of S, E, I, R, and P take lower values in case of $\varepsilon = 2$ as compared to $\varepsilon = 1$ because the embedding parameter q belongs to the shorter interval depending upon the value of ε. Thus, it is easy to observe that the population of susceptible individuals S increases less with decreasing β and the populations of E, I, R, and P decrease more with decreasing β in case of $\varepsilon = 2$ as compared to $\varepsilon = 1$. Thus, it proves that the occurrence of the term $\left(\frac{1}{\varepsilon}\right)^{\mu}$ in the obtained solution provides faster convergence. Consequently, the graphical presentations report that the fractional parameter β and an asymptotic parameter ε prove their pivotal role in decreasing of the infected class of individuals while increasing of susceptible individuals, which is impossible to achieve in the case of an integer-order system.

Figures 11.5 and 11.11 clearly show that the density of viral load P varies for different values of β in a very small range of values. Thus, as the fractional order β of the model system (11.3) decreases, susceptible population S increases whereas all the other compartments decrease. The graphical investigation shows dependency of calculated numerical values on β. It is also worth noting that the fractional HEV model Eq. (11.3) recovers its integer-order form for $\beta = 1$. Figures 11.6 and 11.12 show the mutual variations of $S, E, I, R,$ and P with regard to t for $\beta = 1$ and for $\varepsilon = 1$ and $\varepsilon = 2$, respectively. These figures indicate that the values of $S, E, I, R,$ and P decrease in case of $\varepsilon = 2$ as compared to $\varepsilon = 1$.

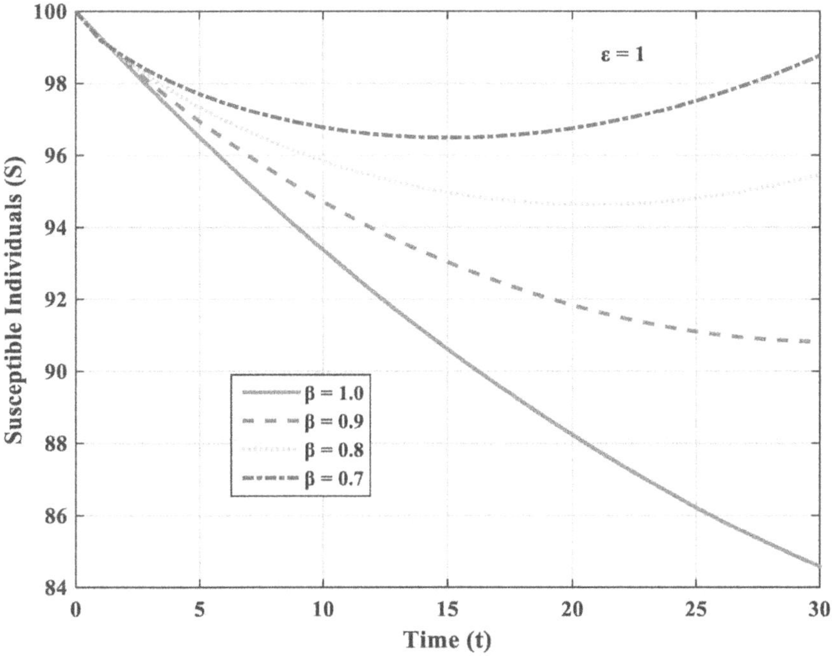

Figure 11.1 Variation of susceptible population S with regard to β and time t.

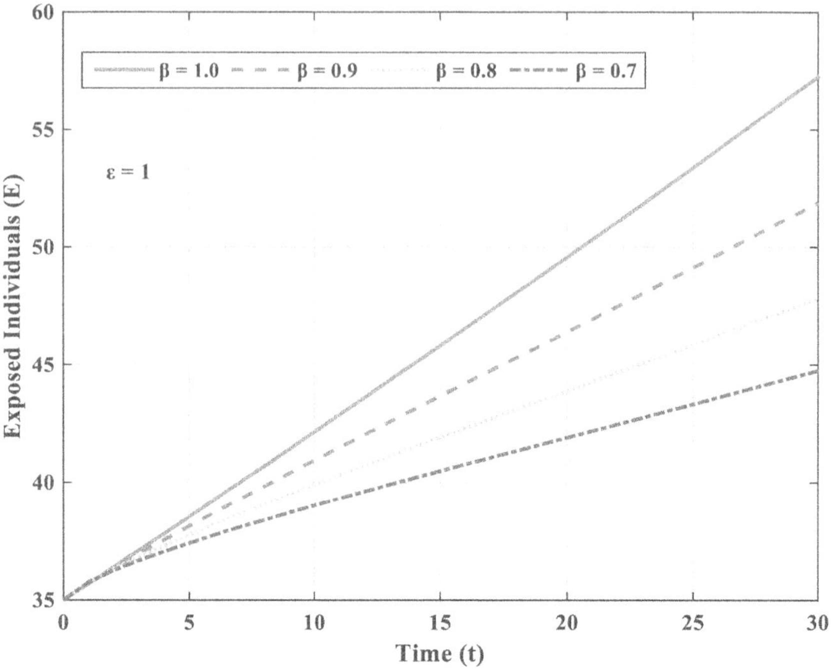

Figure 11.2 Variation of exposed population E with regard to β and t.

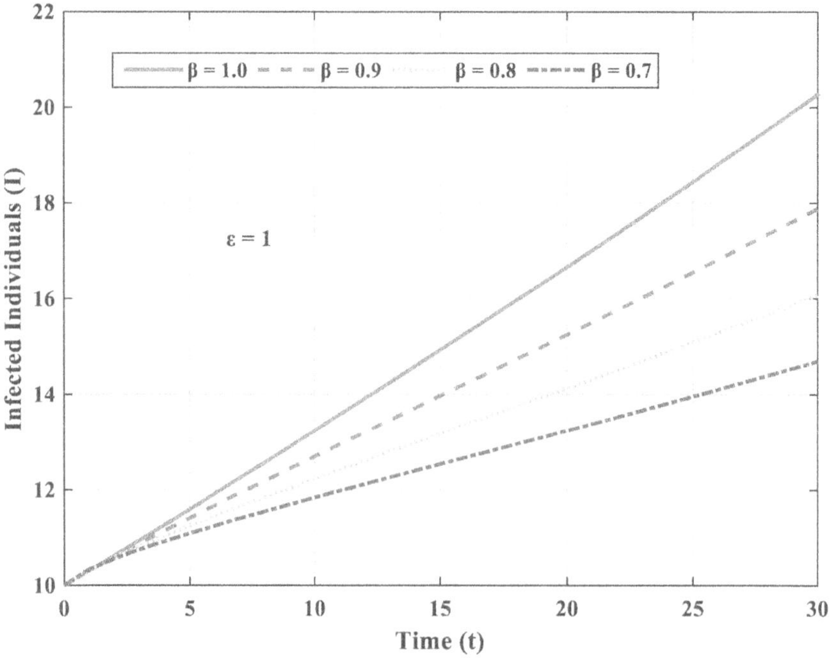

Figure 11.3 Behaviour of infected population I with regard to β and time t.

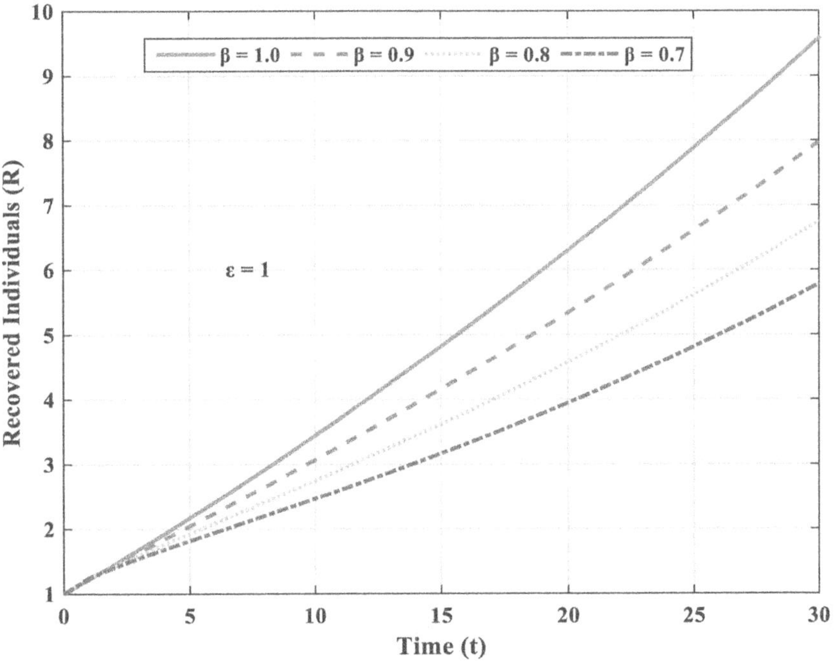

Figure 11.4 Behaviour of recovered population R with regard to β and time t.

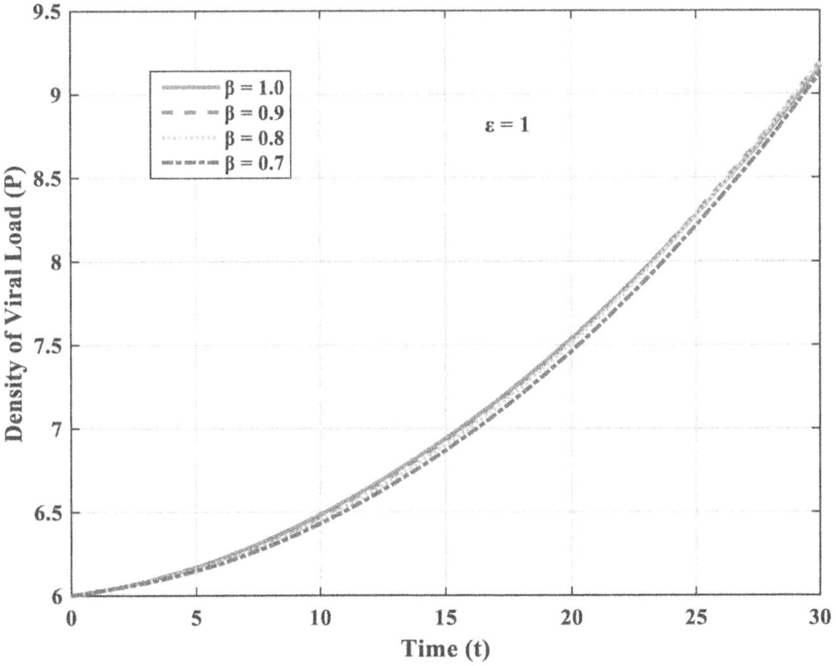

Figure 11.5 Plots of viral load P with regard to β and time t.

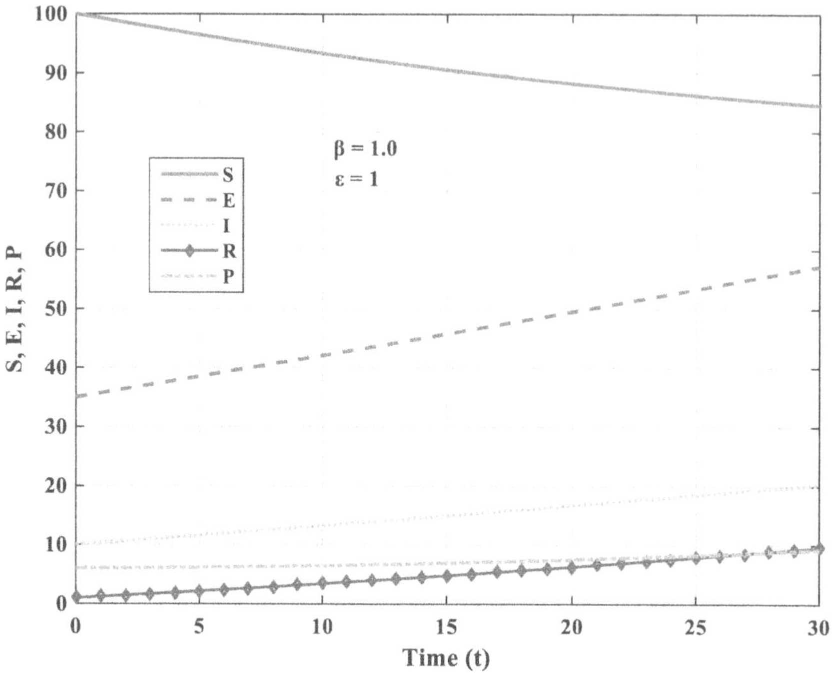

Figure 11.6 Behaviour of S, E, I, R, P with regard to time t for $\beta = 1$ and $\varepsilon = 1$.

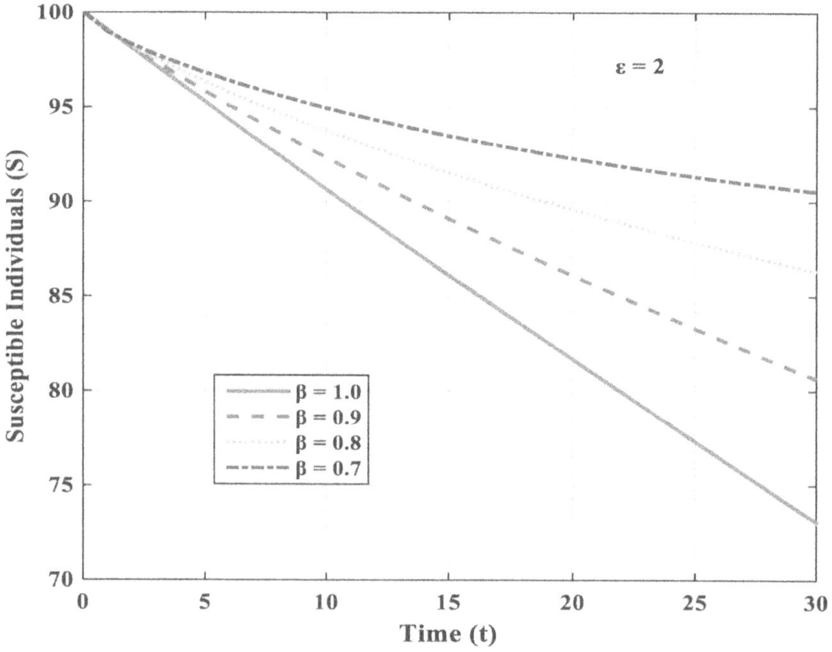

Figure 11.7 Plots of susceptible population S with regard to β and time t.

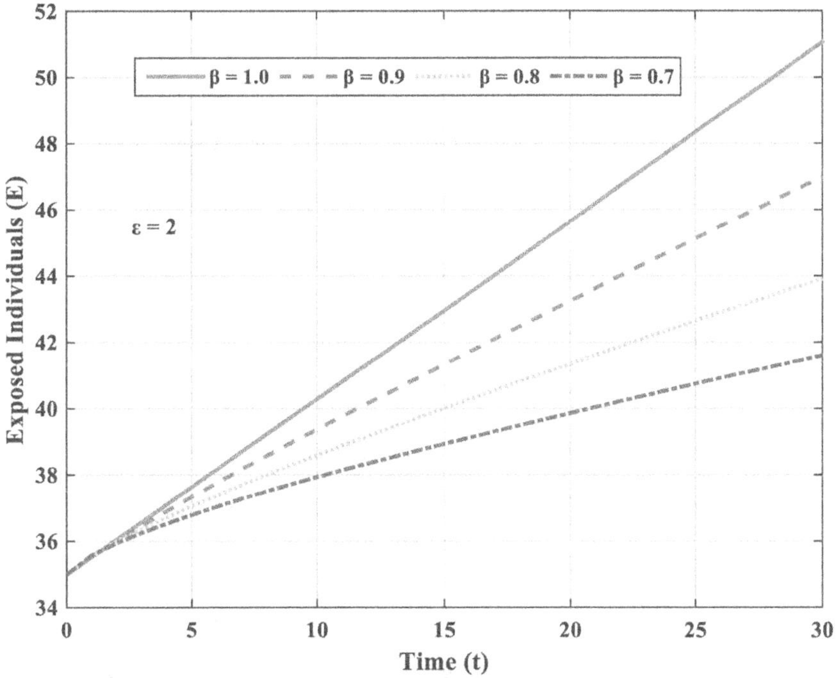

Figure 11.8 Variation of exposed population E with regard to β and time t.

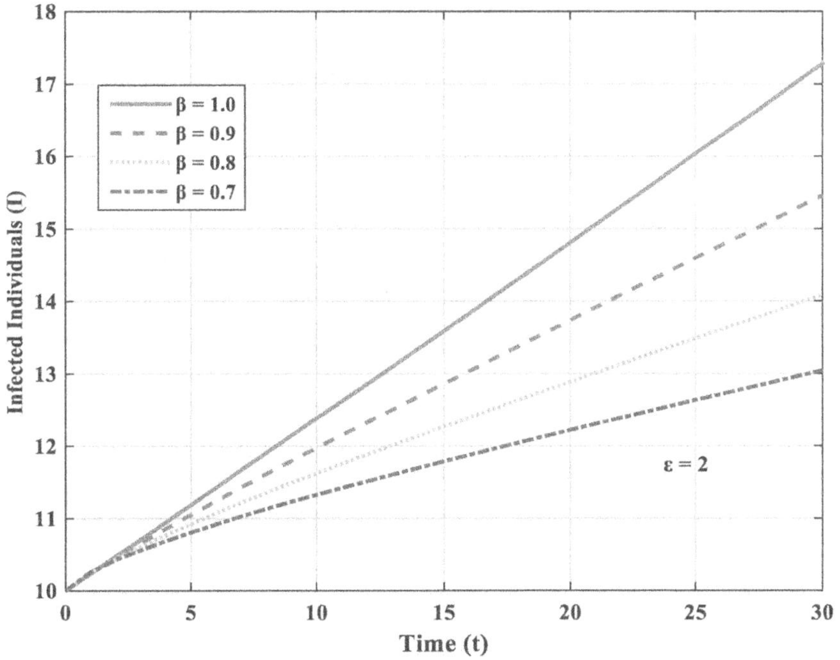

Figure 11.9 Variation of infected population I with regard to β and time t.

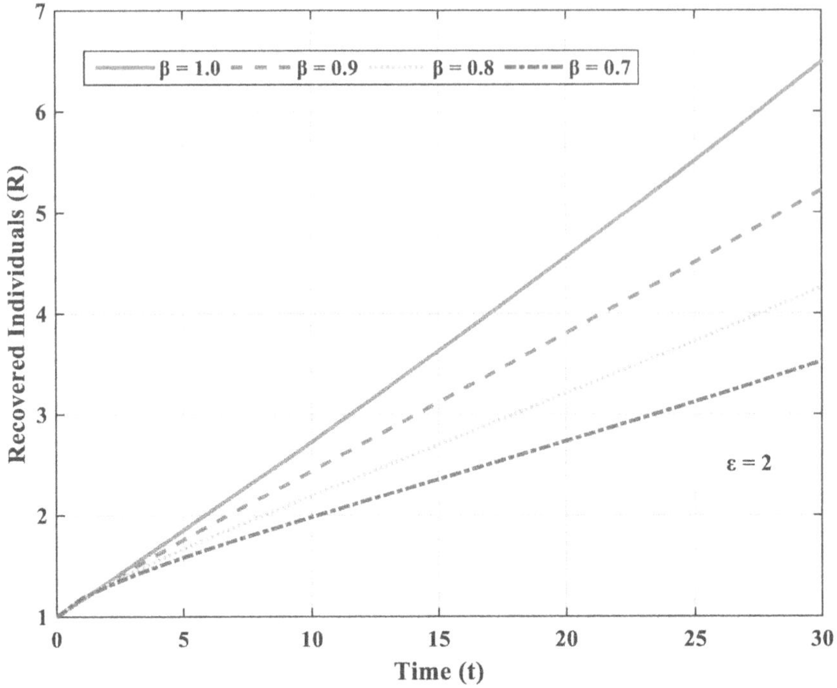

Figure 11.10 Variational behaviour of recovered population R with regard to time t and β.

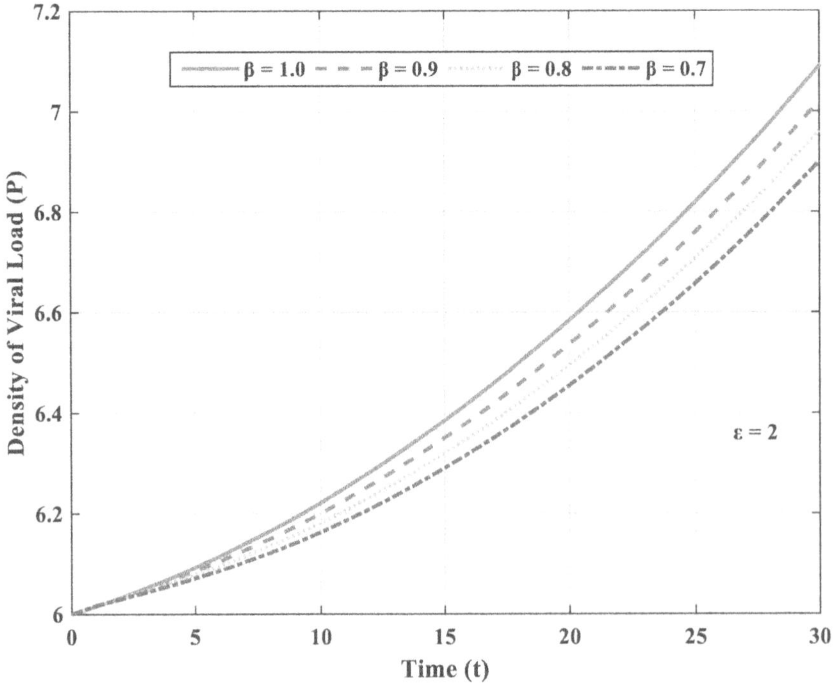

Figure 11.11 Plots of viral load P with regard to time t and β.

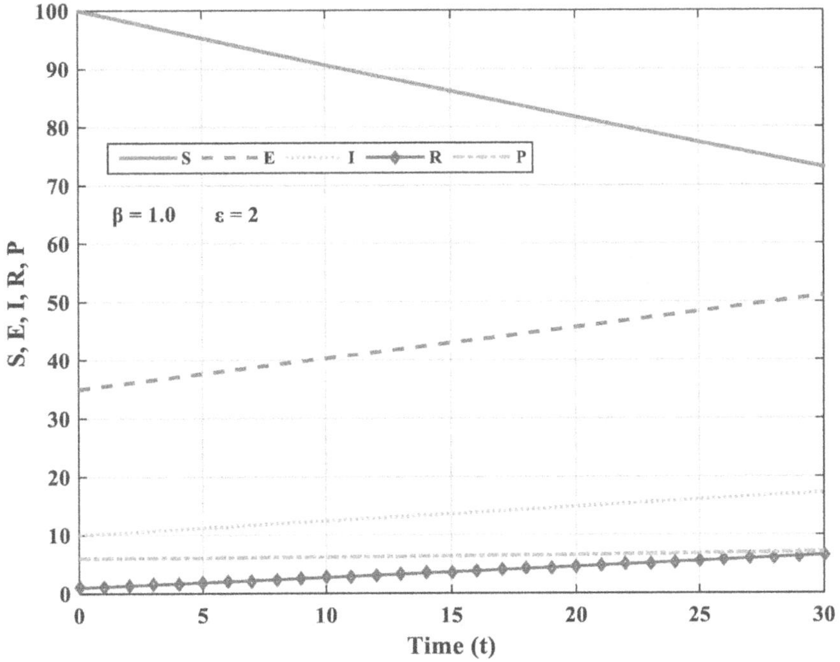

Figure 11.12 Behaviour of S, E, I, R, P with regard to time t for $\beta = 1$ and $\varepsilon = 2$.

11.8 CONCLUDING REMARKS AND OBSERVATIONS

In this chapter, the numerical scheme q-HASTM has been applied to investigate the fractional HEV model describing the viral dynamics of hepatitis E with the Caputo time-fractional derivative. The computational procedure indicates that the q-HASTM solution is more general as compared to the other numerical methods such as the homotopy perturbation method (HPM), HAM, q-HAM, and HASTM, and includes their results as a specific case. The numerical results of the fractional HEV model are achieved for different values of the fractional order $\beta = 1, 0.9, 0.8, 0.7$. This study explores the consequences of the variations of fractional order of a time derivative and time t on susceptible, exposed, infected, and recovered populations through graphical presentations. The numerical results through figures indicate that the population of susceptible individuals S increases, while the populations of exposed, infected, recovered, and density of virus load decrease with decreasing β and increasing t. The numerical results depicted in the form of graphs authenticate the accuracy of the q-HASTM. Hence, the implemented method is very appropriate and relevant to handle various nonlinear fractional models exhibiting real physical phenomena. Moreover, the uniqueness and convergence analysis of the method are also established with the aid of fixed point theory of Banach spaces and some theorems of pure mathematics. Caputo fractional modelling of an HEV model can be helpful in eliminating the spread of hepatitis E disease and can provide a solid base for control related strategies.

REFERENCES

1. I. Nimgaonkar, A. Ploss, A porcine model for chronic hepatitis E, Hepatology 67(2) (2018) 787–790.

2. R. Aggarwal, S. Naik, Epidemiology of hepatitis E: current status, J. Gastroen. Hepatol. 24(9) (2009) 1484–1493.

3. T. Ahmad, J. Hui, T.H. Musa, M. Behzadifar, M. Baig, Seroprevalence of hepatitis E virus infection in pregnant women: a systematic review and meta-analysis, Ann. Saudi Med.40(2) (2020) 136–146.

4. G.N. Mercera, M.R. Siddiqui, Application of a hepatitis E transmission model to assess intervention strategies in a displaced persons camp in Uganda, in 19th International Congress on Modelling and Simulation, Perth, Australia (2011) 12–16.

5. J.A. Backer, A. Berto, C. McCreary, F. Martelli, W.H.M. van der Poel, Transmission dynamics of hepatitis E virus in pigs: Estimation from field data and effect of vaccination, Epidemics 4(2) (2012) 86–92.

6. E.O. Alzahrani, M.A. Khan, Modeling the dynamics of Hepatitis E with optimal control, Chaos, SolitonsFractals 116 (2018) 287–301.

7. B. Nannyonga, D.J.T. Sumpter, J.Y.T. Mugisha, L.S. Luboobi, The dynamics, causes and possible prevention of hepatitis E outbreak, PLoS One 7(7) (2012) 1–8, e41135.

8. H. Ren, J. Li, Z.-A. Yuan, J.-Y. Hu, Y. Yu, Y.-H. Lu, The development of a combined mathematical model to forecast the incidence of hepatitis E in Shanghai, China, BMC Infect Dis 13 (421) (2013) 1–6.

9. D.B. Rein, G.A. Stevens, J. Theaker, J.S. Wittenborn, S.T. Wiersma, The global burden of hepatitis E virus genotypes 1 and 2 in 2005, Hepatology 55(4) (2012) 988–997.

10. J. Gouttenoire, D. Moradpour, A mouse model for hepatitis E virus infection, J. Hepatol. 64 (2016) 1003–1005.

11. D.G. Prakasha, P. Veeresha, and H. M. Baskonus, Analysis of the dynamics of hepatitis E virus using the Atangana-Baleanu fractional derivative, Eur. Phys. J. Plus 134(241) (2019) 1–11, DOI 10.1140/epjp/i2019-12590-5.

12. M.A. Khan, Z. Hammouch, D. Baleanu, Modeling the dynamics of hepatitis E via the Caputo-Fabrizio derivative, Math. Model. Nat. Phenom. 14(2019) 1–19.

13. A. Atangana, D. Baleanu, New fractional derivatives with nonlocal and non-singular kernel: Theory and application to heat transfer model, Therm. Sci. 20(2) (2016) 763–769.

14. M. Caputo, M. Fabrizio, A new definition of fractional derivative without singular kernel, Prog. Fract. Differ. Appl. 1 (2015) 73–85.

15. M. Sholikah, C. Alfiniyah, M. Miswanto, Stability analysis and optimal control of mathematical model for the spread of hepatitis E, Commun. Math. Biol. Neurosci. 2020 (37)(2020) 1–24.

16. S.G. Samko, A.A. Kilbas, O.I. Marichev, Fractional Integrals and Derivatives: Theory and Applications, Gordon and Breach, Yverdon (1993).

17. J. Singh, D. Kumar, M. Al-Qurashi, D. Baleanu, A novel numerical approach for a nonlinear fractional dynamic model of interpersonal and romantic relationships, Entropy 19(7) (2017) 1–17.

18. M.A. El-Tawil, S.N. Huseen, The q-homotopy analysis method (q-HAM), Int. J. Appl. Math. Mech. 8(15)(2012) 51–75.

19. M.A. El-Tawil, S.N. Huseen, On convergence of the q-homotopy analysis method, Int. J. Contemp. Math. Sci. 8(10)(2013) 481–497.

20. S.J. Liao, Proposed Homotopy Analysis Techniques for the Solution of Nonlinear Problems. Ph.D. Thesis, Shanghai Jiao Tong University, Shanghai (1992).

21. S.J. Liao, Beyond Perturbation: Introduction to Homotopy Analysis Method, Chapman and Hall/CRC Press, Boca Raton, FL (2003).

22. V.P. Dubey, R. Kumar, J. Singh, D. Kumar, An efficient computational technique for time-fractional modified Degasperis-Procesi equation in propagation of nonlinear dispersive waves, J. Ocean Eng. Sci. DOI: 10.1016/j.joes.2020.04.006.

23. I. Podlubny, Fractional differential equations, Academic Press, New York (1999).

24. G.K. Watugala, Sumudu transform– a new integral transform to solve differential equations and control engineering problems, Math. Eng. Indust. 6(4)(1998) 319–329.

25. V.B.L. Chaurasia, J. Singh, Application of Sumudu transform in Schrödinger equation occurring in quantum mechanics, Appl. Math. Sci. 4(57)(2010) 2843–2850.

26. F.B.M. Belgacem, A.A. Karaballi, Sumudu transform fundamental properties investigation and applications, J. Appl. Math. Stoch. Anal. 2006(2006) 1–2310.1155/JAMSA/2006/91083.

27. F. Gao, H.M. Srivastava, Y.-N. Gao, X.-J. Yang, A coupling method involving the Sumudu transform and the variational iteration method for a class of local fractional diffusion equations, J. Nonlinear Sci. Appl. 9(11) (2016) 5830–5835.

28. H.M. Srivastava, A.K. Golmankhaneh, D. Baleanu, X.-J. Yang, Local fractional Sumudu transform with applications to IVPs on Cantor sets, Abstr. Appl. Anal. 2014(2014) 1–7.

29. S.K. Choi, B. Kang, N. Koo, Stability for Caputo Fractional differential systems, Abstr. Appl. Anal. 2014(2014) 1–6.

30. G.M. Mittag-Leffler, Sur l' integrable de Laplace-Abel, Comptes Rendus de l'Académie des Sciences Series II 136 (1903) 937–939.

31. Z. Odibat, S.A. Bataineh, An adaptation of homotopy analysis method for reliable treatment of strongly nonlinear problems: construction of homotopy polynomials, Math. Meth. Appl. Sci. 38(5)(2014) 991–1000.

32. I.K. Argyros, Convergence and Applications of Newton-type Iterations, Springer-Verlag, New York (2008).

33. A.A. Magrenan, A new tool to study real dynamics: the convergence plane, Appl. Math. Comput. 248(2014) 215–224.

Index

Note: Locators in *italics* represent figures and **bold** indicate tables in the text.

For Product Safety Concerns and Information please contact our EU representative GPSR@taylorandfrancis.com
Taylor & Francis Verlag GmbH, Kaufingerstraße 24, 80331 München, Germany

www.ingramcontent.com/pod-product-compliance
Lightning Source LLC
Chambersburg PA
CBHW060813220326
41598CB00022B/2608
9780367564803